Characterization
of Proteins

Biological Methods

Characterization of Proteins, edited by **Felix Franks,** *1988*

Methods of Protein Microcharacterization: A Practical Handbook, edited by **John E. Shively,** *1986*

Experiments in Molecular Biology, edited by **Robert J. Slater,** *1986*

Methods in Molecular Biology, edited by **John M. Walker,** *1984*
    Volume I: ***Proteins***
    Volume II: ***Nucleic Acids***

Liquid Chromatography in Clinical Analysis, edited by **Pokar M. Kabra** and **Laurence J. Marton,** *1981*

Metal Carcinogenesis Testing: Principles and In Vitro Methods, by **Max Costa,** *1980*

Characterization of Proteins

edited by
Felix Franks
Pafra Ltd.
Cambridge, UK

Humana Press · Clifton, New Jersey

Library of Congress Cataloging-in-Publication Data
Main entry under title:

Characterization of Proteins.

   (Biological methods)
   Includes bibliographies and indexes.
   1. Proteins—Analysis. I. Franks, Felix.
II. Series. [DNLM: 1. Proteins—isolation.
2. Proteins—physiology.   QU 55 C469]
QP551.C486   1988   574.19'245   87-29874
ISBN 0-89603-109-8

© 1988 The Humana Press Inc.
Crescent Manor
PO Box 2148
Clifton, NJ 07015

All rights reserved.

No part of this book may be reproduced, stored in a retrieval system, or transmitted in any form or by any means, electronic, mechanical, photocopying, microfilming, recording, or otherwise without written permission from the Publisher.

Printed in the United States of America

Preface

Proteins are the servants of life. They occur in all component parts of living organisms and are staggering in their functional variety, despite their chemical similarity. Even the simplest single-cell organism contains a thousand different proteins, fulfilling a wide range of life-supporting roles. Their production is controlled by the cell's genetic machinery, and a malfunction of even one protein in the cell will give rise to pathological symptoms. Additions to the total number of known proteins are constantly being made on an increasing scale through the discovery of mutant strains or their production by genetic manipulation; this latter technology has become known as protein engineering.

The in vivo functioning of proteins depends critically on the chemical structure of individual peptide chains, but also on the detailed folding of the chains themselves and on their assembly into larger supramolecular structures. The molecules and their functional assemblies possess a limited in vitro stability. Special methods are required for their intact isolation from the source material and for their analysis, both qualitatively and quantitatively. Proteins are also increasingly used as "industrial components," e.g., in biosensors and immobilized enzymes, because of their specificity, selectivity, and sensitivity. This requires novel and refined processing methods by which the protein isolate can be converted into a form in which it can be utilized.

This book provides a unified approach to several essential aspects of protein technology: classifications of protein function, both *in situ* and in the form of isolates, criteria that govern in vitro protein stability limits, methods of detection, isolation and separation from nanogram to tonne quantities, analysis *in situ* and in the isolated state, and the technology of protein processing for the production of stable preparations for clinical, agricultural, or industrial use.

This book is based on material initially prepared for several postexperience courses: "Characterization of Proteins—Their Analysis and Isolation," taught by the authors on behalf of the Center for Professional Advancement in the US and Europe. It is intended for academic, medical, and a wide range of industrial research and development scientists who are actively involved in protein isolation, purification, characterization, and the setting and measuring of protein standards of performance.

A general survey of the chemical and functional properties of proteins and the economics of protein isolation and processing is followed by a survey of the parameters that circumscribe protein stability in vitro. The next section deals with in vivo characterization, detection, and localization of several important groups of proteins (e.g., peptide hormones and receptors). This is followed by several chapters describing chromatographic, electrophoretic, and immunological methods employed in the analytical and preparative scale separation, isolation, and characterization of proteinaceous materials, from amino acids through peptide fragments and intact proteins. The final section is devoted to aspects of large-scale processing and quality control methods.

This volume differs from most monographs on proteins in that it treats these molecules as valuable but labile chemical raw materials that are easily inactivated by conventional processing techniques. Assuming some previous knowledge of protein analytical, physical, and biochemistry, the book combines these disciplines to produce a synthesis of the parameters that are essential for the successful isolation, separation, characterization, and stabilization of proteins.

Felix Franks

Contents

Preface v
List of Contributors xix
Introduction 1

CHAPTER 1
Description and Classification of Proteins
Felix Franks

1. Chemical Constitution 9
   1.1. Amino Acid Classification 10
   1.2. Conjugated Proteins 12
   1.3. Amino Acid Profile 15
2. Protein Size 15
3. Classification by Function 15
   3.1. Versatility and Limitations 15
   3.2. Biological Function 18
   3.3. Technological Function 22

CHAPTER 2
Internal Structure and Organization: Relationship to Function
Felix Franks

1. Structural Hierarchy of Proteins 25
   1.1. Experimental Approaches to the Characterization of Protein Structure 25
2. Primary Structure 26
   2.1. General Principles 26
   2.2. Nomenclature 27
   2.3. Ribonuclease 28
   2.4. Insulin 28
   2.5. Homology 28

3. Secondary Structure 31
   3.1. The α-Helix 32
   3.2. The β-Pleated Sheet 34
   3.3. Polyproline Helices 34
   3.4. Other Regular Structures 36
4. Tertiary Structure 37
   4.1. Classification of Tertiary Structures 38
   4.2. Spatial Disposition of Structured Domains 38
   4.3. Tertiary Structure and Homology 40
5. Quaternary Structure 42
   5.1. Motile Structures 44
   5.2. Structural Proteins 45
   5.3. Immunoglobulins 48
   5.4. Multisubunit Enzyme Complexes 49
   5.5. Nuclear Proteins 49
   5.6. Fibrinogen 50
6. Structural Integrity and Technological Function 51

CHAPTER 3
Solution Properties of Proteins
Felix Franks

1. Conformational Stability and Solubility 53
2. Solubility 54
   2.1. Electrostatic Factors 54
   2.2. Classification of Proteins by Solubility 57
   2.3. Solubility as Index of Viability/Denaturation 58
   2.4. Solubility and Processing Conditions 60
   2.5. Solubility and Nonelectrostatic Factors 60
3. Solution Properties of Large Molecules 61
   3.1. Thermodynamics of Ideal and Nonideal Solutions . 61
   3.2. Solution Properties and Molecular Shapes 63
   3.3. Quality of Solvent 64
4. Experimental Techniques 65
   4.1. Membrane Equilibria 65
   4.2. Diffusion and Sedimentation 67
   4.3. Viscosity 70
   4.4. Scattering of Radiation 72
   4.5. Optical Spectroscopy 75
   4.6. Optical Activity, Circular Dichroism (CD), and
        Optical Rotatory Dispersion (ORD) 79

Contents ix

    4.7. Magnetic Resonance 81
References 94

CHAPTER 4
Conformational Stability: *Denaturation and Renaturation*
Felix Franks

1. Origin of the Native State 95
2. Two-State Model of Protein Stability 97
    2.1. Further Analysis of "All-or-None" Denaturation . 101
3. Thermal Denaturation 103
4. Low-Temperature Denaturation 104
    4.1. Freeze Denaturation 107
5. Pressure Denaturation 107
6. Shear Denaturation 108
7. Chemical Denaturation and Stabilization 110
    7.1. pH Denaturation 111
    7.2. Salting In/Out and Protein Stability 112
    7.3. Characterization of Protein D-States 118
8. Surface Denaturation 120
9. Multistate Denaturation 121
10. Protein Engineering: The Means of Affecting
    Protein Stability 123
References 125

CHAPTER 5
Protein Hydration
Felix Franks

1. Historical 127
2. Classification of Protein Hydration Interactions 128
    2.1. Ion Hydration 129
    2.2. Hydrophilic Hydration 130
    2.3. Hydrophobic Hydration and Hydrophobic
        Interactions 131
3. Medium Effect on Protein Stability 135
4. Hydration Sites in Protein Crystals 139
5. Dynamic Aspects of Water in Protein Solutions:
    Protein Hydration 142

6. Protein Hydration and Bound Water: Some Popular Misconceptions 144
7. Proteins at Low Moisture Content: Sequential Hydration 146
8. Computer Modeling of Protein–Water Interactions 151
References 153

CHAPTER 6
Characteristics of Proteins and Peptides *In Situ:* An Overview
P. J. Thomas

1. Proteins and Peptides as Functional Units 155
   1.1. Carrier Proteins 157
   1.2. Hormones and Prohormones 157
   1.3. Neuropeptides and Cotransmitters 163
   1.4. Receptors 165
   1.5. Enzymes 172
   1.6. Proteins Associated with Genetic Material 175
   1.7. Intracellular Structural Proteins 177
   1.8. Storage Proteins 179
   1.9. Toxins 179
2. Multiple Functions of Individual Proteins 180
3. Cyclical Interactions 181
   3.1. The Multicellular Level 181
   3.2. The Intracellular Level 182
   3.3. Long-Term Interactions 183
4. Conclusions 186
References 186

CHAPTER 7
Methods of Measuring Binding, Some Extracellular Carrier Proteins, and Intracellular Receptor Proteins: *A Selective Introduction*
P. J. Thomas

1. Introduction 189
   1.1. Measurement of Binding 189

1.2. Effects of Impurities 191
1.3. Conformational Changes 191
1.4. The Hill Plot 192
1.5. Rate Constants 193
1.6. Specificity 194
1.7. Practical Approaches 196
2. Examples of Carrier Proteins 196
   2.1. Rodent α-Fetoprotein 198
   2.2. Steroid Binding Proteins in Adult Blood 200
3. Intracellular Receptors 204
   3.1. Steroid Receptors 204
   3.2. Aftermath of Steroid-Receptor Binding 208
   3.3. Catecholoestrogens 212
4. Other Intracellular Receptors 212
   4.1. Vitamin D (Cholecalciferol and Ergocalciferol)
      Receptors 212
   4.2. Thyroid Hormones and Their Receptors 213
5. Summary 213
References 213

Chapter 8
Membrane Receptors
P. J. Thomas

1. Measurement of a Hitherto Unknown Cell Surface Receptor 217
   1.1. Choice of Ligand 219
   1.2. Receptor Subclass 219
   1.3. Separation of Bound from Free Receptors 220
2. Purification and Isolation of Receptors 220
   2.1. Dopamine Receptors 221
   2.2. Effects of Use Upon Dopamine Receptors 222
   2.3. Nicotinic Acetylcholine Receptors 222
   2.4. Insulin Receptors 223
   2.5. Other Receptors 224
3. Intracellular Messengers 226
   3.1. Cyclic Nucleotides 226
   3.2. Phosphodiesterases 228
   3.3. Cyclic Nucleotide-Dependent Protein Kinases 229
References 230

CHAPTER 9
Protein and Peptide Hormones
P. J. Thomas

1. Introduction 233
2. Release and Release-Inhibiting Factors Secreted by the Hypothalamus 233
3. Protein Hormones of the Anterior Pituitary (Adenohypophysis) 235
   3.1. Multiple Functions of Individual Anterior Pituitary Hormones 236
4. The Posterior Pituitary (Neurohypophysis) 238
   4.1. Neurophysins 239
   4.2. Some Effects of Neurohypophyseal Hormones Upon the Brain 240
5. Hormones of the Periphery 240
   5.1. Insulin 240
   5.2. Insulin in the Brain 241
6. Conclusions 242
References 242

CHAPTER 10
Analysis of Amino Acids and Small Peptides
Colin Simpson

1. Introduction 245
2. Amino Acid Analysis 246
   2.1. Paper Chromatography 246
   2.2. High-Voltage Paper Electrophoresis 247
   2.3. Combined Paper Chromatography and Electrophoresis 249
   2.4. Thin-Layer Chromatography 249
   2.5. Gas Chromatography 252
   2.6. Amino Acid Analyzers 257
   2.7. Liquid Chromatography 258
   2.8. Separation of Racemic Amino Acids 269
3. Electrophoretic Separation of Amino Acids in Capillaries 275
   3.1. Isotachophoresis 275
   3.2. High-Voltage Zone Electrophoresis 278
References 282

CHAPTER 11
Analytical Liquid Chromatography of Proteins
Colin Simpson

1. Introduction 285
2. Adsorption Chromatography 286
3. Partition Chromatography 287
4. Bonded Phase Chromatography 288
   4.1. Mechanism of Separation 291
5. Hydrophobic Interaction Chromatography 296
6. Ion Exchange Chromatography (Electrostatic
   Interaction Chromatography, EIC) 296
   6.1. The Nature of the Stationary Phase 298
   6.2. Operating Parameters in IEC 299
   6.3. Various Other Parameters That Control
        Separation by IEC 301
   6.4. Other Operating Methods 304
   6.5. Metal Chelate Interaction Chromatography 307
7. Chromatofocusing 309
   7.1. Factors Controlling Resolution 310
8. Affinity Chromatography 311
   8.1. Support Matrices and Coupling 314
   8.2. Spacers 314
   8.3. Operation 316
9. Gel Permeation Chromatography 316
   9.1. The Mobile Phase 318
   9.2. Dispersion Mechanisms 320
10. Field Flow Fractionation 324
    10.1. Theory and Mechanism of FFF 326
    10.2. Types of Field 328
    10.3. Some Separations by Various Modes of FFF 328
    10.4. Advantages and Disadvantages of FFF 332
References 336
Suggested Further Reading 337

CHAPTER 12
Analytical Electrophoresis
Colin Simpson

1. Introduction 339
2. Isoelectric Focusing 340

2.1. The pH Gradient 340
2.2. pH Drift 342
2.3. The Anticonvective Medium 342
2.4. Dependence of Zone Width on Operating
     Parameters 343
2.5. Ultra-Thin-Layer IEF 344
2.6. The Immobiline System 345
3. Zone Electrophoresis 348
4. Isotachophoresis 352
5. Two-Dimensional Separations 353
6. Visualization and Detection 358
7. Blotting 360
References 361

CHAPTER 13
Preparative Liquid Chromatography
Colin Simpson

1. Introduction 365
   1.1. Surface Hardness 365
   1.2. Temperature Effects 366
   1.3. Shear Forces 366
2. Rationale for Sample Overloading (1) 367
   2.1. Column Radius, r 368
   2.2. Distribution Coefficient, K 368
   2.3. Packing Density, d 368
   2.4. Surface Area, A_s 368
   2.5. Column Length (l) and Particle Size, dp . 369
3. Column Overload 369
4. Recycle Technique 373
5. Equipment for Preparative Liquid Chromatography ... 374
   5.1. The Column 374
   5.2. Sample Application 378
   5.3. Column End Fittings 380
6. Technique in Preparative Chromatography 380
   6.1. Developing a Separation 380
References 383

CHAPTER 14
Preparative Scale Electrophoresis
Colin Simpson

1. Scale 385
2. Microscale 386
    2.1. Conventional Analytical Methods (Zone EP and Isoelectric Focusing) 386
3. Semimicro Scale 389
    3.1. Flat-Bed Preparative Isotachophoresis 389
    3.2. Column Isotachophoresis 391
    3.3. Annular Isotachophoresis 392
4. Continuous Preparative Electrophoresis 395
    4.1. Flat-Bed—Trop's Method 395
    4.2. Free-Flow Electrophoresis—Wagner and Mang ... 398
5. Large-Scale Preparative Electrophoresis 406
    5.1. Preparative Isoelectric Focusing 406
    5.2. Large-Scale Free-Flow Electrophoresis 408
References 412

CHAPTER 15
Use and Production of Polyclonal and Monoclonal Antibodies
A. W. Schram

1. Introduction 415
2. Polyclonal Antibodies 416
    2.1. Immunization Procedure 416
    2.2. Detection of Specific Antibodies 417
3. Monoclonal Antibodies 419
    3.1. Principle of the Procedure 419
    3.2. In Vivo Immunization and Fusion 420
    3.3. Cloning Procedure 420
    3.4. Screening of the Clones 421
    3.5. Large-Scale Production of Monoclonal Antibodies 421
    3.6. In Vitro Immunization 422
    3.7. Epitope Analysis 423
4. Characterization and Purification of Antibodies 424
5. Application of Antibodies 424
Suggested Reading 425

Chapter 16
Large-Scale Methods for Protein Separation and Isolation
Peter J. Lillford

1. Aims of Large-Scale Processes 427
2. Description of Raw Materials 428
3. Extraction 430
4. Precipitation and Fractionation 432
   - 4.1. Precipitation Methods 432
   - 4.2. Formation of Precipitates 439
5. Solid/Liquid Separation 440
   - 5.1. Filtration 440
   - 5.2. Centrifugation 441
6. Liquid/Liquid Separation 444
   - 6.1. Ultrafiltration 444
   - 6.2. Chromatography 446
7. Concentration and Drying 447
   - 7.1. Concentration 447
   - 7.2. Drying 448
References 450

Chapter 17
Large-Scale Processing of Plant Proteins
Chester Myers

1. Raw Materials 453
   - 1.1. Species Variation 453
   - 1.2. Seasonal Variation 454
   - 1.3. Preprocessing 455
2. Refining Processes 455
   - 2.1. Mechanical Separation—Air Classification and the Liquid Cyclone 456
   - 2.2. Mixed Solvents 457
   - 2.3. Refinement by Aqueous Extraction 458
3. Protein Isolate Preparation 458
   - 3.1. Isoelectric Precipitation 462
   - 3.2. Separation by Ionic Strength Manipulation 463
   - 3.3. Separation by Temperature Control 464
References 466

CHAPTER 18
Characterization and Functional Attributes of Protein Isolates: *Biochemical Applications*
Peter J. Lillford

1. Introduction	467
2. Hierarchy of Characterization Methods	468
2.1. Molecular Identity	468
2.2. Subunit Composition	469
2.3. Size and Shape Measurement	469
2.4. Secondary Structure by Spectroscopy	470
2.5. Detailed Structure by X-Ray Diffraction	471
2.6. Application of Nuclear Magnetic Resonance to Protein Structure and Dynamics	472
3. Serum Albumin	477
4. Insulin	477
5. Oxytocin	478
5.1. Function and Primary Structure	478
5.2. NMR Studies	478
6. Antibody Structure and Function	479
6.1. Chemical Studies	479
6.2. Crystal Structure	479
6.3. NMR Studies	481
7. Lysozyme	481
7.1. Biological Function	481
7.2. Molecular Identity	482
7.3. X-Ray Crystallography	482
7.4. NMR Mapping of the Active Site	482
7.5. Catalytic Mechanism	482
8. Interferons	483
8.1. General Description	483
8.2. Isolation and Separation of Natural Interferons	484
8.3. Characterization of Interferons	485
8.4. Action of Interferons	487
References	488

CHAPTER 19
Functional Attributes of Protein Isolates: *Foods*
Chester Myers

1. General Considerations for Evaluation of Protein Isolates	491

1.1. Solubility 491
1.2. Electrostatic Charge/pH Titration 493
1.3. Criteria of Purity and Denaturation 495
1.4. Amino Acid Profile 518
1.5. Nutritional Evaluation 519
2. Proteins as Food Ingredients 525
2.1. Protein Gelation/Aggregation 525
2.2. Surface Activity 529
2.3. Texturization and Rheology 541
References 546

Protein Index 551
Subject Index 555

Contributors

FELIX FRANKS · *Department of Botany, Botany School, University of Cambridge, Cambridge and Biopreservation Division, Pafra Ltd., Cambridge, UK*

PETER J. LILLFORD · *Unilever Research, Bedfordshire, UK*

CHESTER MYERS · *St. Lawrence Reactors Ltd., Missisauga, Ontario, Canada*

A. W. SCHRAM · *Vakgroep Biochemie, Universiteit van Amsterdam, Academisch Medisch Centrum, Amsterdam, The Netherlands*

COLIN SIMPSON · *Department of Chemistry, Chelsea College, University of London, London, UK*

P. J. THOMAS · *Department of Pharmacology, Chelsea College, University of London, London, UK*

Introduction

The total international protein literature could fill a medium-sized building, and is growing at an ever-increasing rate. The reader might be forgiven for asking whether yet another book on proteins, their properties, and functions can serve a useful purpose. An explanation of the origin of this book may serve as justification. The authors have participated as tutors for an intensive postexperience course, entitled Characterization of Proteins: Their Analysis and Isolation, organized by the Center for Professional Advancement, East Brunswick, New Jersey, an educational foundation. The course was first mounted in Amsterdam in 1982 and has since been repeated several times, in both Amsterdam and the US, with participants from North America and most European countries. The course has been well received by the participants, as has the course manual, from which this book has been developed.

Although most of the subject matter contained in this volume can be found in other publications, its combination in one volume has probably not been attempted before. The viewpoint taken here is that methods developed for the isolation, purification, processability, analysis, and characterization of proteins rely on our present knowledge of their structures, stabilities, and performance under a variety of physical and chemical conditions. These methods are described and related to economic, clinical, and technological functions and applications. In other words, we are here concerned with the methodologies of relevance to the "protein business." The protein-converting industry is a young, but rapidly growing, industry; some of the quantitative estimates here presented are therefore based on extrapolations of presently available data, which may turn out to be wide off the mark, but this is surely the general nature of industrial and economic forecasts. One may recall the early 1970s when many companies (including the oil majors) built large-capacity plants for the conversion of mineral oil into single-cell proteins, allegedly for human and animal nutrition. After the oil crisis of 1973, some of the same companies, together with others, began to investigate the conversion of proteinaceous material into mineral oil!

In any discussion of the technology of protein isolation and processing from a business perspective, the value of the final product needs to be considered, since this will probably determine the method chosen for the extraction and purification of the protein. Another factor that needs to be considered is the scale of the operation.

Isolated protein powders for food use were virtually unknown prior to 1940, except for dried "natural" products such as milk and egg. The restrictions of the war led to the first protein replacer in Europe—a chemically hydrolyzed milk powder, marketed as a partial egg white replacer. Today, isolated proteins for food use are extracted from yeasts, fungal sources, legumes, oilseeds, cereals, and leaves. Large-scale isolation methods have been developed and are operated on the basis of cost and efficiency. Any improvement in the product is measured against the cost of achieving such an improvement, and this cost features largely in the marketing equation along with other factors, such as the perceived (by the consumer) benefit of such an improvement. This latter attitude is of considerable importance in protein isolation technology. For instance, it is easy to recover fish muscle protein from fish offal by the use of extreme pH treatments, followed by solvent extraction and drying. The product so obtained is a polypeptide, hopefully with the same amino acid sequence as the starting material. It is, however, likely to be a bland, insoluble powder. Even if its nutritional quality is identical to that of intact muscle protein (and this is by no means sure), its appearance does not make it very attractive to the consumer.

At the other end of the quantity scale, an increased understanding of the biochemical origins of disease, coupled with developments in recombinant DNA technology, has led to massive commercial interest in the production (or creation) of highly specific proteins, such as interferon, hormones, and blood factors, and to the current boom in biotechnology. Here again the emphasis is on profitability (and a quick return on capital) so that allowable costs incurred in extraction and recovery must be determined by the value of the product in the market place. The enzyme business appears to be particularly buoyant: in the early 1980s, sales of enzymes in North America and Western Europe totaled $300 million with an estimated rise to $500 million by 1985.

When a protein is produced for "biochemical" purposes, i.e., when its specific biological activity is required, such as enzymic cleavage, the control of diabetes by insulin, and so on, the choice

Introduction

of source material may not be negotiable. If, on the other hand, the protein is isolated for a more general functional property (e.g., gel promotion, emulsification, rheological modification of a substrate), then there may be a wide choice of starting material, so that economic considerations become the determining factor. For example, oilseeds represent an important source of proteins mainly because the oil recovery pays for the crop: meal left after oil extraction is cheap and may be used as animal feed, fertilizer, or fuel.

Estimated production costs of a range of proteins are summarized in Table 1. The allowable limits are set by the product value and hence by the market demand. Each step in a recovery process must be optimized for yield, and the number of steps (n) must be cut to a minimum. For example, with a 90% recovery at each individual stage, the final yield after n steps is $90 \times 10^{[2-n]}$, which, for $n = 5$, is 60%. Present scales of demand for various classes of protein products are shown in Table 2; Table 3 provides an indication of the trade values per kg of protein isolate.

Table 1
Estimate of Production Costs in 1980

	$/g
Interferons from leucocytes	25,000,000
Blood clotting factor VIII from plasma	200,000
Urokinase from cell culture	75,000
Synthetic calcitonin and β-endorphin	10,000
Human growth hormone from donated glands	1,500
Synthetic peptides to 20 amino acids	<1,000
Synthetic ACTH fragments (24 amino acids)	500
Asparaginase from *E. coli* fermentation	100
Insulin from slaughterhouse glands	<50
Gentamicin antibiotic from fermentation	<10
Human serum albumin from blood	1.5

Market information on protein materials is hard to come by and, in any case, soon becomes out of date. The most comprehensive data refer to the production and marketing of enzymes. Tables 4–7 summarize available information. One can conclude that there is a growing demand for carbohydrate-modifying enzymes. Most of the commercially available enzymes belong to the class of hydrolases, although a few isomerases and oxidases can also be produced economically. Once it becomes possible to isolate and stabilize synthases, the demand for such products will probably increase dramat-

Table 2
Scales of Requirements for Purified Proteins

		Annual demand
Fine chemicals	Medical/clinical research enzymes	g–kg
Intermediate	Industrial enzymes, plasma proteins, pharmaceuticals/cosmetics	kg–tons
Large	Industrial enzymes, food specialities	>10 tons
Commodities	Food/feed (nutritional)	>100,000 tons

Table 3
Relative Values of Proteins in 1982

	US $/ton
Antithrombin III	750 million
Cohn I (human fibrinogen)	45 million
Cohn II (human γ-globulins)	6 million
Cohn III (human β-globulins)	6 million
Cohn IV (human α-globulins)	6 million
Cohn V (human albumin)	4.5 million
Soya lectin	6,000 million
Soya trypsin inhibitor	9 million
Soya glycinin	4.5 million
Soya crude 11S (ultrafiltrate)	45 thousand
Soya isoelectric isolate	3 thousand
European EEC cheese	3,500
EEC skim milk powder	1,260
EEC whey powder	360
EEC whey concentrate	2,400
EEC casein powder	3,500

ically. There is also a remote possibility that enzymes (or enzyme-like synthetic catalysts) may eventually supersede heterogeneous catalysts in chemical manufacturing processes. Their sensitivity to temperature, pH, and salt concentrations makes this impractical at the present time, but the exploitation of their catalytic efficiency and high degree of specificity would make such a development extremely attractive.

At the present time it is too early to tell what impact will result from the commercial exploitation of so-called protein engineering;

Table 4
The Japanese Market for Enzymes in 1972

Enzyme	Industrial application	Quality, tons	Value, million yen
Bacterial α-amylase	Textiles (desizing)	10,000	600
	Starch (liquefying)	850	280
	Alcohol (liquefying)	50	15
Fungal glucoamylase	Starch (liquefying)	400	500
	Sake (liquefying)	16	128
Malt	Starch	350	35
Malt β-amylase	Starch	170	40
Streptomyces gluc. isomerase	High fructose syrup	200	220
Bacterial and fungal (amylase + protease)	Digestive (pharmaceutical)	—	520
	Cleaning	80	70
Bacterial proteases	Detergent	30	30
	Soy	40	40
	Leather	900	90
Egg white lysozyme		5	500
Fungal glucose oxidase		0.15	100
Fungal catalase		0.30	140
Fungal cellulase		20	80
Lipases		3	36
Other		—	100

that is, the chemical modification of a primary peptide chain by genetic means. Presumably mutant proteins so produced should have some desirable attribute, e.g., thermal stability, pH tolerance, substrate specificity, and so on. Although mutants can certainly be engineered almost at will, our present knowledge of the factors that determine the desired attribute is so sketchy and qualitative that purposeful protein engineering must remain a development for the future. One need only be reminded of the fact that very minor changes in the amino acid composition (or sequence) can produce pathological conditions (e.g., sickle cell hemoglobin), whereas at the other extreme, fairly major chemical changes in the peptide chain do not appear to affect the biological function of the protein (e.g., cytochrome *c*). The noncovalent forces that determine protein conformation and function are so small and the balance between stabilizing and destabilizing effects is so delicate that it will require a much higher level of understanding before predictive methods can be applied to protein engineering.

Table 5
Microbial Enzymes Produced in Japan in 1977

	Tons	Million yen
beta-Amylase (*Asp. oryzae, Asp. niger, B. subtilis, B. amyloliquifaciens*)	11,100	910
Glucoamylase (*Asp. niger, Rhiz. neveus, Rhiz. delemar, Endomycopsis sp.*)	400	620
Protease (*Asp. oryzae, Asp. niger, Asp. saitoi, Mucor. pussilus, B. subtilis, Strept. griseus*)	1,050	750
Lipase (*Asp. niger, Rhiz. sp., C. cylindracea*)	3	36
Cellulase (*Asp. niger, T. viride*)	45	170
Pectinase (*Coniosthyrium diplodiella*)	13	13
Glucose oxidase (*Pen. amagasakiense*)	0.5	100
Others	140	400
Invertase (*S. cerevisiae*)		
Naringinase (*Asp. niger*)		
Melibiase (*Mortierella vinacea*)		
Aspartase (*E. coli*)		
Aminoacylase (*Asp. oryzae*)		
Autocyanase (*Asp. niger*)		
Glucose isomerse (*Streptomyces sp.*)		
Catalase (*Asp. niger*)		

In this book an attempt is made to provide a unified approach to the problems of protein characterization, stability, analysis, isolation, and processing, while still bearing in mind economic considerations of scale and values. Differences are drawn between in vivo and in vitro methods of characterization and between analytical procedures that are appropriate to the quantities of material to be handled. Certain unifying physical and chemical principles exist, whatever the scale of the operation, and these principles are summarized in the opening chapters.

Finally, a word of warning: historically, much of our knowledge regarding protein structure, stability, and performance has been derived from in vitro studies on isolated proteins acting on isolated substrates. An increasing number of such proteins is becoming available in bottled form, obtainable from laboratory suppliers, a prime example being lysozyme, on which much of the pioneering work was done. Whatever useful information can be obtained about lysozyme from model studies in aqueous solutions of the enzyme, it must be remembered that, in vivo, a lysozyme molecule is most unlikely ever to "see" another lysozyme molecule. It is questionable, therefore, whether a study of protein–protein interactions (such as give rise to crystals) of globular molecules has any relevance

Introduction

Table 6
Growth of Enzyme Markets (US and Europe)[a]

	Million US $		
	1971	1975	1980
Amylases			
Amyloglucosidase	1.70	2.10	2.60
Fungal amylase	2.01	2.40	3.10
Bacterial (α-amylase)	4.60	8.00	8.50
Total	8.31	12.50	14.20
Proteases			
Fungal	0.76	0.82	0.91
Bacterial	0.83	1.00	1.30
Detergent	1.00	2.00	4.70
Pancreatin	0.80	0.75	0.75
Rennins:			
Animal + microbial	7.50	8.10	8.90
Pepsin	2.75	2.80	2.85
Papain	3.58	3.87	4.27
Total	17.22	19.34	23.68
Other			
Glucose oxidase	0.35	0.60	0.90
Cellulase	0.10	0.15	0.20
Invertase	0.10	0.10	0.10
Glucose isomerase	1.00	3.00	6.00
Pectinases	1.56	1.75	2.10
Medical diagnostics, research, and so on	5.50	7.30	9.80
Total US markets			
Amylases	8.31	12.50	14.20
Proteases	18.34	20.77	24.51
Others	8.51	12.15	19.00
Total	35.26	45.42	57.71
World-wide markets			
Pectinases	2.81	3.12	3.66
Cellulase	1.80		
Pancreatin	12.80		
Bromelain	0.8	2.4	
Papain	6.0	6.49	7.17
Fungal protease	1.37	1.60	1.95
Pepsin	4.30		
Bacterial protease	2.83	5.0	10.0
Glucose oxidase	0.62	0.908	1.464
Glucose isomerase	1.00	4.00	8.00
Rennin	23.0	27.53	35.12

[a]From Katchalski-Katzir and Freeman, *TIBS* **7**, 427–431 (1982).

Table 7
The Most Important Industrial Enzyme Preparations

Source	Name	1900	1950	1976	Current annual production, tons of enzyme protein
Animal	Rennet	x			2
	Trypsin		x		15
	Pepsin		x		5
Plant	Malt amylase	x			10,000
	Papain		x		100
Microbial	Koji	x			?
	Bacillus protease		x		500
	Amyloglucosidase			x	300
	Bacillus amylase		x		300
	Glucose isomerase			x	50
	Microbial rennet			x	10
	Fungal amylase	x			10
	Pectinase		x		10
	Fungal protease	x			<10

to physiological situations. Such interactions are of course of great importance in isolation and purification technology. Another graphic example of such differences is the much-quoted figure for the free energy of ATP hydrolysis. In the chemistry laboratory the experiment is carried out in dilute solution (infinite dilution) under standard conditions of pH and ionic strength. The in vivo reaction takes place in membranes, in which the concentration of water is likely to be extremely low, so that the reaction kinetics and energetics can bear little resemblance to those that govern the process in dilute aqueous solution.

We emphasize, therefore, that despite some common physicochemical principles, clear distinctions must be drawn between physiological and technological function and performance and between results obtained from in vivo and in vitro characterization and analysis.

Chapter 1

Description and Classification of Proteins

Felix Franks

1. Chemical Constitution

Proteins are linear condensation polymers of amino acids, formed by the reaction:

$$H_2NCHR_1COOH + H_2NCHR_2COOH \rightarrow H_2NCHR_1CONHCHR_2COOH + H_2O$$

The amino acids that occur in proteins (some 20) differ only in the nature of the residue R. The residues are linked by *identical* bonds (peptide bonds)

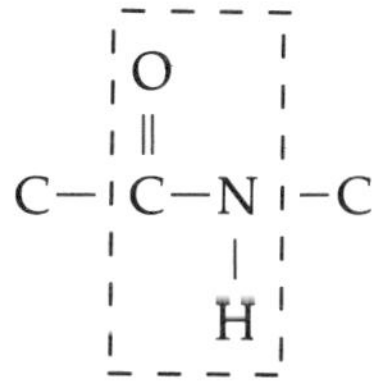

$$C-\underset{\underset{H}{|}}{\overset{\overset{O}{\|}}{C}}-N-C$$

the geometry of which is unique (*see* chapter 2). In this way proteins resemble synthetic homopolymers with identical backbone structures, but variable side chains, e.g., vinyl, acrylic, polyether derivatives.

A comparison of proteins with other biopolymers is shown below:

	Proteins	Carbohydrates	Nucleotides
Monomer variety	>20+	Usually 1–3	4
Linkage	Identical	Variable	Identical
Chain branching	Never	Common	Never

Proteins derive their chemical variety from the amino acid composition and distribution in the peptide chain. Carbohydrates derive their variety mainly from different types of linkages between sugar residues in the polymer chain (namely, cellulose and starch, which are both homopolymers of glucose, but with different linkages between glucose residues).

1.1. Amino Acid Classification

This type of classification is somewhat arbitrary, but in practice three groups of amino acids are distinguished, as shown in Fig. 1. Hydrophobic amino acids contain substantial apolar side chains, e.g., three or four carbon alkyl or benzyl radicals. Acidic/basic amino acids contain side chains capable of ionization, i.e., terminal COOH or NH_2 groups. Polar amino acids contain $-OH$ or $>NH$ groups capable of participating in hydrogen bonding.

1.1.1. Ambiguities

The following amino acids do not fall easily into the above classification:

Glycine (R = H) and alanine (R = Me),
Proline and hydroxyproline are imino acids.
Cysteine, which contains the $-SH$ group and is responsible for providing the only chemical cross links ($-S-S-$) found in proteins.

1.1.2. Rare Amino Acids That Occur in Proteins

5-Hydroxy lysine and some other lysine and glutamate derivatives occur, as does desmosine (in elastin), which is a condensation compound in which four lysine molecules form a pyridine ring.

1.1.3. Nonprotein Amino Acids

Many amino acids occur naturally, e.g., in plants and fungi and in the free state, but not in proteins, although they are often involved as intermediates in amino acids synthesis or metabolism (see Table 1).

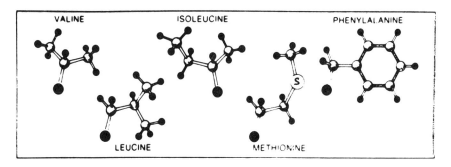
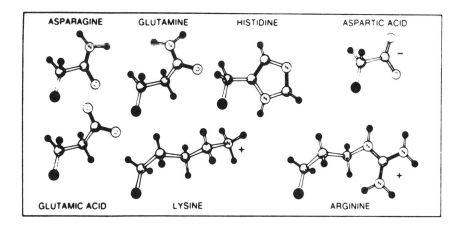
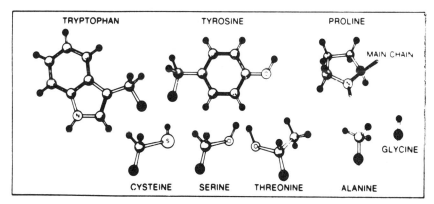

Fig. 1. Three groups of amino acids reproduced with permission from A.C.T. North, in *Characterization of Protein Conformation and Function* (1979) (F. Franks, ed.) Symposium Press, London.

Table 1
Nonprotein Amino Acids

Amino acid	Function
α-Aminobutyric acid	Accumulates under stress conditions
β-Alanine	Coenzyme A synthesis
β-Cyanoalanine	Neurotoxin
Ornithine $H_2N(CH_2)_3CH(NH_2)COOH$	Arginine biosynthesis
Homoserine $HO(CH_2)_2CH(NH_2)COOH$	Threonine, isoleucine, and methionine biosynthesis
O-Acetylserine	Carbon source for the production of sulfur amino acids
Albizzine $NH_2CONHCH_2CH(NH_2)COOH$	Nitrogen storage in legumes
Mimosine	Plant toxin

$$O=\underset{\underset{OH}{|}}{\bigcirc}NCH_2CH(NH_2)COOH$$

1.2. Conjugated Proteins

Many (most?) proteins are linked, often covalently, to other chemical groupings, organic or inorganic, known as prosthetic groups (see Table 2).

1.2.1. Glycoproteins

Most proteins in vivo are glycoproteins, in the sense that they contain small amounts (3% in lysozyme) of carbohydrate that are often removed inadvertently during the isolation and purification stages.

Under in vivo conditions the sugar residues are added to the wholly or partly preformed peptide by the action of enzymes (transferases). This leads to a degree of irregularity in the resulting glycoprotein.

Two types of peptide–sugar links can be distinguished (Table 3): N-linked sugars, via asparagine, and O-linked sugars, via serine, threonine, hydroxylysine, or hydroxyproline.

Description and Classification of Proteins

Table 2
Compositions of Some Conjugated Proteins

Class	Prosthetic group	Approximate percentage of weight
Nucleoprotein systems		
Ribosomes	RNA	50–60
Tobacco mosaic virus	RNA	5
Lipoproteins		
Plasma β_1-lipoproteins	Phospholipid, cholesterol, neutral lipid	79
Glycoproteins		
γ-Globulin	Hexosamine, galactose, mannose, sialic acid	2
Plasma orosomucoid	Galactose, mannose, N-acetylgalactosamine, N-acetylneuraminic acid	40
Seed lectin (potato)	β-Arabinose, galactose,	50
(soya bean)	N-acetylgalactose	6
Fish antifreeze protein	N-acetylgalactosamine, galactose	50
Immunoglobulins	N-acetylglucosamine, mannose, fucose, and so on	15
Collagen (tendon)	Galactose, glucose	0.5
Phosphoproteins		
Casein (milk)	Phosphate esterified to serine residues	4
Hemoproteins		
Hemoglobin	Iron protoporphyrin	4
Cytochrome c	Iron protoporphyrin	4
Flavoproteins		
Succinate dehydrogenase	Flavin nucleotide	2
D-Amino acid oxidase	Flavin nucleotide	2
Metalloproteins		
Ferritin	$Fe(OH)_3$	23
Tyrosine oxidase	Cu	0.2
Alcohol dehydrogenase	Zn	0.3
Pyruvate carboxylase	Mn	0.02
Nitrate reductase	Mo	0.2
Glutathione	Se	3

Table 3
Carbohydrate–Peptide Linkages in Glycoproteins

Linkage	Anomeric type	Occurrence	Stability to acid	Stability to alkali
N-Glycosydic				
N-Acetylglucosaminyl-asparagine	β	Widespread	+	+
O-Glycosidic				
N-Acetylgalactosaminyl-serine or threonine	α	Mucins, blood group substances, fetuin, antifreeze glycoproteins	+	−
Xylosyl-serine	β	Proteoglycans	+	−
Galactosyl-serine	α	Plant cell walls, earthworm cuticle	+	−
Galactosyl-5-hydroxylysine	β	Collagens	++	++
L-Arabinofuranosyl-O-hydroxyproline	β	Plant cell walls	−	

1.3. Amino Acid Profile

Table 4 summarizes the amino acid profiles of some representative proteins. Apart from the classification in Fig. 1, amino acids fall into two groups: common, e.g., alanine, glycine, and leucine, and rare, e.g., histidine and tryptophan. Few patterns emerge from a study of amino acid profiles:

> Globular proteins have a fairly uniform amino acid profile with a higher proportion of acidic/basic residues than fibrous proteins.
> Fibrous proteins have high proportions of relatively few amino acids, e.g., glycine and proline.
> Histones are very basic and pepsin is exceptionally acidic.
> Most proteins have few cysteine residues, most of them being involved in providing disulfide links.

Amino acid analytical methods are dealt with in chapter 10.

2. Protein Size

Proteins vary in size from approximately 20 to >500,000 amino acid residues. The limiting number of residues per peptide chain appears to be about 1000. Any increase in size beyond this is by means of mainly noncovalent aggregation of individual peptide chains to form homo- or heterooligomeric structures. In some cases, $-S-S-$ linkages join the individual peptide chains (known as subunits), making up such a supermolecular aggregate (e.g., immunoglobulins).

Table 5 summarizes size and degree of aggregation data for some representative proteins.

3. Classification by Function

Distinction must be made between in vivo biological function and in vitro technological function (often referred to as functionality in the technical jargon of the day).

3.1. Versatility and Limitations

Even a simple microorganism, *Escherichia coli*, contains more than 1000 different proteins. It is believed that the total number of proteins is of the order of 10^{10} to 10^{12}. This is only a small frac-

Table 4
Amino Acid Profiles of Representative Classes of Proteins

	Lysozyme	Chymotrypsinogen	Carbonic Anhydrase	Alcohol dehydrogenase	Pepsin	Cytochrome c	β-Lactoglobulin	Serum albumin	Ferredoxin	Myoglobin (human)	TMV protein	α-Casein	Soya bean trypsin inhibitor	Histone H2A	Myosin	Silk fibroin	Collagen	Elastin	Wool keratin
Ala	12	22	19	28	18	6	30	3	9	12	14	9	8	18	78	334	107	58	46
Val	6	23	17	39	18	3	18	45	7	7	14	11	14	7	42	31	29	118	40
Leu	8	19	20	25	21	6	44	58	8	17	12	17	15	16	79	7	28	56	86
Ile	6	10	10	24	28	6	17	9	4	8	9	11	14	6	42	8	15	26	
Pro	2	9	17	20	15	4	17	31	4	5	8	17	10	5	22	6	131	136	83
Phe	3	6	11	18	14	4	9	33	2	7	8	8	9	1	27	20	15	29	22
Trp	6	8	6	2	6	1	3	1	1	1	3	2	2	0	4	0	0		9
Met	2	2	2	9	5	2	8	6	0	3	0	5	2	0	22		5	0	5
Gly	12	23	16	38	38	12	7	15	6	15	6	9	16	11	39	581	363	376	87
Asx	21	23	31	25	44	8	32	46	13	11	18	15	26	7	85	21	47	4	54
Glx	5	15	22	29	27	12	48	80	13	21	16	39	18	10	155	15	77	22	96
Lys	6	14	18	30	1	19	29	58	4	20	2	14	10	13	85	5	31	3	19
Arg	11	4	7	12	2	2	6	25	1	2	11	6	9	12	41	6	49	6	60
Ser	10	28	13	26	44	0	14	22	7	7	16	16	11	4	41	154	32	9	95
Thr	7	23	14	24	28	10	16	27	8	4	16	5	7	55	41	13	19	10	54
Cys	8	10	1		4	2	6	20	5		1	0	4	0	4		0		

Description and Classification of Proteins

Tyr	3	4	8	4	18	4	8	18	4	2	4	10	4	3	18	71	5	8	36
His	1	2	11	7	1	3	4	16	1	9	0	5	2	4	15	2	5	0	7
Total	129	245	260	274	513	104	316	513	97	153	158	199	181	129	840	1274	598	861	799
Percent non polar	46	49	45	44	50	44	50	55	36	49	46	40	50	50	29	78	72	93	48
No. positively charged	17	20	36	49	4	24	39	99	6	31	13	32	20	36	141	13	85	9	86
No. negatively charged	26	33	27	38	71	20	80	126	20	32	34	25	46	8	240	36	124	26	150
% positively charged	13	8	14	13	1	23	13	20	6	20	8	16	1	28	18	1	10	2	11
% negatively charged	20	13	11	10	21	19	25	24	20	21	21	13	25	6	29	3	13	3	19

Table 5
Molecular Weights and Subunit Structures of Proteins

	Molecular weight	No. of residues	No. of subunits
Insulin (bovine)	5,733	51	2
Ribonuclease (bovine pancreas)	12,640	124	1
Lysozyme (egg white)	13,930	129	1
Myoglobin (horse heart)	16,890	153	1
Chymotrypsin (bovine pancreas)	22,600	241	3
Hemoglobin (human)	64,500	574	4
Serum albumin (human)	68,500	~550	1
Hexokinase (yeast)	96,000	~800	4
Hemerythrin	107,000	~890	8
Tryptophan synthetase (*E. coli*)	117,000	~975	4
Soyabean lectin	120,000	1,000	4
γ-Globulin (horse)	149,900	~1,250	4
Cholesterol esterase	400,000	~3,300	6
Glycogen phosphorylase (rabbit muscle)	495,000	~4,100	4
Myosin	620,000	~5,160	3
Glutamate dehydrogenase (bovine liver)	1,000,000	~8,300	~40
Fatty acid synthetase (yeast)	2,300,000	~20,000	~21
Tobacco mosaic virus	40,000,000	~336,500	2,130

tion of the total possible number of polypeptides, as illustrated by two examples: For a peptide containing one each of 20 *different* amino acid residues, there are 20! = 2×10^{18} possible amino acid sequences. For a protein of molecular weight 34,000 containing equal numbers of 12 different amino acid residues, there are 10^{300} sequence isomers.

3.2. Biological Function

Table 6 provides a small selection of representative functions. Of the groups included in Table 6, the enzymes have received the most intensive study. They are the in vivo catalysts of biochemistry and act by perturbing the stability of the substrate molecule(s), weakening covalent bonds.

Storage proteins are chemically very stable, with a low turnover rate. They are usually concentrated in so-called protein bodies under

Description and Classification of Proteins

Table 6
Classification of Proteins by Biological Function

Type and example	Occurrence or function
Enzymes	
Ribonuclease	Hydrolyzes RNA
Cytochrome c	Transfers electrons
Trypsin	Hydrolyzes some peptides
Storage proteins	
Ovalbumin	Egg-white protein
Casein	Milk protein
Ferritin	Iron storage in spleen
Gliadin	Seed protein of wheat
Zein	Seed protein of corn
Transport proteins	
Hemoglobin	Transports O_2 in blood of vertebrates
Hemocyanin	Transports O_2 in blood of some invertebrates
Myoglobin	Transports O_2 in muscle
Serum albumin	Transports fatty acids in blood
β-Lipoprotein	Transports lipids in blood
Iron-binding globulin	Transports iron in blood
Ceruloplasmin	Transport copper in blood
Transducer proteins	
Myosin	Stationary filaments in myofibril
Actin	Moving filaments in myofibril
Dynein	Cilia and flagella
Opsin	Photoreceptor
Defense proteins	
Antibodies	Form complexes with foreign proteins
Complement	Complexes with some antigen–antibody systems
Fibrinogen	Precursor of fibrin in blood clotting
Thrombin	Component of clotting mechanism
Interferon	Antiviral agent
Lectins	Seeds; agglutinating agents
Toxins	
Clostridium botulinum toxin	Causes bacterial food poisoning
Diphtheria toxin	Bacterial toxin
Snake venoms	Enzymes that hydrolyze phosphoglycerides.
Ricin	Toxic protein of castor bean

(continued on next page)

Table 6 (*continued*)
Classification of Proteins by Biological Function

Type and example	Occurrence or function
Hormones	
Insulin	Regulates glucose metabolism
Adrenocorticotrophic hormone (ACTH)	Regulates corticosteroid synthesis
Growth hormone	Stimulates growth of bones
Structure proteins	
Viral-coat proteins	Sheath around chromosome
Glycoproteins	Cell coats and walls
α-Keratin	Skin, feathers, nails, hoofs
Sclerotin	Exoskeletons of insects
Fibroin	Silk of cocoons, spider webs
Collagen	Fibrous connective tissues (tendons, bone, cartilage)
Elastin	Elastic connective tissue (ligaments)
Mucoproteins	Mucous secretions, synovial fluid
Chromosomal proteins	
Histones	Eukaryotes
Protamines	Fish sperm, some mammals
Nonhistone proteins	For example, enzymes for DNA replication.
HMG proteins (structural?)	Thymus, kidney, liver
Ubiquitin	Removal of damaged proteins (?)

conditions of low moisture. Transport proteins can be localized, as in membranes, where they regulate the flow of material in and out of the cell, e.g., ions, molecules, or electrons, or they can themselves be transported, supplying nutrients (e.g., oxygen) to distant parts of the organism and removing waste products.

Transducer proteins interconvert different types of energy. Thus, actomyosin and flagellin convert chemical into mechanical energy and opsin converts electromagnetic radiation into electrical impulses as part of the photoreceptor system of the eye. Structural proteins are load-bearing systems, usually characterized by a high tensile strength.

Defense proteins recognize and inactivate foreign microbes and materials (antigens), rendering them harmless. They are often quite resistant to proteolytic attack. Toxins act by chemical attack or by blocking mechanisms. They are usually small and highly mobile.

There is no sharp dividing line in terms of size between oligopeptides and proteins. Even dipeptides can have pronounced physiological activity, as illustrated by the artificial sweetener aspartame. It has recently been reported that a dipeptide that is removed from the blood during hemodialysis is implicated in promoting schizophrenia.

3.2.1. Enzyme Types

Common bioreactions that are catalyzed by enzymes are summarized in Table 7. Of these, hydrolysis is the simplest, involving only the cleavage of a covalent bond by water. Hydrolases are the simplest enzymes and the easiest to immobilize. They usually contain only one peptide chain (e.g., lysozyme, ribonuclease) of a relatively low molecular weight.

Complex chemical reactions, especially those involving transfer of chemical groups, require multifunctional enzymes, usually oligomers composed of several different subunits, e.g., amino acid synthetases or transferases.

Table 7
Classification of Enzymes and Subclassification of Proteolytic Enzymes

Class	Group affected	Example
Oxidoreductases	CHOH	Lactate dehydrogenase
	CH—CH	Acyl CoA dehydrogenase
	CH—NH$_2$	Amino acid oxidase
Transferases	Methyl	Guanidinoacetate methyltransferase
	Hydroxymethyl	Serine hydroxymethyltransferase
	Acyl	Choline acetyltransferase
	Amino	Transaminase
Hydrolases	Carboxylic ester	Esterase, lipase
	Phosphoric monoester	Phosphatase
	Phosphoric diester	Ribonuclease
	Glycosidic bond	Amylase
	Peptide	See below
Lyases	C—C	Aldolase
	C—O	Fumarate hydratase
Isomerases	Aldose-ketose	Phosphoglucoisomerase
Ligases	Forms C—O bonds	Amino acid activating enzyme
	Forms C—N bonds	Glutamine synthetase
	Forms C—C bonds	Acetyl CoA carboxylase

(continued on next page)

Table 7 (continued)
Proteolytic Enzymes

Source	Enzyme	Specificity
Mammalian	Trypsin	R_1 = arg or lys
	Chymotrypsin	R_1 = phe or tyr (preferred)
	Pepsin	R_2 = phe, tyr, or trp
	Prolidase	R_1 or R_2 = glycine (preferred)
	Carboxypeptidase A	R_2 = Any amino acid except arg, lys, and pro
	Carboxypeptidase B	R_2 = lys or arg
	Iminopeptidase	Liberates proline
	Leucine amino peptidase	R_1 = Any amino acid except pro and gly
	Aminopeptidase M	Nonspecific[b]
Plant	Papain	Nonspecific[b]
	Ficin	Nonspecific[b]
	Carboxypeptidase C	Unknown
Bacterial	Subtilisin[a]	Nonspecific[b]
	Nagarase[a]	Nonspecific[b]
	Pronase[a]	Nonspecific[b]

[a]Commercial commonly used names.
[b]Most proteolytic enzymes are inactive toward the imide link of proline and attack glycine peptides very slowly.

3.3. Technological Function

In the case of technological function, a protein is regarded as a chemical material with desirable technological attributes, to be exploited in fabricated products, such as foods, textiles, chemicals, or pharmaceutical preparations. Some of the specific characteristics are summarized in Table 8 (nutritional attributes are not included).

The exploitation of the technological attributes (except where the in vivo properties are to be exploited, as in enzymes) requires the destruction and subsequent modification of the natural protein structure (and function). Processing stages can include some or all of the following: extraction, isolation, fractionation and purification, chemical modification, and reassembly into the new structure. Usually the end product will still be susceptible to microbial attack and should be suitably protected.

Description and Classification of Proteins

Table 8
General Classes of Attributes of Proteins Important in Food Applications

General property	Special attribute	Examples
Organoleptic	Color, flavor, odor, texture, mouth-feel, smoothness, grittiness, turbidity, and so on	Beverages, creams, pastes
Hydration	Solubility, dispersibility, wettability, water sorption, swelling, thickening, gelling, water-holding capacity, syneresis, viscosity, dough formation, and so on	Dough, meat products, desserts, frozen/thawed products
Surface	Emulsification, foaming, aeration, whipping, protein/lipid film formation, lipid binding, stabilization, and so on	Processed meat, coffee whitener, whipped desserts
Structural, textural, rheological	Elasticity, grittiness, cohesion, chewiness, viscosity, adhesion, network cross-binding, aggregation, stickiness, gelation, dough formation, texturizability, fiber formation, extrudability, elasticity, and so on	Beverages, toppings, textured and frozen products
Other	Compatibility with additives, enzymatic activity/inertness, modification properties	

As will be seen, the physical properties of proteins are extremely sensitive to environmental factors such as the solvent, pH, ionic strength, concentration, temperature, pressure, and shear stresses. The processor should allow for the labile nature of the macromolecules if he is to optimize the desirable attributes of the end product. Normal chemical engineering practice must thus be modified, and chapters 16–19 deal with the processing and quality assessment of proteinaceous materials in the light of their physical properties, as described in chapters 2–5.

Chapter 2

Internal Structure and Organization

Relationship to Function

Felix Franks

1. Structural Hierarchy of Proteins

There are several levels of protein structure and different types of interactions that maintain such structures. Thus, the primary structure is the amino acid residue sequence in the peptide chain. The secondary structure is the intramolecular rearrangement wherein domains of high symmetry are produced in sections of the peptide chain. The tertiary structure is the intramolecular arrangement of the above domains with respect to each other. Quaternary structures are formed by the aggregation of individual peptide chains (subunits) to form ordered supramolecular structures.

1.1. Experimental Approaches to the Characterization of Protein Structure

Experimental techniques can be divided into chemical and physical methods.

1.1.1. Chemical Methods

Chemical methods include the following:
(1) Amino acid sequence, location of −S−S− bonds, and quantitative carboxyl and amino terminal analysis define the primary structure and the number of peptide chains.
(2) Chemical modification with monofunctional and polyfunctional reagents identifies "reactive" and "masked" residues in the peptide chain, which themselves are determined by the microenvironment (secondary and/or tertiary structure).
(3) Susceptibility to proteolytic enzymes and other chain cleaving reagents (e.g., cyanogen bromide). Native (functional) proteins are resistant or almost resistant to proteolytic attack, whereas denatured (inactive) proteins tend to be rapidly hydrolyzed.

1.1.2. Physical Methods

Hydrodynamic properties monitor the size, shape, and charge of the macromolecule in solution. These are discussed further in chapter 3.

Scattering and absorption of radiation:
(i) Small angle scattering by macromolecules (X-rays, neutrons, visible light) is a function of size and shape.
(ii) Absorption (visible and UV) and fluorescence measurements provide information about the microenvironment of chromophores (usually tyr, trp, and phe).
(iii) Optical rotatory dispersion (ORD) and circular dichroism (CD) properties are characteristic of particular ordered (secondary structure) domains.
(iv) Nuclear magnetic resonance (NMR) provides both structural (short-range) and dynamic information.
(v) X-ray and neutron diffraction of *crystalline* proteins provides the complete three-dimensional structure in terms of the atomic coordinates. It requires previous knowledge of the amino acid sequence. X-ray diffraction relies on electron density and cannot, therefore, "see" hydrogen atoms, whereas neutron techniques can locate hydrogen (preferably deuterium) atoms.

2. Primary Structure

2.1. General Principles

Proteins are assembled on the ribosome. The primary amino acid sequence therefore contains *all* the basic information that is

Structure and Function

required for the subsequent conformational rearrangements leading to structures that convey biological activity to the polypeptide chain. Because the amino acid sequence is the most important property of the protein, in principle it should be possible to calculate a protein's higher-order structures from the primary sequence. Despite many attempts, this has not yet been achieved. In a few cases the primary structure can be related to function, e.g., casein, which behaves as a giant surfactant molecule, with phosphoserine residues (charged) at one end of the chain and a long hydrophobic sequence at the other. This arrangement promotes the formation of micelles.

2.2. Nomenclature

The following abbreviations for the amino acids are commonly used. The one-letter code is gaining in popularity.

Amino acid	Conventional code	One-letter code
Valine	val	V
Leucine	leu	L
Isoleucine	ile	I
Methionine	met	M
Phenylalanine	phe	F
Asparagine	asn	N
Glutamine	gln	Q
Histidine	his	H
Lysine	lys	K
Arginine	arg	R
Aspartic acid	asp	D
Glutamic acid	glu	E
Tryptophan	try or trp	W
Tyrosine	tyr	O
Proline	pro	P
Hydroxyproline	hyp or hypro	
Cysteine	cys	C
Serine	ser	S
Threonine	thr	T
Glycine	gly	G
Alanine	ala	A
Asparagine or aspartic acid	asx	B
Glutamine or glutamic acid	glx	Z

Two typical examples of primary structures of commonly studied simple proteins are described below.

2.3. Ribonuclease

The amino acid sequence of ribonuclease is shown in Fig. 1. This important protein consists of a single peptide chain containing 124 residues, including eight cysteine residues. Chemical analysis indicates disulfide links between the following pairs of cys residues: 4-5, 1-6, 3-8, and 2-7. The corresponding positions in the peptide chain are: 65-72, 26-84, 58-109, and 40-95.

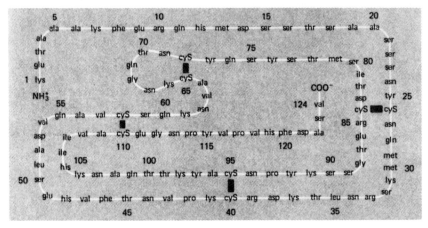

Fig. 1. Primary structure of bovine pancreatic ribonuclease, showing the positions of the four S—S bridges [from F. Wold, in *Macromolecules: Structure and Function* (1971) Prentice-Hall, New Jersey].

2.4. Insulin

Insulin in the active state is composed of two peptide chains that are linked by disulfide bonds in two positions, as shown in Fig. 2. The A chain of 21 residues also contains an intrachain disulfide link between residues 6 and 11. The B chain is composed of 30 residues (*see* also proinsulin, section 5).

2.5. Homology

Some proteins exhibit species differences in the amino acid composition, although *not* in the function of the particular protein. Table 1 summarizes amino acid replacements in positions 8, 9, and 10 of the A chain of mammalian insulins. Vertebrate cytochrome *c* contains 100 residues, of which only 36 are completely conserved in

Structure and Function

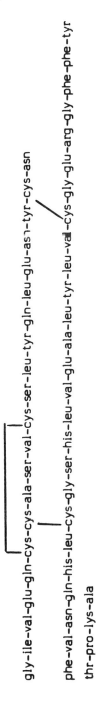

Fig. 2. Primary structure of human insulin.

all species. These include a complete sequence of 11 residues in positions 70-80 (—NPKKOIPGTKM—). Table 2 shows amino acid differences between species. Despite the large degree of chemical variability, the protein fulfills the same function in all species. Table 3 gives details of the degree of conservation of the rarer amino acids

Table 1
Amino Acid Substitutions in A Chains of Insulins from Different Species

	Position		
	8	9	10
Beef	ala	ser	val
Pig	thr	ser	ile
Sheep	ala	gly	val
Horse	thr	gly	ile
Sperm whale	thr	ser	ile
Man	thr	ser	ile
Dog	thr	ser	ile
Rabbit	thr	ser	ile

Table 2
Amino Acid Differences in Cytochrome c Between Different Species

	Number of differences
Man–monkey	1
Man–horse	12
Man–dog	10
Pig–cow–sheep	0
Horse–cow	3
Mammals–chicken	10–15
Mammals–tuna	17–21
Vertebrates–yeast	43–48

Table 3
Conservation of Rarer Amino Acids in Cytochrome c

Amino acid	Position
phe	10, 82
cys	14, 17
his	18
lys	27, 72, 73, 79, 87
pro	30, 71, 76
tyr	48, 67, 74
trp	59

that are thought to be important to the structure and function of cytochrome c.

In general, if a protein plays a central role in many different forms of life (e.g., glycolytic enzymes), then complete homology is observed. Thus, nucleohistones are 100% conservative.

In some cases single amino acid substitutions can give rise to pathological conditions, e.g., sickle cell hemoglobin. On the other hand, mutant proteins with minor amino acid changes can enable organisms to survive under adverse conditions, e.g., the alcohol dehydrogenase enzymes in halophilic bacteria contain more acidic residues (glu and asp) than do the corresponding mesophiles. This enables the bacteria to survive under conditions of high salt concentrations, as in the Dead Sea. In fact, being obligate halophiles, they cannot exist under normal salt conditions (0.15M).

3. Secondary Structure

Compared to the covalent bond energies involved in the primary structure (>200) kJ/mol), the secondary structure relies on weak, hydrogen-bonding interactions ($\approx$ 10 kJ/mol). It originates from the stereochemistry of the peptide bond, shown in Fig. 3. The six atoms within the shaded area lie in one plane because the C—N bond has some double bond character. Rotation can occur about

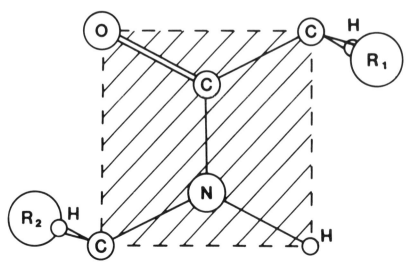

Fig. 3. Stereochemistry of the *trans*-peptide bond, as found in proteins.

the C—C and C—N bonds, and successive residues in the polypeptide chain are characterized by a pair of torsional angles, as shown in Fig. 4. By convention, $\phi = 0$ when the four atoms C—N—C_α—C lie in one plane, and $\psi = 0$ when N—C_α—C—N are coplanar. If successive residues take up similar orientations (same values for ϕ and ψ) to one another, a symmetrical structure results. Such symmetry is most likely in cases in which the amino acid sequence shows repeating patterns.

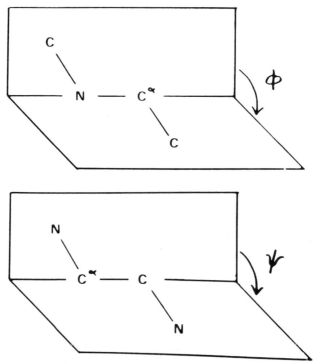

Fig. 4. Torsional angles ϕ and ψ that determine the secondary structures of proteins [from F. Wold, in *Macromolecules: Structure and Function* (1971) Prentice-Hall, New Jersey].

3.1. The α-Helix

Apart from possible interactions between amino acid side chains, the α-helix is the form of lowest free energy. It therefore forms spontaneously, especially in fibrous proteins (keratins in hair, wool, feather, and skin), which usually contain repeating patterns in their primary sequences.

Structure and Function

The helix is described in terms of a symmetry index η_m, where η is the number of residues that form a perfect repeat after m turns. For the α-helix, $18_5 = 3.6$ peptides per turn. The helix is right-handed with a pitch (rise per turn) of 0.54 nm and an average rise per residue of 0.15 nm. The representation of the α-helix in Fig. 5 shows that hydrogen bonds are formed between NH groups and

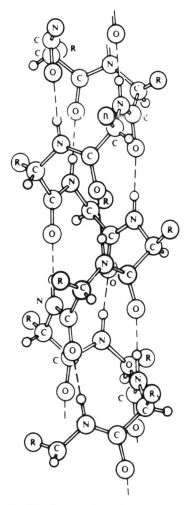

Fig. 5. The α-helix. Hydrogen bonds are indicated by broken lines. Amino acid side chains are designated by R [from B. W. Low and J. T. Edsall, in *Currents in Biochemical Research* (1968) (D. E. Green, ed.) Wiley Interscience, New York].

the CO groups of the *fourth* residue along the chain. The hydrogen bonds are almost parallel to the helix axis; this makes for maximum bond strength and macroscopic elasticity. The intrinsic stabilizing forces can be enhanced or diminished by interactions between amino acid side chains. Proline and hydroxyproline are incapable of inclusion in the helix because they do not contain NH groups. Glycine is also considered to be an α-helix structure breaker.

3.2. The β-Pleated Sheet

Another high-symmetry form is the so-called β-structure, which is adopted by peptide chains that cannot easily form an α-helix. It is an extended, corrugated structure with $\eta_m = 2_1$ in which alternate side chains project above and below the axis of the chain, respectively. The chains are linked laterally by hydrogen bonds to form sheets, as shown in Fig. 6. Two possible configurations exist, depending on whether neighboring chains run parallel or antiparallel. In fibrous proteins, only antiparallel structures are observed. This structural arrangement provides high tensile strength, but renders the fiber brittle.

3.3. Polyproline Helices

Collagen is the most abundant protein in nature. It contains high proportions of gly, pro, and hypro residues and is therefore incapable of forming regular arrays of intramolecular hydrogen bonds. Two symmetrical structures exist for polyproline: polyproline I with $\eta_m = 10_3$, a right-handed helix in which successive imino groups are in a *cis*-configuration; and polyproline II, with $\eta_m = 3_1$, a left-handed helix with *trans*-imino groups. (Although there is a possibility of *cis* and *trans* isomerism across the peptide bond with all pairs of amino acids, the *trans* form is much more stable, so that *cis* isomers have never yet been identified. The energies of the two diproline isomers are very similar, so that they are both capable of existence. In the absence of intramolecular hydrogen bonds, the helices are presumably stabilized by solvation interactions, and the choice of solvent determines the particular isomer that will be formed. The polyproline II chain with the imide groups in the *trans*-planar configuration is shown in Fig. 7.

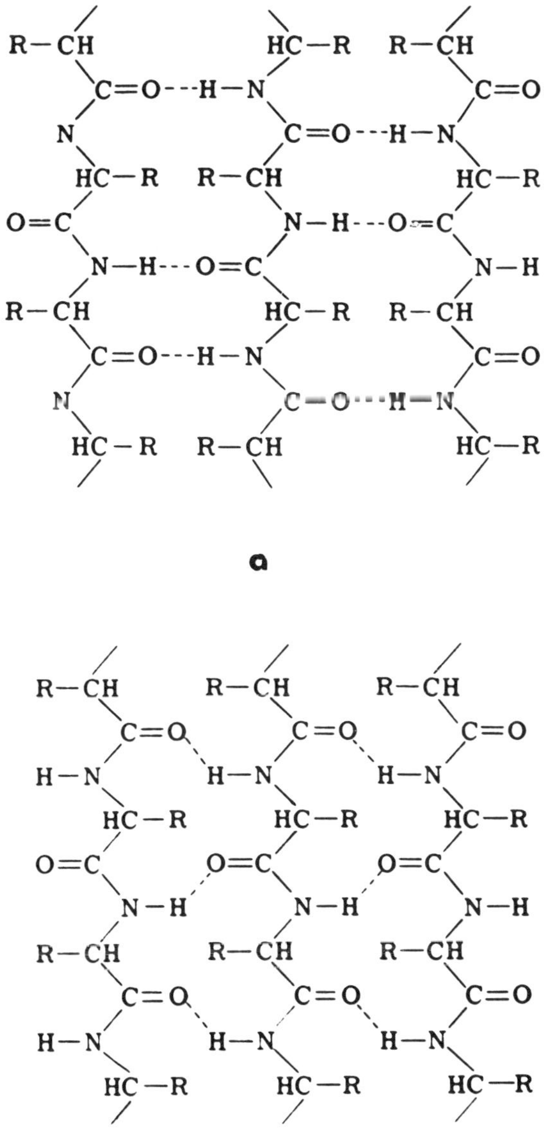

Fig. 6. The β-sheet structures commonly found in proteins: (a) antiparallel, and (b) parallel.

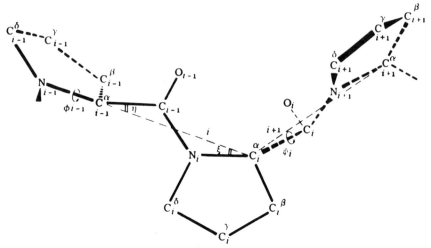

Fig. 7. Stereochemical details of the polyproline-II structure [from A. G. Walton and J. Blackwell, in *Biopolymers* (1973), Academic, Florida].

3.4. Other Regular Structures

Table 4 provides details of some standard polypeptide conformations. In globular proteins α- and β-structures predominate. Domains of high symmetry are linked by amino acid sequences that do not possess long-range order. These can take the form of long flexible loops or a series of so-called reverse turns, as shown in Fig. 8 for a sequence from lysozyme.

Table 4
Regular Secondary Peptide Structure, With Physical Parameters

Structure	Symmetry index, ζ_m	Peptides per turn, ζ/m^a	Residue repeat, nm
α-Helix	18_5	3.60	0.15
β-Sheet (parallel)	2_1	2.0	0.325
β-Sheet (antiparallel)	2_1	2.0	0.35
ω-Helix	4_1	−4.0	0.1325
Polyglycine II	3_1	±3.0	0.31
Polyproline II	3_1	−3.0	0.312
Polyproline I	10_3	3.33	0.19

^{a}Positive values for right-hand, negative values for left-hand helices.

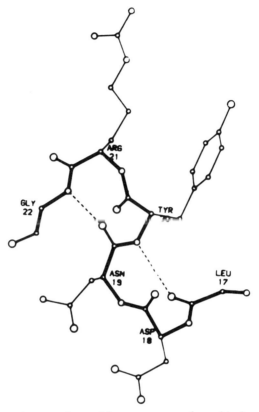

Fig. 8. Reverse turn amino acid sequence, as found in lysozyme [from A. C. T. North, in *Characterization of Protein Conformation and Function* (1979) (F. Franks, ed.) Symposium].

4. Tertiary Structure

The tertiary structure is the spatial disposition of various ordered domains with respect to each other; it is also the origin of the *native* or *functional* state of the protein in vivo. The interactions that maintain tertiary structures are weak (<5 kJ/mol), short-range (*see* Chapters 4 and 5), and sensitive to environmental factors, such as pressure, temperature, pH, ionic strength, and the composition of the solvent medium. The tertiary structure is therefore labile and easily disrupted.

4.1. Classification of Tertiary Structures

Many attempts are on record to classify known tertiary structures and to calculate the generation of the tertiary structure from the primary sequence. There is, however, little understanding of the relationships between tertiary structure and protein function.

Tertiary structures can be grouped in terms of secondary structure sequences in the peptide chain, as shown in Fig. 9. We can distinguish between sequences of α-helices, sequences of β-structures, segregation of α- and β-structures, and alternating α- and β-structures. The structured domains are connected by hinges. Examples of each of the four groups are given in Fig. 10. In α- or β-structures, adjacent structural domains are antiparallel. In α/β structures, the β-sheets tend to lie parallel to each other (*see* Fig. 6), separated by α-helices that run in the opposite direction.

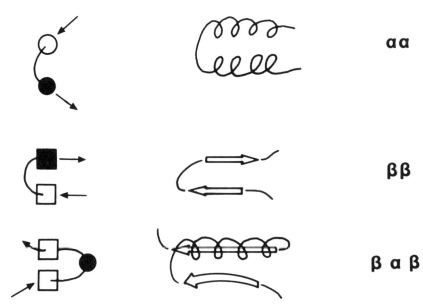

Fig. 9. Classification of tertiary structures according to secondary structure sequences.

4.2. Spatial Disposition of Structured Domains

X-ray and neutron diffraction techniques provide ever more detailed information on three-dimensional structures of proteins

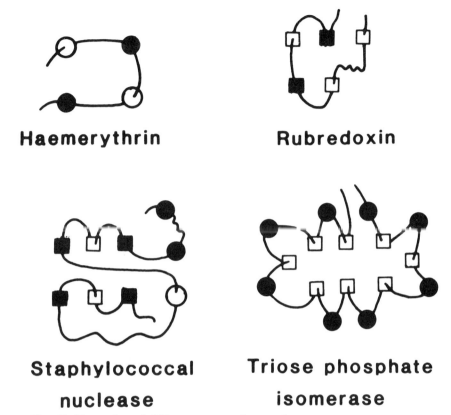

Fig. 10. Examples of different types of secondary structure sequences.

in the crystalline state, and it is generally believed that the secondary and tertiary structures in solution resemble those in the crystal in the essential features, i.e., those that determine the biological effectiveness of the molecule.

The projections in Fig. 10 can now be further elaborated to illustrate the topology of the molecule as a whole. Examples are shown in Fig. 11. Hemerythrin and rubredoxin are simple examples of the $\alpha\alpha$ and $\beta\beta$ types, respectively. In both cases the structural units are arranged in pairs. Staphylococcal nuclease is an example of the $\alpha\alpha + \beta\beta$ type, in which two triplets of antiparallel β-sheets are joined to three helical segments. In the α/β arrangement of triose phosphate isomerase, alternating α and β sequences resemble a barrel arrangement in which the parallel β-strands run in one direction and the α-helices in the opposite direction. The folding of the protein gives rise to a hollow cylinder.

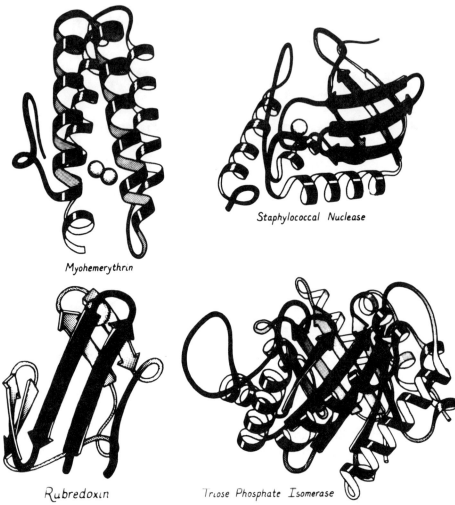

Fig. 11. Different types of secondary structure sequences, folded to adopt tertiary structures [from C. Chothia and A. M. Lesk, in *Protein Folding* (1980) (R. Jaenicke, ed.) Elsevier, Netherlands].

4.3. Tertiary Structure and Homology

Natural selection produces protein families with very different amino acid sequences that nevertheless have very similar tertiary structures. For example, the globins occur in many widely differing organisms, where they store and transport oxygen. They contain a heme group in which the iron can reversibly bind oxygen.

Structure and Function

Each molecule (or subunit) contains 136–154 amino acid residues, of which only 16% are conserved. Despite this variability in composition, the tertiary structures are remarkably similar: seven or eight α-helices are assembled, as shown in Fig. 12. In an extreme case, a comparison of nine different globins has established that the same folded structure is obtained, but only two residues are conserved!

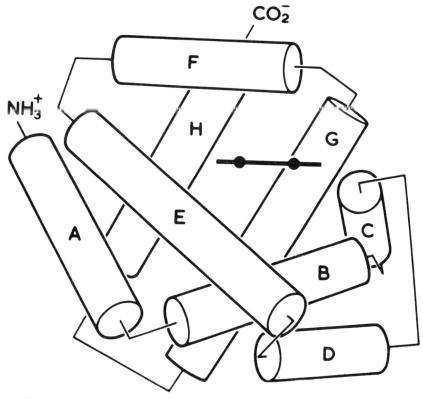

Fig. 12. Spatial disposition of the α-helices in globins; the heme group is shown as a bold line [from J. S. Richardson, in *Protein Folding* (1980) (R. Jaenicke, ed.) Elsevier, Netherlands].

On the other hand, a comparison of pancreatic trypsin inhibitor (PTI), which contains 58 residues with a number of different toxins having the same number of residues, but totally different functions, shows that although only 19 residues are the same, yet the folding pathway and folded structure of the different proteins is identical. Finally, a survey of several plastocyanins (small Cu-containing proteins) indicates that of the 21 amino acid residues buried in the interior of the proteins, 20 are conserved, whereas none of the 46 surface exposed residues are fully conserved. It appears that the crucial feature for identical tertiary structures is the maintenance of the hydrophobic interior of the protein, whereas a greater degree of variability can be tolerated in those residues that lie on the periphery, exposed to the solvent. This is not to say, however, that *all* mutant proteins have the same degree of biological activity.

5. Quaternary Structure

As summarized in chapter 1, Table 5, many proteins in their active state are composed of aggregates of identical or different peptide subunits. Aggregation numbers range from 2 (insulin) to several thousands, e.g., 2100 for tobacco mosaic virus coat protein. The supermolecular structure is formed by a process known as self-assembly, from the correctly folded individual subunits. The driving forces for self-assembly are believed to be mainly hydrophobic interactions (*see* chapter 5), although contributions from hydrogen-bonding and charge–charge effects cannot be ruled out.

Self-assembly is highly specific: if two tetrameric enzyme complexes (aldolase and glyceraldehyde-3-phosphatase) in a mixture are dissociated and then allowed to reassemble, this process results in a complete regeneration of the two individual enzymes, despite the fact that many random collisions must occur between peptide chains belonging to the *different* enzyme. There is no evidence that dissociation and reassembly occur via the same pathway.

The complexity and specificity of self-assembly is well illustrated by the construction of T-even bacteriophages, as portrayed in Fig. 13. Three different, but converging, pathways have been identified by which the head, the tail, and the tail fibers are assembled. For instance, in the construction of the tail the first event is the aggregation of subunits to form the base plate. The tail tube is then built up with an α-helical arrangement of subunits. Around this tail tube a second array of polypeptides that acts as the contractile element in the sheath is then built up. Finally the tail assembly is completed. Experiments on mutants have shown that there is a sequential information transfer during the assembly process, so that subunits cannot become misassembled before the right structure has appeared for their incorporation.

Structure and Function

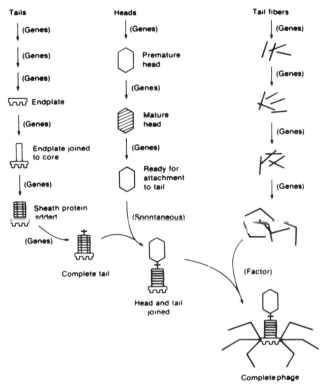

Fig. 13. Self-assembly scheme of T-even bacteriophage [from R. N. Perham, in *Characterization of Protein Conformation and Function* (1979) (F. Franks, ed.) Symposium Press, London].

This assembly information is contained in so-called propeptides (also known as signal peptides or teleopeptides), short amino acid sequences that are essential for the correct folding/assembly of the subunits, but that are subsequently cleaved off by a proteolytic enzyme. It is thus impossible to dissociate such a structure in vitro and to reassemble it.

One of the simplest cases of spontaneous in vivo folding and assembly is provided by insulin. The protein that is synthesized on the ribosome is proinsulin, the primary structure of which is shown in Fig. 14. It consists of a single chain and has no hormonal activity. It is activated by trypsin to produce insulin and a peptide of molecular weight 4000 that contains residues 31–63 of the original molecule. The propeptide is therefore required to ensure the cor-

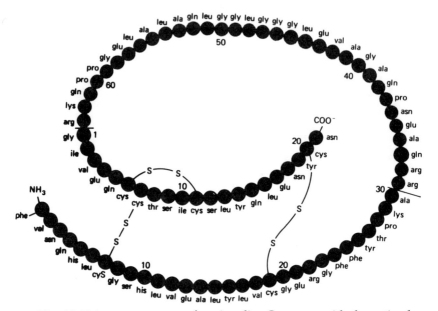

Fig. 14. Primary sequence of proinsulin. Compare with the active hormone sequence in Fig. 2.

rect folding of the chain, but until the removal of part of the chain, the protein remains inactive. The active protein now has two chains, but once the tertiary structure is disrupted, it cannot refold correctly because the necessary information has been removed by the cleavage of the C-chain.

5.1. Motile Structures

Cilia and flagella are the simplest motile organelles. A characteristic structure in eukaryotic cells involves 11 protein filaments of which nine (paired microtubules) are peripherally located, whereas the other two are usually found at the center of the flagellum, as shown schematically in Fig. 15. The filaments are composed of the globular protein tubulin, which is the basic subunit of microtubules. The flagella or cilia of prokaryotes are composed of flagellin that resembles tubulin, but is of a simpler, helical design. For instance, bacterial flagellin protein takes up an α-helical pattern, as shown in Fig. 16.

Compared to the flagellum, muscle is a very complex proteinaceous contractile assembly of specialized fibers (myofibrils).

Structure and Function

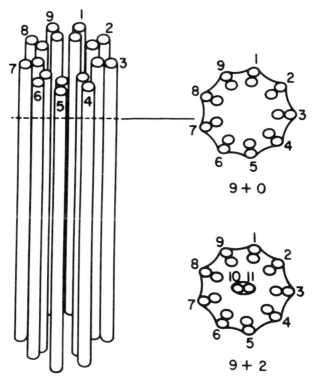

Fig. 15. Schematic diagrams showing the pairwise organization of microtubules in a flagellum of eurkaryotes [from G. H. Brown and J. J. Wolken, in *Liquid Crystals and Biological Structures* (1979) Academic, Florida].

The two major components, actin and myosin, self-aggregate and then associate with one another to form actomyosin, the active contractile system. Two additional globular proteins, tropomyosin and troponin, are associated with the actomyosin structure and regulate the contractile motion. The total supermolecular assembly is shown in Fig. 17. The myofibril is composed to a hexagonal array of these assemblies.

5.2. Structural Proteins

Collagen is one of the most studied proteins. It forms the connective tissue of the bulk of metazoan animals, but also occurs in immunoglobulins, bone, skin, and cornea of vertebrate eyes. The collagen molecule is a single polypeptide chain:

Fig. 16. Flagellin (related to tubulin) α-helical assembly, as found in the flagella of prokaryotes. The protein forms a single strand, compared to the double-stranded tubulin assembly [from G. H. Brown and J. J. Wolken, in *Liquid Crystals and Biological Structures* (1979) Academic, Florida].

$$NH_2-(20\ residues)-(gly-X-Y)_{300}-(20\ residues)-COOH$$

where 30% of the X or Y residues are pro or hypro. The conformation resembles the left-handed polyproline-II helix, and three

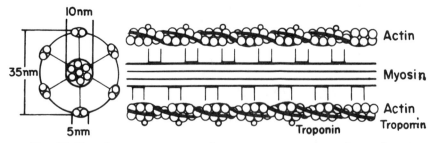

Fig. 17. Protein components and quaternary structure of muscle actomyosin fibrils [from G. H. Brown and J. J. Wolken, in *Liquid Crystals and Biological Structures* (1979) Academic, Florida].

Structure and Function

molecules aggregate to form a tropocollagen triple helical structure, as illustrated in Fig. 18a. The individual triple structures are laid down in the collagen fiber to produce the staggered array shown in Fig. 18b. Their exact positions in the fibril are determined by the 20-residue propeptide sequence that is subsequently removed, leaving the 67 nm displacements. The regular occurrence of glycine allows the three chains to approach sufficiently closely to produce the collagen superhelix. The molecules are also linked by covalent bonds and hydrogen bonds, all of which enhance the tensile strength of the fibrils.

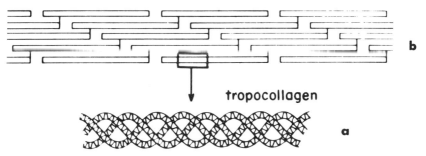

Fig. 18. Three-stranded tropocollagen molecule (a) and their method of assembly in the collagen fiber (b), showing overlap and gap regions [from G. H. Brown and J. J. Wolken, in *Liquid Crystals and Biological Structures* (1979) Academic, Florida].

Table 5 summarizes other quaternary protein assemblies that give rise to load-bearing structures. A common feature is the chemical periodicity of certain amino acids, usually glycine. An interesting

Table 5
Predominant Amino Acid Residues (Chemical Rhythm) of Fibrous Proteins

Protein	Source	Predominant residues, percent of total residues
Fibroin	Silk moth	gly 45, ala 29, ser 12
Fibroin	*Chrysopa* egg stalk	ser 43, gly 25, ala 21
Elastin	Lung tissue	gly 32, apolar 60
Resilin	Insect exoskeleton	gly 38, ala 11
Keratin	Whole wool	ser 11, cys/2 11, glu 11
Myosin	Rabbit muscle	glu 18, lys 11
Actin	Rabbit muscle	glu 11
Fibrinogen	Human blood	glu 15, asp 13

variant of the common helical and β-sheet structures is the protein found in the egg stalk of the lacewing fly *Chrysopa flava*. The stalk has to support the weight of the egg and must therefore possess a rigid structure, unlike most other fibrous proteins that fulfill functions in which high tensile strength is required. The egg stalk protein is characterized by high proportions of serine (43%), glycine (25%), and alanine (21%). The rigidity is achieved by short stretches of β-structures that double back on themselves by means of reverse turns, with a glycine residue at each corner. The structure is referred to as a cross-β-structure.

5.3. Immunoglobulins

Immunoglobulins are multisubunit proteins responsible for providing adaptive immunity in vertebrates. The unique characteristics of the adaptie immune response are specificity, dissemination, amplification, and memory.

All immunoglobulins have a common basic structure consisting of two "heavy" and two "light" polypeptide chains linked by —S—S— bridges, as shown in Fig. 19. The general formula for all immunoglobulins is therefore $(H_2L_2)_n$. Their diversity derives from the different types of heavy and light chains.

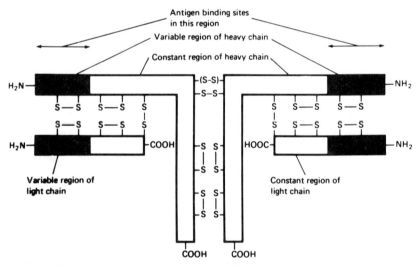

Fig. 19. Basic structure of immunoglobulin molecule, showing the two light and two heavy peptide chains and constant and variable regions.

5.4. Multisubunit Enzyme Complexes

The aggregation of several different subunits facilitates the performance of a series of sequential reactions, in which the product of one such reaction serves as the substrate of the next. The reaction sequence can thus be performed without the need for random diffusion of intermediates and the possibility of decomposition of reactive species. An example is provided by yeast pyruvate dehydrogenase. Under anaerobic conditions, pyruvate is decarboxylated, eventually to yield ethanol. In aerobic yeast pyruvate is converted to acetyl-CoA by a complex system of three individual enzyme functions, each associated with a different polypeptide chain. The three enzymes have molecular weights of 100,000, 80,000, and 56,000, respectively, and the particle of the multisubunit complex has a diameter of 30–40 nm. The complex is held together by noncovalent forces and can easily be dissociated. Enzyme E2 is itself composed of 24 subunits, clustered in trimers at the corners of a cube. In the active complex the polypeptide chain stoichiometry (E1:E2:E3) is 2:1:1, which indicates a total of 96 peptide chains and a molecular weight of 8×10^6. The subunits will self-assemble in vitro, but in the absence of the peptide chains belonging to E2, no assembly occurs. E2 forms the core to which E1 and/or E3 can be added independently. E2 therefore has two roles: it has an enzyme function and it also forms the structural core around which the complex is built.

All enzymes that perform complex chemical functions, such as synthases or kinases, are multisubunit structures of greater or lesser complexity.

5.5. Nuclear Proteins

Higher cell DNA is organized in a specific manner into chromosomes of a fibrillar appearance (chromatin fibers) that contain equal amounts by weight of DNA and proteins (histones). Several different protein fractions can be isolated. The histones exist as discrete structures of two or four subunits, each aggregate being associated with 200 base pairs of DNA such that the charges on the phosphate groups are largely neutralized by the basic residues of the histones (*see* Table 3 in chapter 1). Histones are the most conserved of all proteins, especially the parts of the peptide chains that possess globular tertiary structures and are involved

5.6. Fibrinogen

In the process of blood clotting, the soluble protein fibrinogen is converted into the fibrous protein fibrin by the action of the enzyme thrombin. The fibrous structure is then rendered insoluble by yet another protein, plasmin. Fibrinogen is a dimeric structure, composed of pairs of three different peptide chains, and is rich in aspartic acid (12% of the total amino acids). Its three-dimensional structure is still uncertain, but in the electron microscope fibrinogen has the appearance of three globules linked by a thread. The assembly of the dimer is shown in Fig. 20. The molecular weights of the Aα, Bβ, and γ-chains are 67,000, 58,000, and 47,000, respectively, and the chains are linked by multiple $-S-S-$ bonds, of which five are permanent and eight labile. During the formation of the blood clot, thrombin cleaves the four A and B chains at arg–gly linkages, thus activating the polymerization and crosslinking steps by an as yet unidentified mechanism.

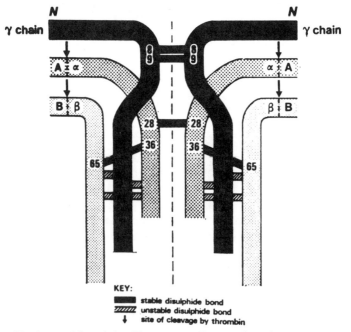

Fig. 20. Assembly of the fibrinogen dimer, showing permanent and labile S—S linkages.

6. Structural Integrity and Technological Function

Self-assembly of large multisubunit structures takes place spontaneously in vivo, but can sometimes also be induced in vivo under carefully controlled conditions. The criterion of success is then the degree of biological activity achieved. In vitro, technological applications of proteins usually require the isolation of a protein from its source, followed by concentration and, possibly, further manipulation and processing stages.

It is highly desirable (or even essential) to maintain the reversible quaternary structures during the isolation and concentration stages and to modify such structures in a predictable and reproducible manner during the final processing stages. Just as the biological properties of proteins are determined by the quaternary structure (e.g., fibrous vs globular proteins), so the physical properties of a processed proteinaceous material are very sensitive to the type of aggregation induced during processing. This is illustrated in Table 6, which compares the rheological behavior of three products, all derived from globular proteins. The response to an applied shear stress τ is expressed in terms of an empirical power law:

$$\tau - \tau_0 = kd^n$$

where τ_0 is the shear stress corresponding to the yield point, d is the shear rate, and k and n are fitting parameters, called consistency index and flow behavior index, respectively. It is seen that at low concentrations there is a general similarity between the three preparations, but that divergent behavior becomes pronounced with increasing concentration, i.e., where the rheological response is determined mainly by particle–particle interactions.

Table 6
Rheological Parameters of Protein Products
Derived From Globular Proteins

Product	Shear stress at yield point	Flow behavior index	Consistency index
Soya protein isolate	Zero at $c = 6\%$, increases rapidly; 1670 at $c = 16\%$	Decreases with increasing c; 0.52 at $c = 16\%$	0.11 at $c = 4\%$ 62 at $c = 16\%$
Calcium caseinate	Zero	0.9 (constant)	0.04 at $c = 4\%$ 3.9 at $c = 20\%$
Whey protein concentrate	Zero	0.9 at $c = 4\%$; 0.77 at $c = 16\%$	0.03 at $c = 4\%$ 3.9 at $c = 20\%$

Chapter 3

Solution Properties of Proteins

Felix Franks

1. Conformational Stability and Solubility

The conformational stability of a protein is determined by intramolecular factors and solvent interactions (hydration); to obtain information about the latter, dilute solutions have to be employed. The solubility is determined primarily by intermolecular effects (protein/protein), but proteins are solvated, so that hydration effects are also involved in changes in solubility. Protein/water interactions are discussed in detail in chapter 5.

Conformational changes (changes in function/activity) can be induced by changes in temperature, pressure, and the solvent medium. Technological performance critically depends on conformation, hydration (water-holding capacity), and solubility. Biological performance is normally restricted to narrow physiological ranges of pressure, temperature, and medium composition. There are many examples, however, of adaptation (genetic) to extreme physical and chemical environments: thermophiles, psychrophiles, halophiles, acidophiles, and so on (1). Such adaptations are usually achieved through minor changes in the amino acid composition (2).

Like other polymers, proteins can be characterized by their chain conformations. However, the methods of polymer statistics cannot be applied to most proteins because they adopt *specific* (native) conformations under physiological conditions.

2. Solubility

From a physiological viewpoint the solubility of a protein is unimportant because its in vivo concentration is governed by biochemical factors. When a protein is isolated from a source material, and concentrated as part of a subsequent processing chain, solubility becomes one of the most important attributes. Since proteins are polyelectrolytes, their solubility behavior is governed largely by electrostatic (ionic) interactions.

2.1. Electrostatic Factors

In determining charge/charge interactions, the pK_a and pK_b values of the individual amino acids play an important role. They are summarized in Table 1. Only those amino acids with acidic or basic side chains affect the solution behavior (except the end groups). Histidine is the only amino acid with a buffering capacity in the physiological pH range of 6–8.

A pH titration of a protein provides a useful fingerprint of the number of ionic groups that are accessible to the solvent medium. Dissociation equilibria are expressed in terms of $\bar{r}$, the number of

Table 1
Dissociation Constants (pK_a) of Amino Acids

Amino acid (or residue)	pK, —COOH Free acid	pK, —NH$_2$	pK, Side chain free acid	pK, Side chain in protein
Glycine	2.3	9.6		
Alanine	2.3	9.7		
Leucine	2.4	9.6		
Serine	2.2	9.2		
Threonine	2.6	10.4		
Glutamine	2.2	9.1		
Aspartic acid	2.1	9.8	3.9	4
Glutamic acid	2.2	9.7	4.3	4
Histidine	1.8	9.2	6.0	6
Cysteine	1.7	10.8	8.3	10
Tyrosine	2.2	9.1	10.1	10
Lysine	2.2	9.0	12.5	10
Arginine	2.2	9.0	12.5	12
Terminal —NH$_2$		8		
Terminal —COOH	3			

dissociated protons per molecule (3). Figure 1 is the titration profile of ribonuclease that should be read in conjunction with the amino acid composition/sequence in Fig. 1 in chapter 2. Not all of the 36 potentially titratable groups are in fact titrated in the range 2 < pH < 12. Also the shape of the curve depends on the ionic strength (I), reflecting electrostatic effects on ion binding/dissociation.

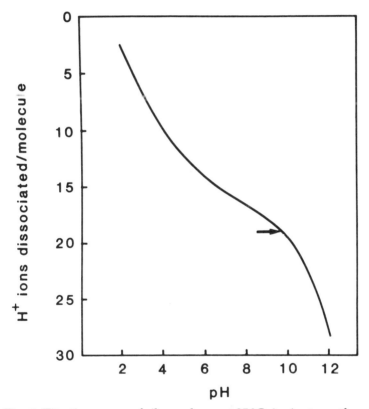

Fig. 1. Titration curve of ribonuclease at 25°C, ionic strength = 0.01; see ref. 3.

Individual groups cannot be resolved from the titration profile, but the curve can be seen to be divided into several main regions, depending on the respective pK values of the amino acids. More detailed information can be obtained from a spectrophotometric titration curve of, say, tyrosine at 295 nm. In the case of ribonuclease this indicates that of the six tyr residues, three titrate normally, with pK ≈ 10, but three have abnormally high pK values.

A semi-empirical, but very useful, method for analyzing titration profiles makes use of an "intrinsic" constant, $K_{i,in}$ and an electrostatic free energy of interaction $\Delta G°(\bar{z})$, such that

$$K_i = K_{i,in} \exp[\Delta G°(\bar{z})/RT] \quad (1)$$

where $\bar{z}$ is the average charge on the protein molecule. It can then be shown that for any single type of residue i,

$$\text{pH} - \log \frac{x_i}{1 - x_i} = pK_{i,in} - 0.868\, w\bar{z} \quad (2)$$

where x_i is the fraction of class i that have been titrated, and w is an empirical constant (3). The success of Eq. (2) is shown in Fig. 2. Note the effect of I on the slope (w). The linearity increases as

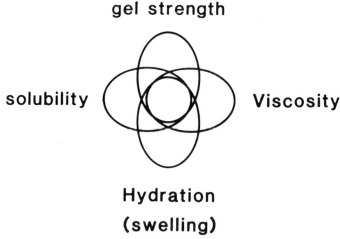

Fig. 2. Venn diagram showing the interdependence of the parameters that govern the "quality" of meat products.

I is reduced. Equation (2) also gives the intrinsic pK of the protein (the pH value for which $\bar{z}$ = 19). Note that the isoelectric point, pI, is a function fo the ionic strength, the nature of the buffer used, and any other solutes present. It is *not* dependent on protein concentration. Table 2 provides a summary of pI values and ranges for a variety of proteins. It must be emphasized that even at the isoelectric point the protein molecule carries a charge and it is only the *net* charge that is zero. Since proteins are polyelectrolytes, many of their properties depend on their net charge, the charge distribution, ionic strength of the medium, and its dielectric permittivity. Technologically, protein–protein interactions determine viscoelasticity, plasticity, yield value, gelation, emulsification, foaming, texturization, and so on, and hence electrostatic effects are probably of paramount importance in protein processing technology.

Table 2
Isoelectric Points (pI) of Proteins[a]

Protein	pI	Protein	pI
Salmine	12.1	Collagen	6.7
Thymohistone	10.8	Gelatin	4.7–5.0
Ovalbumin	4.7	α-Casein	4.5
Conalbumin	7.1	Lysozyme	11.1
Serum albumin	4.9	Myoglobin	7.0
Myogen A	6.3	Hemoglobin (human)	7.1
β-Lactoglobulin	5.0	Hemocyanins	4.6–6.4
Livetin	4.9	Hemerythrin	5.6
γ_1-Globulin (human)	6.6	Cytochrome c	10.6
γ_2-Globulin (human)	8.2	Rhodopsin	4.5
Myosin A	5.3	Chymotrypsinogen	9.5
Tropomyosin	5.1	Urease	5.0
Thyroglobulin	4.6	α-Lipoprotein	5.5
Fibrinogen	4.8	Bushy stunt virus	5.3
α-Crystallin	4.8	Vaccinia virus	5.3
β-Crystallin	6.0	Prolactin	5.7
Arachin	5.1	Insulin	5.3
Keratins	3.7–5.0	Pepsin	1.0

[a]Values given refer to measurements at low ionic strength (<0.01).

2.2. Classification of Proteins by Solubility

A purely phenomenological classification system makes use of the solubility:

Globulins	Insoluble in water, but dissolve in aqueous salt solutions.
Albumins	Soluble in water and in salt solutions.
Prolamins	Insoluble in water, but dissolve in 50–90% aqueous ethanol. Prolamins have only been detected in plants.
Glutelins	Insoluble in all the above solvents, but dissolve in dilute solutions of acid or base; they only occur in plants and are not well characterized.
Scleroproteins	Insoluble in most ordinary solvents; they swell in salt solutions.

2.3. Solubility as Index of Viability/Denaturation

Technological performance often depends on the complex interplay of several physical properties. The Venn diagram in Fig. 2 shows such interrelationships in model meat systems; they determine technological attributes such as storage life, moisture loss, and response to additives, such as salts.

Solubility is a good index of the extent of denaturation, and the pH/ionic strength/solubility profile is often the best practical index of technological usefulness. Figure 3 shows the effects of pH and NaCl concentration on the solubility of β-lactoglobulin, and Fig. 4 shows the solubility of isoelectric carboxyhemoglobin as a function of *I* for various electrolytes. Two important principles are illustrated: First, the solubility undergoes a minimum in the neighborhood of the isoelectric pH, and second, the effect of salt on the solubility is complex with *qualitatively* similar behavior being pro-

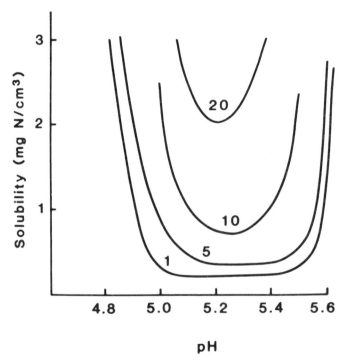

Fig. 3. The effects of pH and NaCl concentration on the solubility of β-lactoglobulin; concentrations in mol/L.

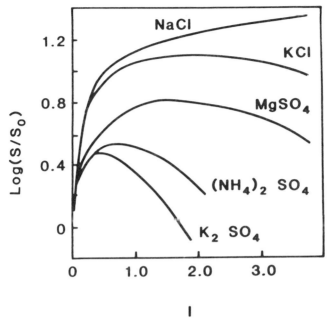

Fig. 4. Influence of different electrolytes at the same ionic strength on the solubility of isoelectric carboxyhemoglobin. Note the differences in salting in/out behavior, indicating that the electrolyte effects are not of a purely electrostatic nature.

duced by all salts. The quantitative effects differ widely between salts. Thus, besides the general ionic strength effect, there is another effect that is ion specific. At low salt concentrations the ionic strength effect predominates, and *all* salts raise the solubility. At $I > 0.15$, ion-specific effects are first observed, with some ions reducing the solubility, and others enhance it further. These phenomena are referred to as salting out and salting in, respectively. The solubility at a given pH can be expressed by

$$\log S = b - K_s I \qquad (3)$$

where S is the solubility (g/L), b is the logarithm of the hypothetical solubility at $I = 0$, and K_s is known as the salting-out constant. It is found that K_s varies markedly with the nature of the salt, but hardly depends on the nature of the protein. On the other hand, b is very dependent on the nature of the protein. At pH > pI, K_s becomes independent of temperature and pH.

The ion specificity is expressed in terms of the lyotropic series [also known as Hofmeister series (4)], according to which the ions are arranged in terms of their salting-in potential, as follows (5–7):

Salting out　　　　　　　Neutral　　　　Salting in
$SO_4'' > HPO_4'' > OAc' > F' >$ citrate $> Cl' < NO_3' < I' < CNS' < ClO_4'$
$Li^+ > Na^+ < K^+ < Rb^+ < NH_4^+$
$Me_4N^+ < Et_4N^+ < Pr_4N^+ << Bu_4N^+$

Solubility determinations are complicated by the fact that ions also affect the stability of the native states of proteins and their degree of association. Thus, salting-out ions enhance the stability of native states, whereas salting-in ions give rise to denaturation (8). (There is as yet no convincing explanation for the lyotropic series, although it has been "rediscovered" several times since Hofmeister's original report. It has also been found to influence processes such as the micelle formation of surfactants and the flocculation behavior of colloidal systems.)

2.4. Solubility and Processing Conditions

The total soluble nitrogen content of a protein is markedly influenced by several factors related to the processing conditions, such as isolation, precipitation, concentration, and so on. This aspect is treated in more detail in chapters 20 and 21. Here we only cite a few examples. Alfalfa protein is very sensitive to heat precipitation, as can be seen in Fig. 5. On the other hand, acid-precipitated material retains its original solubility/pH profile. Similarly, fish muscle protein is susceptible to freeze/thaw treatment, which renders it largely insoluble (see Fig. 6). The effects of freeze concentration are discussed in chapter 19.

2.5. Solubility and Nonelectrostatic Factors

Proteins can also be "salted" in/out by nonelectrolytes; e.g., urea enhances the solubility of most proteins (8), and ethanol at low temperature and low concentration salts out, but at higher temperatures at all concentrations salts in (9). Sugars and sugar alcohols generally salt out globular proteins (10–13). Here the effect does not depend on dielectric permittivity and therefore does not originate from electrostatic interactions. Solubility determinations are made difficult by the effects of organic solvents on the stability of the native state. The same is true for the effect of temperature: frequently a rise in temperature enhances protein solubility up to a point beyond which precipitation will occur.

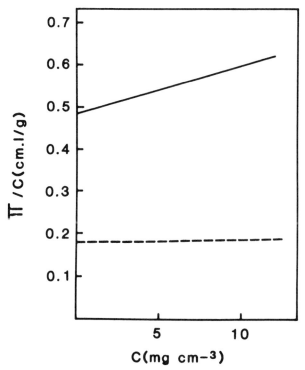

Fig. 5. Osmotic pressure data for native aldolase at neutral pH (broken line) and the individual subunits of the protein in 6M GuHCl (solid line).

3. Solution Properties of Large Molecules

Apart from the charge, the main determinants of the solution behavior of large molecules are size and shape. Since most measurement techniques are based on solution thermodynamics, we summarize briefly the main features of binary solutions.

3.1. Thermodynamics of Ideal and Nonideal Solutions

The chemical potential μ_i of component i in a solution is referred to the ideal solution (Raoult's law) by

$$\mu_i - \mu_i^o = RT \ln x_i \tag{4}$$

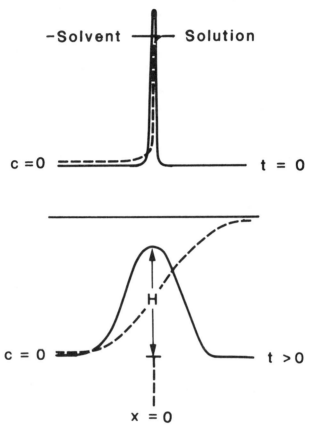

Fig. 6. The broadening of an initial sharp boundary ($t = 0$) as a function of time (t) during free diffusion. The broken line represents the concentration as a function of the position in the cell (x). The solid line shows the gradient, as given in Eq. (17).

where x_i is the mol fraction concentration, and μ_i^o is the chemical potential of the *pure* component $i(x_i = 1)$. More practical units of concentration are mol/L or g/L (c_i). In the latter case μ_i^o refers to $c_i = 1$ g/L.

Nonideal behavior is taken into account by the activity coefficient f_i by which the concentration must be multiplied for Eq. (4) still to hold. For a binary solution in the limit of infinite dilution $f_1 \to 0$ as $x_1 \to 1$ (or $c_2 \to 0$). Usually $f_1 < 1$. Also in a binary solution $x_1 = (1 - x_2)$ and $\ln(1 - x_2) \approx -x_2 - x_2^2/2 \ldots$, so that

$$\mu_i - \mu_i^o = -RT(x_2 + \ldots) + RT \ln f_1$$

Solution Properties of Proteins

and $x_2 \approx c_2 V_1^o / M_2$, where V_1^o is the molar volume of the solvent (water) and M_2 is the molecular weight of the solute. Making this substitution for x_2, we obtain

$$\mu_i - \mu_i^o = -(RTV_1^o c_2)M_2 + RT \ln f_1$$

Expressing $\ln f_1$ in terms of a power series in c_2:

$$\ln f_1 = -(\alpha c_2^2 + \beta c_2^3 + \ldots)$$

so that Eq. (4) now becomes

$$\mu_i - \mu_i^o = -RTV_1^o \left\{ \frac{c_2}{M_2} + Bc_2^2 + \ldots \right\} \quad (5)$$

$$= -RTV_1^o c_2 / M_2 [1 + BM_2 c_2 + \ldots] \quad (6)$$

B is known as the second viral coefficient and serves as a convenient measure of nonideality. Equation (6) serves as the basis for the determination of molecular weights.

3.2. Solution Properties and Molecular Shapes

The second viral coefficient B and many other solution properties of macromolecules are sensitive to the shape (conformation) of the large molecule. A linear polymer in which there is little or no resistance to rotation around the bonds adopts a *random* conformation (random coil), the average dimensions of which can be calculated (*14*). The *average* radius of gyration, $<r_g>$, is defined in terms of the *average* squared distance $<r_i^2>$ of residue i from the center of mass, by

$$r_g^2 = (\Sigma <r_i^2>)/N \quad (7)$$

where N is the number of residues. For a true random coil, r_g is proportional to $N^{1/2}$. The structural proteins in which the predominant secondary or quaternary structure is helical adopt a rod-like conformation in solution. On the other hand, globular proteins can often be quite adequately described as spheres or ellipsoids. Nonideal behavior in solutions of macromolecules arises primarily from their size. The center of each molecule is excluded from a volume determined by the volumes occupied by all the other molecules. B therefore depends on this *excluded volume* (u):

$$B = Nu/2M_2^2 \quad (8)$$

where N is the Avogadro number. The actual value of u is calculated from the molecular dimensions:

for spheres: $u = 8M_2v/N$ and $B = 4v/M_2$
for rods: $u = 2LM_2v/Nd$ and $B = Lv/dM_2$

where v is the specific volume of the polymer and L and d are the length and diameter of the rod.

3.3. Quality of Solvent

The thermodynamics of mixing a solute and a solvent are determined by the balance of the interactions between solvent molecules, solute molecules, and cross-interactions between solute and solvent. If the sum of the self-interactions balances the cross-interactions, the solvent is known as a theta (θ) solvent and the solution exhibits a pseudoideal behavior, with $B = 0$. When preferential solvation occurs (good solvent), $B > 0$, whereas in a poor solvent the polymer molecules tend to aggregate and $B < 0$. In the extreme case the polymer will precipitate. The concept of good and poor solvents is related to the phenomenon of salting in/out, already discussed in section 2.3. Table 3 summarizes some calculated nonideality parameters for solutions of macromolecules at a concentration of 5 mg/cm³, a typical protein concentration in solution experiments

Table 3
Solution Nonideality Corrections, as Expressed in Eq. (6)[a]

Particle	B	$(1 + BM_2c)$
Sphere, $M = 10^5$	3×10^{-5}	1.015
Rod, $L/d = 1000$, $M = 10^5$	7.5×10^{-4}	1.375
Random coil, $M = 10^5$	5×10^{-4}	1.250
Random coil, θ solvent	0	1.000

[a]Units: B in cm³/g²/mol, $c = 5$ mg/cm³.

(15). The excluded volume effect *always* lowers μ_1, but a poor solvent may cause an increase in μ_1. In mechanistic terms, the excluded volume effect always pushes molecules apart, but this may be compensated by a desolvation in a poor solvent, which will promote solute aggregation. The θ-condition corresponds to an exact balance between the two effects.

4. Experimental Techniques

We here describe methods for the determination of molecular weight, shape, and charge of proteins in solution.

4.1. Membrane Equilibria

Semipermeable membranes find multiple applications in biochemistry and biotechnology, e.g., dialysis, membrane filtration, osmosis, and reverse osmosis. The membrane acts as barrier that retains some substances and allows others, of a lower molecular weight, to pass. In practice, cellophane membranes that are commonly used have a molecular weight cutoff of $M < 10,000$, but membranes can be prepared that select at much higher or lower molecular weights.

4.1.1. Osmotic Pressure

The separation of a protein solution from the pure solvent by means of a semipermeable membrane leads to a flux of solvent into the solution that gives rise to an excess pressure (osmotic pressure). Equation (5) can then be written as

$$\mu_1 - \mu_1^o = -RTV_1^o(c_2/M_2 + Bc_2^2 + \ldots) + V_1^o\pi \tag{9}$$

where π is the osmotic pressure. Equation (9) applies at atmospheric (low) pressure and low c_2. Now

$$\pi = RT(c_2/M_2 + Bc_2^2 + \ldots) \tag{10}$$

In the limit of $c_2 \to 0$

$$\pi = RTc_2/M_2 \tag{11}$$

analogous to the ideal gas equation, from which the molecular weight of the protein can be obtained. In practice, molecular weights are calculated from the relationship

$$\pi/c_2 = RT/M_2 + BRTc_2 + \ldots \tag{12}$$

by a plot of π/c_2 vs c_2 and extrapolation to $c_2 = 0$. The method has several disadvantages. (1) The possibility of protein aggregation/denaturation while the determination is in progress, (2) the measurement should be performed at net zero charge, but the pI may not be known, (3) the method gives no indication of the purity of the sample, i.e., a *mean* molecular weight may be obtained. This is a well-known effect in polymer chemistry.

Table 4 summarizes some molecular weights obtained from osmotic pressure measurements, and Fig. 5 is a typical osmotic pressure plot for native aldolase and its dissociated subunits. The slopes of the lines are proportional to B, according to Eq. (12); note the larger B for the denatured subunits.

Table 4
Molecular Weights Obtained From Osmotic Pressure Measurements

Polymer	M_n
Ovalbumin	44,600
Hemoglobin	66,500
Aldolase	156,500
Aldolase subunit	42,400
Amylose	32,000–150,000

4.1.2. Equilibrium Dialysis

Dialysis can be used to purify proteins in solution by the removal of low molecular weight impurities, but also to measure the binding of ions or molecules. The protein solution is contained in a membrane bag (phase α) that is suspended in a solution containing the low molecular weight substance component 3, (phase β). Component 3 will distribute itself between phases α and β, until, at equilibrium, $\mu_3^\alpha = \mu_3^\beta$. Neglecting pressure differences, then

$$RT \ln c_3^\alpha f_3^\alpha = RT \ln c_3^\beta f_3^\beta$$

or

$$c_3^\alpha f_3^\alpha = c_3^\beta f_3^\beta$$

If we now assume that both bound and free component 3 exist in phase α, and that the environments on both sides of the membrane are sufficiently similar, so that $f_3^\alpha = f_3^\beta$, then

$$c_3^\alpha \text{ (free)} = c_3^\beta$$

and

$$c_3^\alpha \text{ (bound)} = c_3^\alpha - c_3^\beta \tag{13}$$

Unless binding is strong, the results obtained from dialysis equilibrium measurements must be treated with caution.

4.1.3. Donnan Equilibrium

When the macromolecule carries a charge (e.g., protein), the membrane equilibrium is further complicated (16). Let the protein be designated by PX_z, which dissociates according to

$$PX_z \rightarrow P^{+z} + zX^-$$

where X^- is a monovalent counterion. In the presence of a salt BX, which dissociates into B^+ and X^-, and at equilibrium, we have

$$c_B^\alpha c_X^\alpha = c_B^\beta c_X^\beta$$

In addition, electroneutrality demands that

$$c_B^\alpha - c_B^\beta = -zc_P^\alpha c_B^\alpha/(c_B^\alpha + c_B^\beta) \tag{14a}$$

$$c_X^\alpha - c_X^\beta = zc_P^\alpha c_X^\alpha/(c_X^\alpha + c_X^\beta) \tag{14b}$$

In other words, at equilibrium, the counter-ion X will be more concentrated on the polymer side of the membrane, whereas the concentration of the cation B will be lower. These concentration differences will contribute to the osmotic pressure, and it can be shown that

$$\pi = RT \left\{ \frac{c_P}{M_P} + \frac{z^2 c_P^2}{4M_P^2 c_{BX}} \right\} \tag{15}$$

In other words, the solution behaves as if it were nonideal, with a virial coefficient proportional to the charge/mass ratio. This effect is minimized at the isoelectric point, but can become large at pH values far removed from pI.

It is evident from Eq. (15) that π will approach its ideal value as c_{BX} becomes large. Analogs of the Donnan effect also exist in other physical measurements (sedimentation, light scattering), so that it is advisable to use high salt concentrations for molecular weight determinations, especially when measurements are made at pH values far removed from pI.

4.2. Diffusion and Sedimentation

The basic equation governing the free diffusion of particles in a continuous medium is

$$(\partial c/\partial t)_t = D(\partial^2 c/\partial x^2)_t \text{ (Fick's law)} \tag{16}$$

where the left-hand side is the rate of change of concentration and $\partial c/\partial x$ is the concentration gradient; D is the proportionality constant (diffusion coefficient). By imposing certain boundary conditions, Eq. (16) can be integrated:

$$(\partial c/\partial t)_x = \frac{c_0}{2(\pi Dt)^{1/2}} \exp[-x^2/4Dt] \qquad (17)$$

where c_0 is the initial concentration. In practice, D is measured by the broadening of an initially sharp interface, as detected by a suitable optical device. Figure 6 is a schematic representation of the diffusion broadening, from which D can be obtained as

$$A/H = 2(\pi Dt)^{1/2} \qquad (18)$$

where A is the area of the diffusion peak and H its height. Provided the molecular weight is known, D can provide information about the shape of the protein in solution through the introduction of f, the frictional coefficient, which is a measure of the asymmetry and hydration of the protein molecule:

$$f/f_o = (f/f_o)_{\text{hydration}} (f/f_o)_{\text{sym}} \qquad (19)$$

The frictional coefficient f_o is that of an unhydrated sphere of the same mass and is given by

$$f_o = 6\pi\eta r \qquad (20)$$

in terms of the viscosity of the solvent (η) and the radius of the sphere (r). The axial ratios, f/f_o, diffusion coefficients, and molecular weights of proteins are summarized in Table 5. If the shape of the molecule is known from some independent measurement, then f/f_o permits the evaluation of the degree of hydration. On the other hand, assuming zero hydration, f/f_o for an ellipsoid can be calculated from its dimensions. Thus, $(f/f_o\;_{\text{asym}}) = 1.5$ corresponds to a prolate ellipsoid with an axial ratio of 10:1. The axial ratio is best determined from scattering measurements (*see* section 4.4).

The development of the analytical ultracentrifuge has facilitated the determination of sizes and shapes of macromolecules with a high degree of precision. Two distinct techniques find application: (1) sedimentation equilibrium and (2) sedimentation velocity. In the former method the speed of rotation is adjusted so that the centrifugal force just balances diffusion and there is no net solute migration. M is then given by

$$M = \frac{RT\left(\frac{d(\ln \gamma)}{dc}\right)}{r\omega^2(1 - \bar{v}\rho)} \frac{1}{c} \frac{dc}{dr} \qquad (21)$$

Solution Properties of Proteins

Table 5
Diffusional Properties and Molecular Weights of Proteins

Protein	Molecular weight	Diffusion coefficient, $D_{20,\omega} \times 10^7$	Frictional ratio, f/f_0	Sedimentation coefficient, $S_{20,\omega}$
Cytochrome c	13,370	11.4	1.19	1.17
Ribonuclease	13,683	11.9	1.14	
Lysozyme	14,100	10.4	1.32	
Chymotrypsinogen	23,200	9.5	1.20	2.54
β-Lactoglobulin	35,000	7.82	1.25	2.85
Ovalbumin	45,000	7.76	1.17	
Serum albumin	65,000	5.94	1.35	4.6
Hemoglobin	68,000	6.9	1.14	4.46
Catalase	250,000	4.1	1.25	11.3
Urease	480,000	3.46	1.20	18.6
Tropomyosin	93,000	2.24	3.22	
Fibrinogen	330,000	2.02	2.34	7.63
Collagen	345,000	0.69	6.8	
Myosin	493,000	1.16	3.53	6.43
Tobacco mosaic virus	40,590,000	0.46	2.03	198

where ω is the angular velocity, $\bar{v}$ the specific volume, and ρ the density. Values of c and dc/dr are obtained from the experimental measurements, which are extrapolated to infinite dilution, where the activity coefficient term $1 + [d(\ln\gamma)/dc]$ approaches unity.

This is the most precise method for the determination of M, and has been further refined to reduce the time required for equilibrium to be established.

In the sedimentation velocity method the centrifugal force on the macromolecule is increased until it just balances the frictional force opposing migration. The solute then moves down the tube at a constant velocity. The molecular weight is obtained by the use of the Svedberg equation

$$M = RTs/D(1 - \bar{v}\rho) \quad (22)$$

where is is known as the sedimentation constant and is given by

$$s = (dx/dt)/\omega^2 x \ s^{-1}$$

x being the distance from the center of the rotor. The basic unit is taken as $10^{-13}s^{-1}$ (one Svedberg). Sedimentation constants for most proteins fall in the range 1–50 Svedbergs. Table 5 includes some sedimentation constant data.

Both s and D depend on the concentration via the frictional coefficient, f, already referred to. The experimental data must therefore be extrapolated to $c \to 0$. They must also be referred to some standard condition, e.g., water at 20°C. If the measurements are performed in a buffer, b, at temperature, T, then

$$s_{20,w} = \frac{(1 - \bar{v}\rho)_{20,w}\eta_{t,b}}{(1 - \bar{v}\rho)_{T,b}\eta_{20,w}} s_{T,b}$$

it being assumed that f is accurately proportional to the viscosity η.

Various refinements have been developed to increase the resolving power of the ultracentrifuge. In particular, the introduction of a density gradient with the aid of an "inert" substance, such as sucrose, makes possible the separation of the components in a mixture of proteins, and the s value of each fraction can be calculated.

4.3. Viscosity

Viscosity data supplement the information derived from diffusion and sedimentation measurements. The contribution of a macromolecule to the viscosity of a solution depends mainly on its volume, in contrast to the frictional coefficient, which depends on the linear dimensions.

A fluid is termed Newtonian when its viscosity is constant, independent of the shear stress. For a dilute solution of solute concentration ϕ (volume fraction)

$$\eta/\eta_0 = 1 + k\phi \tag{23}$$

where η_0 is the viscosity of the solvent and $k = 2.5$ for spherical particles. For other shapes $k > 2.5$, as shown in Fig. 7 for ellipsoids of different axial ratios. Equation (23) is usually extended to take account of solute–solute interactions at higher concentrations:

$$\eta_{sp} = \frac{\eta}{\eta_0} - 1 = \frac{\eta - \eta_0}{\eta_0} = k\phi + k'\phi^2 + \ldots \tag{24}$$

where η_{sp} is known as the specific viscosity and is dimensionless. In most cases it is more convenient to express concentration in terms of g/cm^3. Equation (24) then becomes

$$\eta_{sp}/c = k v + k' v^2 c \tag{25}$$

Solution Properties of Proteins

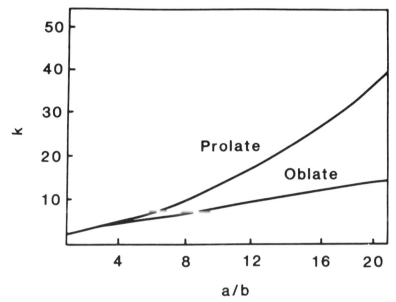

Fig. 7. Constant k in the viscosity Eq. (23) for ellipsoids of different axial ratios.

The limit of η_{sp}/c as $c \to 0$ is called the intrinsic viscosity $[\eta]$. It depends on the properties of the isolated protein molecule, and from Eq. (25)

$$[\eta] = kv$$

In other words, $[\eta]$ is a function only of the shape (through k) and the volume of the protein, and $k \geqslant 2.5$. For many proteins the *anhydrous* specific volume is set at ~ 0.75, so that the minimum value of $[\eta]$ is approximately 2 cm^3/g. The $[\eta]$ data included in Table 5 illustrate that this value is sometimes approached. The slope of the η_{sp}/c versus c plot gives an indication of the protein–protein interactions. Figure 8 shows that neither intact molecules of hemocyanin nor the dissociated, but native, subunits interact strongly. Pronounced aggregation does however take place between the unfolded subunits. Also from Fig. 8 we see that the unfolded subunit appears to be much larger than its native, compact counterpart. A useful expression for *random-coil* polymers (including proteins) is

$$[\eta] = KM^a \qquad (26)$$

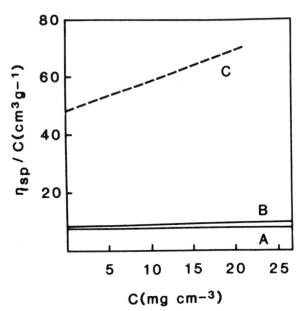

Fig. 8. Determination of the intrinsic viscosity of crab hemocyanin. (A) Native protein ($M = 10^6$); (B) Subunits ($M = 7 \times 10^4$); (C) Denatured (GuHCl) protein.

where $a \geq 0.5$. For a θ-solvent, $a = 0.5$. K is an empirical constant that depends on the degree of hydration of the macromolecule. For compact folded proteins there is no direct relationship between $[\eta]$ and M.

When the viscosity is a function of the shear rate, the fluid is said to be non-Newtonian. Two types of behavior can be distinguished: shear thinning and shear thickening. The former is more common and is observed for elongated molecules (fibrous proteins, nucleotides) that can be aligned by the flow gradient. Non-Newtonian flow can provide problems in the experimental determination of $[\eta]$. For instance, the shear rate in an ordinary capillary viscometer is very high, so that artificially low $[\eta]$ values are indicated. Viscosity measurements, especially on concentrated solutions, should normally be performed over a range of shear rates and extrapolated to zero shear rate.

4.4. Scattering of Radiation

4.4.1. Light Scattering

If an assembly of molecules is irradiated with light of a given wave length λ, their electronic charge distribution will be affected.

Solution Properties of Proteins

This results in a dispersion of the radiant energy in directions other than that of the incident radiation. This is the common basis for *all* scattering processes (light, X-rays, neutrons, and so on) (14). For very dilute solutions of macromolecules in which the solute molecules behave as independent point scatterers and are small compared to λ, the scattering intensity i_θ at an angle θ is expressed by

$$R_\theta = \frac{i_\theta}{I_0} \frac{r^2}{1 + \cos^2 \theta}$$

$$= \left[\frac{2\pi^2 n_0^2 (dn/dc)^2}{N\lambda^4} \right] cM = KcM \quad (27)$$

where R_θ is the Rayleigh scattering ratio, I_0 is the intensity of the incident beam, n is the refractive index, and N is the Avogadro number. For an ideal solution, Eq. (27) reduces to

$$Kc/R_\theta = M^{-1}$$

For a nonideal solution, a second virial coefficient B is introduced, analogously to Eq. (10):

$$Kc/R_\theta = M^{-1} + 2Bc \quad (28)$$

When the greatest dimension of the scattering particle exceeds the wavelength of the radiation, then the scattering experiment becomes more complex, but also more informative. In that case

$$\frac{Kc}{R_\theta} \sim \left\{ 1 + \frac{16\pi^2 r_g^2}{3\lambda^2} \sin^2 \frac{\theta}{2} \right\} \left\{ M^{-1} + 2Bc \right\} \quad (29)$$

It is now necessary to extrapolate Kc/R_θ to $c \to 0$ and $\theta \to 0$. This is achieved by means of a Zimm plot, as shown in Fig. 9. The points along lines such as that marked I refer to measurements at constant c, but different θ. The value of Kc/R_θ that will be approached at $\theta = 0$ and at the particular c value is given by the point at which line I has the abscissa value Kc, where K is some arbitrary constant. The equation of line II is

$$\left. \frac{Kc}{R_\theta} \right|_{\theta=0} = M^{-1} + Bc$$

If a number of determinations is made at constant θ, but varying concentrations (line III), the extrapolation to $c \to 0$ will yield

$$\left.\frac{Kc}{R_\theta}\right|_{c=0} = M^{-1}\left[1 + \frac{16\pi^2}{3}\frac{r_g^2}{\lambda^2}\sin^2\frac{\theta}{2}\right]$$

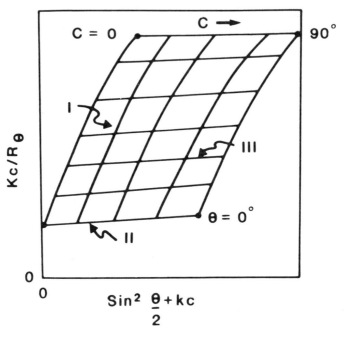

Fig. 9. Typical Zimm plot, as obtained from scattering data at different scattering angles and concentrations (for details, see text).

Both procedures therefore yield an estimate of M and the radius of gyration r_g is also obtained from the experiment. Its interpretation in terms of the dimensions of the particle does, however, depend on the shape of the particle. For instance, for a sphere of radius R, $r_g = (3/5)^{1/2}R$, whereas for a very long rod of length L, $r_g = L/(12)^{1/2}$.

4.4.2. Small Angle X-Ray Scattering

Although light scattering is adequate for the determination of M, the method fails when the dimensions of particles with $r_g < 10$ nm are to be estimated. Unfortunately most materials absorb strongly in the wavelength region < 300 nm, but they become transparent again in the X-ray region. However, the scattering is almost entirely confined to a narrow angular region. Experimentally

Solution Properties of Proteins

this does not matter, because reliable extrapolations to $\theta = 0$ can be performed. For the low angle range we can write

$$-h^2 r_g^2/3 = -\frac{16\pi^4 r_g^2 \sin^2(\theta/2)}{3\lambda^2} \tag{30}$$

with $h = (4\pi/\lambda)\sin(\theta/2)$. A linear relationship exists between $\ln i(\theta)$ and $\sin^2(\theta/2)$, with a slope $-16\pi^4 r_g^2/3\lambda^2$ and an intercept $\ln i(0)$. Even the smallest protein molecules can be investigated by this method.

4.5. Optical Spectroscopy

In biochemistry optical spectroscopy is usually confined to the observation of absorption spectroscopy (except for fluorescence spectroscopy). The requirement for the absorption of energy is that the energy available must be related to the difference between energy levels, according to Planck's law, $E = h\nu$. Usually the determination of absorption spectra is also confined to dilute solutions, where the Beer-Lambert law governs the absorption process:

$$A = \log(I/I_0) = \epsilon l c \tag{31}$$

where A is the absorbance, I and I_0 are the intensities of the transmitted and incident beams, respectively, l is the optical path length, and c is the concentration of the absorbing species (usually in mol/L). Data are sometimes expressed in terms of percent transmission, $100 I/I_0$, and other concentration units can also be used, e.g., g/100 cm^3. This is useful when the molecular weight of the absorbing molecule is not known. The proportionality constant ϵ is known as the extinction coefficient, with the units of L/mol/cm in Eq. (31).

Broadly speaking, four regions can be distinguished in the electromagnetic spectrum, corresponding to different types of energy transitions in the molecule. The microwave and far infrared regions of the spectrum ($\lambda > 10{,}000$ nm) monitor rotational transitions, corresponding to energies of the order of < 10 kJ/mol. Vibrational transitions are observed in the near infrared region, 1000–10,000 nm. Electronic transitions involving outer shell electrons give rise to visible and near ultraviolet spectra at 200–800 nm, whereas the study of inner shell electronic transitions requires X-ray absorption measurements.

Experimentally, the determination of absorption spectra requires a source of radiation of the requisite frequency range, a monochromator, the sample, and a suitable detection system. Most instruments in common use also include a beam splitter, so that the beam that passes through the sample is automatically compared with the beam that is made to pass through a reference standard of some kind. The output is then passed through a difference circuit and an amplifier and the spectrum is either recorded or the data collected on disk for future processing.

4.5.1. Infrared Spectroscopy

Pure rotation spectra can only be obtained from dilute gases, where collisions between molecules are negligible. Protein spectra are obtained from solutions, and therefore the near infrared region becomes of interest. Unfortunately there is such a wealth of detail in the spectra that it is often difficult to assign observed bands to definite bond vibrations. It must be remembered that a nonlinear molecule with n atoms has $3n - 6$ fundamental vibrational modes. A complete analysis of the protein spectrum is therefore out of the question. Nevertheless, certain groups display characteristic vibrational transitions, which are also sensitive to the environment and to hydrogen bonding, and this enables useful information to be derived from such spectra. The >CO group has a fundamental stretching vibration near 1700 cm^{-1} and the >NH group stretch vibration is observed near 3400 cm^{-1}. (Note that the frequency expressed in wave numbers, per cm, is the reciprocal wave length, i.e., the number of wave lengths cm^{-1}.) Both the >CO and >NH vibrations are very sensitive to hydrogen bonding. Their involvement in hydrogen bonding shifts the observed frequencies to lower values. One troublesome interfering factor is the strong absorption of water that occurs in the two regions 3500 and 1500 cm^{-1}. This makes the direct observation of fundamental stretching vibrations difficult, though not impossible with modern instrumentation.

In practice, the two most informative infrared bands are the so-called amide I and amide II bands (17). The former is the >CO band, referred to above, and the latter is a combination of the >NH deformation and the CN stretching vibration. The amide I and II band frequencies are diagnostic of the protein secondary structure, as shown in Table 6. A further series of amide bands (III–VII) have also been identified and assigned to specific vibrational modes involving the peptide groups. Because of the sensitivity of the various amide bands to hydrogen bonding interactions, they can be used

Table 6
Infrared Frequencies of Amide I and II Bands of Polypeptides
in Various Secondary Structures (cm^{-1})

Conformation	Amide I	Amide II
Random coil	1656 (s)	1535 (s)
Helix	1650 (s)	1516 (w)
	1652 (m)	1546 (s)
Pleated sheet (parallel)	1645 (w)	1530 (s)
	1630 (s)	1550 (w)
Pleated sheet (antiparallel)	1685 (w)	1530 (s)
	1632 (s)	

to monitor the interaction of water with the polar residues of the protein. Thus, infrared spectroscopy has become a popular technique for the study of protein hydration.

4.5.2. Visible and UV Spectroscopy

Visible and near UV spectra originate from relatively low-energy electronic transitions. Two main groups of chemical species are involved: compounds containing metals (especially transition metals), such as metalloproteins, and large aromatic ring structures or conjugated double-bond systems. Electronic transitions are usually very broad because they include a large number of closely spaced bands, each corresponding to a vibrational energy change.

Among the proteins, most of the measured absorption in the 260–280 nm region results from the aromatic amino acid residues phe, trp, and tyr. Figure 10 shows the absorption spectra of these three residues. Farther in the UV, other groupings begin to absorb. Thus, at < 230 nm other amino acid side chains absorb, but also the electron displacements in the peptide backbone produce absorption spectra. A strong band is observed at 200 nm and a weaker one at 225 nm. Below 180 nm oxygen, water and all other solvents absorb strongly, so that only dry materials can be studied under vacuum conditions; hence the term vacuum UV. As is the case for most spectroscopic techniques, the details of visible/UV spectra are sensitive to the environment of the chromophore and can therefore be used to monitor conformational changes, during which the absorbing residues become more or less able to perform diffusional motion or to interact with water. Applications of visible/UV spectrophotometry to assess the quality of protein isolates are discussed in Characterization of Isolates, in this volume.

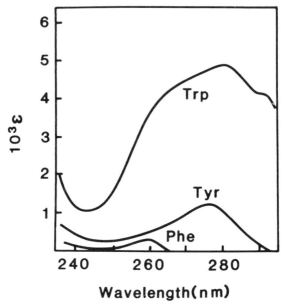

Fig. 10. Absorption spectra of aromatic amino acids.

4.5.3. Fluorescence Spectroscopy

In the case of absorption spectroscopy, one measures the energy absorbed during the excitation of a molecule from the ground state to some excited state, but without considering the eventual fate of the excited state. In a few cases, however, energy is reradiated at a different frequency from that of the exciting radiation. This process is called *fluorescence*. The fluorescence spectrum is independent of the wave length of the exciting radiation, but it appears at a higher wavelength, as seen in Fig. 11 for the case of tyrosine. The fluorescence spectrum is a sensitive fingerprint, for a compound and, since very low intensities of emitted radiation can be detected, it serves as a very sensitive analytical technique.

If the exciting radiation is polarized, then the degree of depolarization of the fluorescent radiation provides a measure of the Brownian diffusion of the chromophoric species.

Under favorable conditions, excitation energy can be transferred from one chromophore to another, so that the fluorescence spectrum of the latter is observed. The efficiency of such an energy transfer is a function of r^{-6}, where r is the distance between the two chromophores involved. For proteins it is usually found that even when excitation is in the phe or tyr bands, the trp fluorescence spectrum is in fact observed.

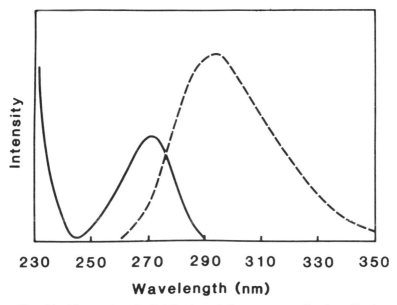

Fig. 11. Absorption (solid line) and fluorescence (broken line) spectra of tyrosine.

4.6. Optical Activity, Circular Dichroism (CD), and Optical Rotatory Dispersion (ORD)

The responses of dissolved molecules to polarized radiation depend explicitly on their structural asymmetry. Hence this response is a useful identifying tool for the type of structure that exists in a protein molecule.

Optical rotation is the consequence of a different refractive index (n) for left and right circularly polarized light, whereas CD results from a corresponding difference in absorption. CD is usually expressed as molar ellipticity [θ], which is equal to 3300 $\Delta\epsilon$, where $\Delta\epsilon = (\epsilon_L - \epsilon_R)$ is the difference in the extinction coefficients. If the dimensions of $\Delta\epsilon$ are L(mol/cm), then [θ] is in deg cm^2/decimol. The ORD is the wavelength dependence of the optical rotation. The Kronig-Kramer equation expresses the relationship between the optical rotation and CD at a particular wavelength λ:

$$[m']_\lambda = 2.303 \frac{9000}{\pi^2} \int_0^\infty \Delta\epsilon_{\lambda'} \frac{\lambda'}{\lambda^2 - (\lambda')^2} \, d\lambda' \qquad (32)$$

In proteins there are three distinct kinds of asymmetry that can give rise to optical activity:

1. The primary structure may be asymmetric. The α-carbon atoms of most amino acids have four different substituents.
2. A helical secondary structure can result in optical activity for electronic transitions in the main chain or in helically organized amino acid side chains.
3. The folded, tertiary structure may lead to the introduction of a symmetric or weakly asymmetric group into an asymmetric environment. Thus, tyr, which in isolation produces a weak optical activity, may exhibit marked activity when the residue is buried within the folded structure.

The three main types of organization in proteins, namely α-helix, β-sheet, and "random" coil, have distinctive CD spectra, as shown in Fig. 12. CD band intensities at 207, 290, or 222 nm have been

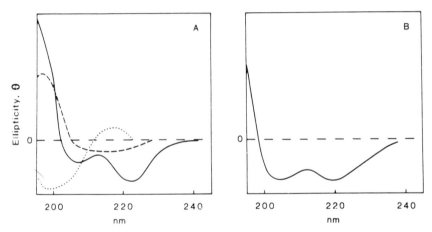

Fig. 12. (A) CD spectra of poly-L-lysine in the α-helical (solid line), β-sheet (broken line), and random coil (dotted line) conformations. (B) Calculated CD spectrum of myoglobin, based on the model spectra in (A).

used to determine the percentage of helix in a protein. An alternative method is by the use of the Moffitt equation:

$$[m'] = \frac{a_0 \lambda_0^2}{\lambda^2 - \lambda_0^2} + \frac{b_0 \lambda_0^4}{(\lambda^2 - l_0^2)^2} + \cdots \qquad (33)$$

where a_0 and b_0 are constants and $[m']$ is the molar rotation. The importance of this equation lies in the fact that it predicts a functional dependence of the ORD of helical polymers on $(\lambda^2 - \lambda_0^2)^{-2}$, where λ_0 is the center of the ORD band. Figure 12 also shows how the "pure" spectra associated with specific secondary structures can be mixed to calculate the percent helix, β-sheet, and random coil in a globular protein. On the whole, the agreement with X-ray diffraction results is satisfactory (see Table 7). Some empirical working rules have been established that help in the interpretation of CD and ORD spectra:

Table 7
Helix Content of Proteins:
Comparison of CD and X-ray Diffraction Results

Protein	Percent α-helix	
	CD	X-ray
Myoglobin	77	77
Lysozyme	29	29
Ribonuclease	18	19
Papain	21	21
Lactate dehydrogenase	31	29
α-Chymotrypsin	8	9
Chymotrypsinogen	9	6

1. The spectra are additive, i.e., they are the simple sum of the spectra of the individual chromophores.
2. The amplitude of the ORD curve or the rotational strength of a CD curve is a direct measure of the degree of asymmetry.
3. A chromophore that is symmetric can become optically active when placed in an asymmetric environment. This may or may not be accompanied by a change in λ_0.
4. The value of λ_0 and the magnitude and sign of $\Delta\epsilon$ at λ_0 allow the chromophore to be identified because λ_0 is always very near the absorption maximum of the molecule.

4.7. Magnetic Resonance

Two groups of techniques have in recent years been developed to the stage where they constitute powerful tools for the probing of macromolecular structures and interactions (18). Both of them are based on the fact that a spinning charged particle behaves as a magnet; it possesses a magnetic dipole moment, which, according to the laws of quantum mechanics, can adopt only certain values, measured as spin quantum numbers. If such a particle has a nonzero spin, then it will interact with an external magnetic field, and different energy levels can be distinguished. The energy levels involved are very closely spaced and correspond to radiation in the microwave region of the spectrum. In practice, the microwave radiation is kept at constant frequency and the magnetic field is swept so as to bring various transitions into resonance; hence the term magnetic resonance.

The two techniques referred to above are electron paramagnetic resonance (EPR), which relies on the existence of unpaired electrons for the nonzero spin, and nuclear magnetic resonance (NMR), in which nuclei with nonzero spin give rise to the spectrum. Magnetic resonance techniques have reached such degrees of refinement that only the briefest accounts can be given here. The interested reader is advised to consult one of the many specialist texts that already exist and more of which appear with regular frequency.

4.7.1. Electron Paramagnetic Resonance (EPR)

With the exception of very few molecular species (e.g., oxygen), unpaired electrons appear only in transition metals with incomplete d or f shells and in free radicals. Few biological molecules contain unpaired electrons, so that a popular device is to attach covalently a group of atoms that contains such an unpaired electron and is also chemically stable. The technique is known as spin labeling.

The frequency ν of the radiation corresponding to the spin transition is

$$\nu = g\beta H/h \qquad (34)$$

where β is the Bohr magneton, H is the magnetic field, h is Planck's constant, and the so-called g factor is the local magnetic field produced by the molecule that contains the unpaired electron. The measurement of g therefore identifies the source of the spectrum. In addition the magnetic moment of the electron interacts with the magnetic fields of nearby spinning nuclei, and this gives rise to the splitting of the line (hyperfine splitting). If the electronic spin is affected by more than one spinning nucleus, the spectrum can become quite complex. In principle, if the spin quantum number of the nucleus is I, then the epr spectrum has $(2I + 1)$ lines of equal intensity.

The spectrum in Fig. 13 illustrates the complex appearance that is the hallmark of many epr spectra. It is the spectrum of the copper–conalbumin complex and is composed of two distinct features: the splitting on the left (low-field region) is caused by the interaction between the d electron with the spinning Cu nucleus. The hyperfine splittings on the right result from the interactions of the electron with nitrogen ligands.

Solution Properties of Proteins

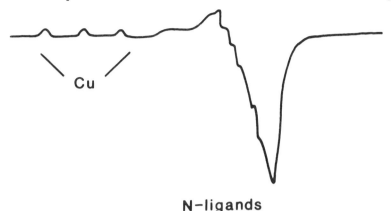

Fig. 13. EPR spectrum of the copper–conalbumin complex, showing the splittings caused by the copper d-electrons and the nitrogen ligands.

Apart from structural information, epr spectra also provide an indication of dynamic properties, e.g., rigidity or flexibility of a residue, and diffusion rates. By attaching a spin label to specific amino acid residues, it then becomes possible to monitor the motional freedom of that residue as affected by certain environmental factors.

4.7.2. Nuclear Magnetic Resonance (NMR)

All spinning nuclei that have a nonzero spin give rise to NMR signals. Nuclei that are frequently used in studies of proteins include ^1H, ^2H, ^{13}C, ^{14}N, ^{15}N, ^{17}O, and ^{31}P. Of these, the proton receives most attention. As is the case with other branches of spectroscopy, there are several distinct features of the spectrum that provide different types of information. The frequency at which a signal appears is closely related to the energy of the process that gives rise to the signal. The intensity of the signal is a measure of the concentration of the species, and the band shape provides information about the type of interaction involved and its dynamics.

The splitting, or clustering, of lines is caused by the interaction of one nucleus with another nucleus to which it is linked covalently. The intensities of such multiplets are related, e.g., 1:1 for a doublet, 1:3:1 for a triplet, 1:3:3:1 for a quartet, and so on.

In NMR it is the fact that the observed resonance frequency of a given nucleus is very sensitive to its environment that makes the technique so valuable. The shift in resonance frequency caused by the chemical environment is known as the *chemical shift*. Chemical shifts are usually measured as displacements from some reference standard and are measured in ppm:

$$\text{Chemical shift (ppm)} = \frac{\nu - \nu_{\text{ref}}}{\nu_{\text{ref}}} \times 10^6$$

Figure 14 summarizes some proton chemical shift values, referred to the protons in $SiMe_4$, a commonly used standard. Compared to other nuclei, the proton shifts are small; they lie within a range of 0–15 ppm. By comparison, ^{13}C shifts cover a range of 350 ppm and ^{31}P shifts, a range of 700 ppm. Nevertheless, because of the common occurrence of hydrogen nuclei in organic compounds, proton resonance is the commonly applied technique.

Reference was made above to the sensitivity of the chemical shift to many environmental factors. Some of these factors are of special importance in protein characterization. *Intramolecular shielding* can produce shifts of up to 20 ppm; it is caused by the movement of electrons by the external magnetic field, thus generating a field at the nucleus opposite in direction to that of the applied field. The effect is particularly pronounced in aromatic molecules and those containing conjugated double-bond systems. The direction of the shift produced (up-field or down-field) depends on the position of the nucleus with respect to the carbon atoms making up the aromatic system.

Paramagnetic shifts are caused by unpaired electrons. They can amount to 20 ppm and can be detected when the unpaired electron is as much as >1 nm distant from the nucleus producing the resonance. The effect is particularly useful because it is related to r^{-3}, where r is the distance between the nucleus and the unpaired electron. Paramagnetic shifts therefore provide information of a structural nature.

Intermolecular effects occur if the electron distribution in one molecule affects the chemical shifts of nuclei in another molecule. In polymers such effects can also be intramolecular, i.e., residues in one part of the chain can influence the chemical shifts of nuclei in a distant part of the chain, if the two regions are placed in close proximity as, for instance, in the tertiary, folded structure of a protein.

If a molecule undergoes chemical or physical changes of a reversible nature or is subjected to reversible changes of environment (*chemical exchange*), the observed chemical shift depends on the rates of such processes. If exchange is very slow, two distinct signals

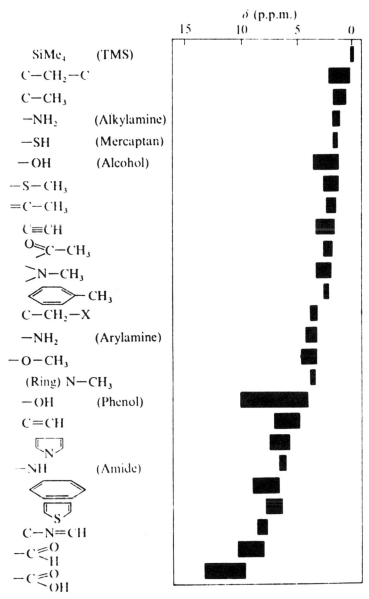

Fig. 14. Ranges of ¹H NMR shifts (referred to SiMe₄) of functional groups that commonly occur in proteins (ref. *18*).

are observed because the molecule will then be in either of the two states, rather than in transition between them. As exchange becomes more rapid, the two lines will broaden and eventually overlap and merge. In the limit of very rapid exchange, only one narrow signal is observed at a position that constitutes an average between those of the two shifts observed for slow exchange. Exchange broadening is a useful phenomenon because the areas of the two resonance lines are proportional to the number of nuclei producing the lines, and hence to the respective populations of the two states. Examples of chemical exchange effects include the binding of substrate molecules to enzymes, *cis–trans* isomerism, and other conformational changes.

There are yet two further important effects to be considered. The magnetic field caused by a spinning nucleus is affected not only by unpaired electrons, but also by the fields caused by other, nearby nuclei. This gives rise to spin–spin interactions that result in line splitting (spin coupling). The appearance of multiplets can be of great help in the assignment of signals to specific protons. The spacing between the components of a multiplet bears a relationship to the distance between the nuclei that give rise to the splitting. The spacing is expressed as a coupling constant, J, which is measured in Hz and is related to the torsional angle ϕ about the $-C_\alpha N$ bond in a peptide group by

$$J = 8.5 \cos^2 \phi, \quad 0 < \phi < 90° \quad cis$$
$$J = 9.5 \cos^2 \phi, \quad 90 < \phi < 180° \quad trans$$

The above equations (Karplus equations) are two examples of relationships between J and torsional angles. Similar equations have been derived for other types of chemical linkages, e.g., $H-C-C-H$, $H-C-O-H$. Their applications are subject to various assumptions and complications. Although much useful structural information can be derived from spin coupling interactions, these processes give rise to very complex spectra. The useful device of spin decoupling allows for a simplification of the proton spectrum. Its comparison with the primary spectrum allows for a more reliable assignment of individual signals to individual protons.

The final NMR effect to be considered in this abbreviated discussion is *nuclear magnetic relaxation*. In the absence of a magnetic field, the magnetic moments of a system of nuclei ($I = \frac{1}{2}$) will have random orientations and the populations of the two magnetic energy levels are equal. When a field is applied, the nuclei must adopt one of two allowed orientations; the populations are given by the Maxwell-Boltzmann distribution law. The equilibrium is attained by a first-order rate process with a rate constant $1/T_1$, where T_1 is

Solution Properties of Proteins

known as the *spin-lattice relaxation time*. It follows that when the field is removed, the nuclei will relax according to the same rate law. The important feature of T_1 is that it can be related to the Brownian diffusion of the molecule that contains the nucleus under study, or to motions of a segment of such a molecule, e.g., the aromatic ring in phenylalanine.

Depending on the plane of magnetization, another relaxation time, T_2, is defined. Here T_2 measures the decay of magnetization in the x and y axes, whereas the decay of magnetization along the z axis is measured by T_1. If the molecular motion is rapid and isotropic, then $T_1 = T_2$. This is the case for many small molecules, but not necessarily for polymers or for systems in which diffusion is inhibited. Two practical results are of importance: T_2 is related to the nmr line width by $\Delta \nu_{1/2} = 1/\pi T_2$, where $\Delta \nu_{1/2}$ is the line width at half-height and T_2, being directly related to molecular motion, becomes long with rapid motion.

As is the case with other spectroscopic techniques, NMR is subject to a number of technical problems and limitations. Many of them, e.g., the powerful water proton signal that dominates aqueous solution spectra, can now be overcome by the use of various refinements and modifications. Chief among them has been the introduction of Fourier transform techniques, although greatly improved spectral resolution resulting from the development of more powerful magnets and the diversification of the nuclei that can be studied have also contributed materially to the ever-growing popularity of NMR.

Certain useful rules can be applied to the interpretation of NMR spectra: (1) the proton shift of a particular group, e.g., $-OH$, is affected by other groups in the same molecule, e.g., alcohol, amino acid; (2) proton shifts are also affected by molecular weight and molecular complexity. Such effects derive from paramagnetic centers, ring currents, or electric fields produced by charged groups; (3) a change in the proton shift as a result of treating the protein with a chemical or physical agent is indicative of a change in the structure of the protein, so that the environment of the proton has changed. Such a change can be of a minor nature, e.g., small shifts in pH; (4) covalent coupling of nuclear spins gives rise to splitting. The number of lines is symptomatic of the nature of the group in question. The coupling constant J is related to the through-space distance of the nuclei involved. In a protein, if some chemical treatment results in a change in J, then a conformational change is indicated; (5) in a system that is subject to chemical exchange, e.g., an enzyme–substrate interaction, the width of the line and/or the number of lines provide information about the population of the two states and the rate of exchange between the two environments; (6) the line width is a measure of the relative mobility of the nucleus. Note, however, that changes in band width can result from exchange phenomena (*see* rule 5 above); (7) the binding of a ligand produces changes in the spectra of the ligand *and* of the macromolecule with which it interacts, particularly of those nuclei that are close to the site of the interaction. Both the position of the shift as well as the line width are usually affected. In the interpretation of protein spectra, the observed signals must first of all be assigned to the constituent amino acids. The latter have been well studied and tabulations exist of the proton and carbon resonances under various conditions of pH. Table 8 lists the ^{1}H resonances of three amino acids, each one characteristic of a different type: apolar, charged, and aromatic.

When the primary sequence is known, fragments can then be obtained and purified by stepwise enzymatic cleavage of the protein. The determination of the NMR spectra of the fragments then allows certain residues to be assigned to a particular region of the macromolecule. However, the chemical shift of a given residue in the fragment is not always identical with that in the intact molecule.

A further device in the assignment of peaks makes use of the ability of certain residues (or groups of residues) to bind ligands. Partial deuteration, when possible, simplifies the spectrum to be analyzed, because proton resonances from the deuterated residues will vanish. Simple organisms can be grown in a completely deuterated medium to which a single protonated amino acid might be added, so that only signals due to this residue will appear in the spectrum. This method is powerful, but expensive. When a full structure is known from X-ray or neutron diffraction, the NMR spectrum can be compared with the predictions of the structure. Many mutant proteins have been found that differ from the mesophilic protein by only one amino acid residue, although performing the same function and, presumably, possessing the same three-dimensional structure. A comparison of the NMR spectra should make possible the identification of the particular residue, but spectral changes can also occur for those residues that interact with the mutant residue. Finally, spin–spin decoupling can be used to simplify the spectrum and help in the assignment of signals to specific residues and their positions in the macromolecule. In recent years, methods employing two-dimensional NMR have also been de-

Table 8
Chemical Shifts (Hz) of ^{1}H Resonances in Amino Acids and Peptides; Measured at 220 MHz With an Internal Standard $CH_3Si(Me)_2(CH_2)_3SO_2Na^+$

Hydrogen type	Equivalent H per residue	Compound	Chemical shift
Leucine			
CH$_3$	6	L-leucine	208
		gly-leu	197
		his-leu	193
β-CH$_2$ + γ-CH	3	L-leucine	374
		gly-leu	350
		his-leu	345
α-CH	1	L-leucine	813
		gly-leu	920
		his-leu	905
Lysine	2	L-lysine	321
γ-CH$_2$	2	poly-L-lysine	315
δ-CH$_2$	2	L-lysine	375
β-CH$_2$	2	L-lysine	412
ε-CH$_2$	2	L-lysine	664
	2	poly-L-lysine	660
α-CH	1	L-lysine	821
	1	poly-L-lysine	947
Tyrosine	1	gly-tyr	628
β-CH$_2$	1	tyr-ala	694
α-CH	1	gly-tyr	972
	1	tyr-ala	906
Aromatic ortho to OH	2	L-tyrosine	1514
	2	gly-tyr	1504
Aromatic meta to OH	2	L-tyrosine	1583
	2	gly-tyr	1572

veloped and are now being used extensively in conformational studies of macromolecules. Of all the instrumental techniques that have been developed and refined during the past two decades to probe the properties of macromolecules in solution, NMR is by far the most powerful. Developments in magnet design, microelectronics, and data processing equipment have further enhanced the potency and versatility of NMR techniques. The following examples serve to illustrate just a few of the applications of NMR techniques to the study of various aspects of protein chemistry and technology. The first five examples describe the most common applications of NMR to the study of proteins.

(1) The proton of the α-C atom of some amino acids is sensitive to the secondary structure to the extent that two characteristic chemical shifts are observed for the α-helix and the random coil. The fraction of the amino acids in the α-helical configuration can thus be calculated. (2) One of the most common applications of NMR is to the study of protein unfolding (denaturation), caused by thermal, chemical, or other means (see Chapter 4). Figure 15 is a diagrammatic proton spectrum of the histidine and tryptophan residues in a protein. The subscripts N and R refer to the native

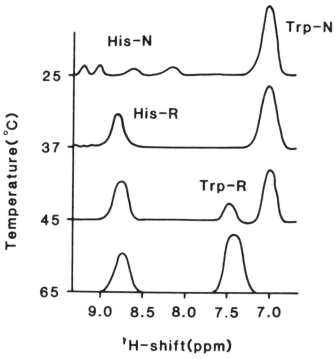

Fig. 15. Histidine and tryptophan proton NMR spectra of a hypothetical protein showing various stages of denaturation (N = native and R = random coil, denatured).

and random coil denatured state of the protein. At 25°C the protein is in its native state and the four histidine signals indicate four distinct chemical environments. The intensity of the tryptophan signal is large compared to that of the individual histidine signals, indicating the presence of several trp residues in the same (average) environment. At 37°C the his signals have merged and the resulting signal is found in the position associated with the R state. However, the region of the protein in which the trp residues are buried is still in its native conformation. At 45 °C the resonance of trp in the R state first appears and denaturation is complete at 65 °C. The area of the subsidiary peak observed at 45 °C is 1/6 of that of the total trp signal, indicating that one trp residue is affected by exposure to 45 °C. By further refinement of the NMR studies it can be established which of the several trp residues is responsible for the signal at 7.5 ppm at 45 °C. A plot of the line positions as a function of temperature, as shown in Fig. 16, provides detailed information about the thermal denaturation of the protein, especially if the peptide sequence and tertiary structure are known. From the

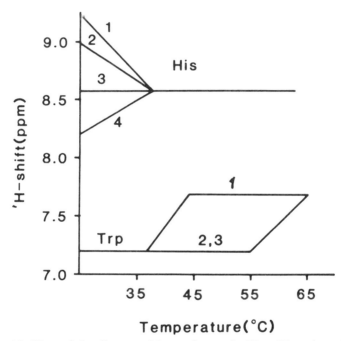

Fig. 16. Plot of the line positions shown in Fig. 15 as function of temperature.

various line widths, the relative mobilities of the residues can be estimated. Thus, the unchanged width of the trp residue before and after denaturation indicates that the trp residues are situated in a flexible region of the protein. Normally, a narrowing of the signal would be expected upon denaturation. (3) The active site of an enzyme can be probed by the change in the NMR signals of certain amino acids as substrate is added. Thus, Fig. 17 shows

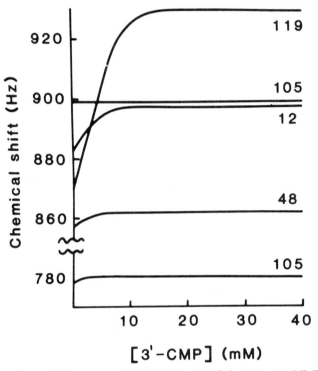

Fig. 17. Effect of 3'-CMP concentration of the proton NMR signals of histidine protons in ribonuclease (for details, see text).

the effect of 3'-CMP concentration on the five histidine signals in ribonuclease (RNase); only three residues (12, 48, and 119) show significant effects. More extensive studies, using also the proton signals of the substrate, as well as imidazole instead of histidine and phosphate instead of 3'-CMP, have shown that his-119 is hydrogen bonded to the phosphate group of 3'-CMP. On the other hand, his-12 is not affected by the binding of inhibitors to the enzyme, indicating that any shift in the his-12 resonance is caused by a more general environmental effect, e.g., a change in pK, and that it must be close to the active site, without taking a direct part in the catalytic function. (4) Binding can be studied in a more direct manner by monitoring the NMR signal of the ligand or its analog. Thus, the binding of Ca^{2+} to calmodulin has been studied by ^{113}Cd

Solution Properties of Proteins

and ^{43}Ca NMR, indicating that calmodulin binds two Ca^{2+} ions, but that there are two distinct kinds of sites with different affinities (19). (5) Great advances have been made in the quantitative determination of protein structure in solution with the aid of techniques such as paramagnetic shift reagents, spin decoupling, and two-dimensional NMR. A classic study of lysozyme illustrates some of the methods and conclusions. By taking a difference spectrum of a solution with and without Gd(II), it was possible to identify those amino acids that are close to the metal binding site. Figure 18 is a map of the results obtained from X-ray crystallography, which

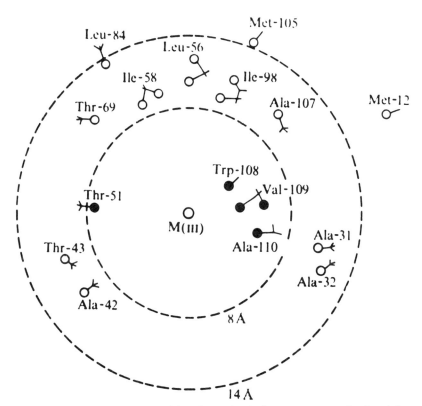

Fig. 18. Map of the metal binding site in lysozyme, as obtained from X-ray diffraction and ^{1}H NMR (ref. 18).

is fully consistent with the NMR findings that three methyl signals and one aromatic signal are primarily affected: the Me signals belong to val-109 and ala-110, whereas the aromatic proton signal belongs to trp-108.

Although the early studies relied on a knowledge of the crystal structure, this is no longer necessary because with the aid of computers the various shifts can be fitted to a variety of possible structural models and refined.

References

1. R. Jaenicke, *Protein Folding* Elsevier, Amsterdam (1980).
2. J. A. Schellmann and R. B. Hawkes, in *Protein Folding* (R. Jaenicke, ed.) Elsevier, Amsterdam (1980), p. 331.
3. C. Tanford, in *Physical Chemistry of Macromolecules* John Wiley & Sons, New York (1961), Chap. 8.
4. F. Hoffmeister, *Arch. Exp. Pathol. Pharmakol.* **24**, 247–260 (1888).
5. P. H. von Hippel and T. Schleich, *Acc. Chem. Res.* **2**, 257–265 (1969).
6. P. H. von Hippel and K. Y. Wong, *J. Biol. Chem.* **240**, 3909–3923 (1965).
7. P. H. von Hippel and A. Hamabata, *J. Mechanochem. Cell Motil.* **2**, 127–138 (1973).
8. F. Franks and D. Eagland, *Crit. Rev. Biochem.* **3**, 165–219 (1975).
9. J. F. Brandts and L. Hunt, *J. Am. Chem. Soc.* **89**, 4826–4840 (1967).
10. S. Y. Gerlsma and E. R. Stuur, *Int. J. Pep. Protein Res.* **4**, 377–383 (1972).
11. T. Arakawa and S. N. Timasheff, *Biochemistry* **21**, 6536–6544 (1982).
12. H. Uedaira and H. Uedaira, *Bull. Chem. Soc. Japan* **53**, 2451–2455 (1980).
13. K. Gekko and T. Morikawa, *J. Biochem.* **90**, 39–50 (1981).
14. E. G. Richards, *An Introduction to Physical Properties of Large Molecules in Solution* Cambridge University Press, Cambridge, UK (1980).
15. K. E. van Holde, *Physical Biochemistry* Prentice-Hall, New Jersey (1971).
16. H. Eisenberg, *Biological Macromolecules and Polyelectrolytes in Solution* Clarendon Press, Oxford (1976).
17. H. R. Mahler and E. H. Cordes, in *Biological Chemistry* Harper & Row, New York (1966), Chap. 3.
18. R. A. Dwek, *Nuclear Magnetic Resonance in Biochemistry* Clarendon Press, Oxford (1973).
19. S. Forsen, E. Thulin, T. Drakenberg, J. Krebs, and K. Scamon, *FEBS Lett.* **117**, 189–194 (1980).

Chapter 4
Conformational Stability

Denaturation and Renaturation

Felix Franks

1. Origin of the Native State

In order to perform its biological function, a protein must possess the correct configuration, i.e., it must be in its *native* state, with the correct secondary, tertiary, and, where applicable, quaternary structure. Under in vivo conditions, this is achieved on the ribosome during the synthesis of the protein. Any mistakes in the folding mechanism are then symptomatic of a pathological condition. Howver, under in vitro conditions of isolation, concentration, drying, and so on, changes may occur that will result in partial or complete inactivation. The protein is then said to be *denatured*. With the vast majority of proteins, the stability of the native (N) state, relative to the denatured (D) state, is highly marginal, amounting to no more than 60 kJ/mol, which is equivalent to the strength of only 3–4 hydrogen bonds (1). Yet, the native structure usually contains several hundred such bonds. From a biological point of view, this marginal stability is required so that proteins can be turned over rapidly, thus avoiding the buildup of, say, immunoglobulins or hormones in the serum. On the other hand, the labile nature of the native state presents problems for the processor who must avoid extremes of pH, ionic strength, temperature, shear, and so on during the various stages of the isolation and concentration process.

The parameters that govern the native state stability have been subject to intensive study, as have the interactions that appear to be mainly responsible for the maintenance of the tertiary and quaternary structures. As a first approximation, any measured property X of a native protein can be divided into three contributions:

$$X = X_{intrinsic} + X_{medium} + X_{interaction}$$

where the first term results from intramolecular forces within the polypeptide(s), the second term describes the properties of the solvent medium, and the third term describes the effects arising from the interactions between the polypeptide and the components of the solvent medium. It is this last term that is responsible for the maintenance of the native state stability, but it only contributes a few percent to the total measured magnitude of X. When the native structure consists of two or more subunits, a further term must be added to account for the interactions *between* subunits.

Although the N structures of many proteins are accurately known, thanks to the important developments in crystallography, little is yet known about the configurational properties of D states. Under physiological conditions proteins exist in their native states, but any clinical application of an isolated protein will also require this state to be maintained throughout the isolation procedure, or to be recovered at the end of the processing treatment. The same is true for industrially used enzymes. The demands made on the processor are not as severe as might be expected: Under carefully chosen conditions many proteins can be precipitated or crystallized without loss of activity.

The effects produced by partial or complete conversion into the D state are difficult to generalize, because the D state itself is not well defined. Denaturation may be indued for one of several reasons. A controlled dissociation of the N state often precedes the eventual reassembly of subunits into a new state, a "technological" N state with predictable attributes that are not the same as the original biological attributes. For instance, a globular plant storage protein may be converted into a fibrous protein with certain desirable properties, such as tensile strength or water-holding capacity. This "technological" state, which we shall call the T state, may be formed irreversibly, e.g., by crosslinking reactions. It is not inconceivable that such a state might possess some form of biological activity of its own, although to the author's knowledge no such cases have been reported. On the other hand, the D state may be a purely transitional stage during the processing, when it is intended to renature the protein and to recover the original activity. For this to be achieved, the denaturation must of course be performed reversibly. It must be borne in mind, however, that denatured proteins are very susceptible to proteolytic attack and liable to irreversible aggregation, the latter being dependent on the protein concentration.

2. Two-State Model of Protein Stability

The simplest model treats protein stability in terms of two populations of states in equilibrium, as depicted in Fig. 1 (*1*). The group of states on the left-hand side (low energy) is identified with the N state and the group on the right-hand side (high energy) with the D state. Treating the equilibrium in terms of two *populations* implies that there is no unique N or D state, but that we are dealing with a group of structures of similar energy. Very sensitive analytical techniques, such as NMR, may well be able to differentiate between the various substructures making up each group. The main point is that the two populations do not overlap and that no intermediate species exist. The experimental study of the N $\rightleftharpoons$ D equilibrium requires that three conditions be met: (1) thermodynamic reversibility, (2) an experimental technique that can detect the transition in the presence of a perturbing influence, and (3) a common conformational state so that the stabilities of different proteins can be compared (*2*).

Thermodynamic reversibility usually requires measurements on very dilute solutions, but in many cases it has been established experimentally (*3,4*). As regards experimental techniques, many of those described in chapter 3 have been applied to the study of denaturation (*5*). The reference state for purposes of stability comparisons is usually taken to be the N state.

If an equilibrium constant can be assigned to the N $\rightleftharpoons$ D transition, then for the limiting case of the infinitely dilute solution

$$\Delta G^0 = -RT \ln K \qquad (1)$$

where ΔG^0 measures the stability of the N state relative to the D state. From Eq. (1), all other thermodynamic functions can be obtained by standard procedures. If we chose some physical parameter P, which is sensitive to the difference between the N and D states, then the effect of any perturbing influence on the N state

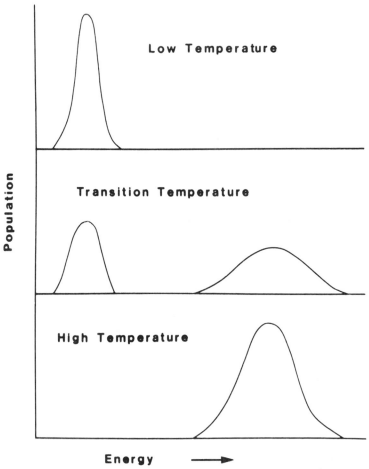

Fig. 1. Schematic representation of the N → D equilibrium, assuming only two populations of states. A change in the temperature alters the relative abundance of these two groups without any other energy state becoming significantly populated.

can be monitored and K can be obtained as a function of the intensity of the perturbing effect (1). This is shown diagrammatically in Fig. 2a, and some actual experimental data for 3-phosphoglycerate kinase are shown in Fig. 2b (6). The method chosen was circular dichroism at 222 nm, which monitors the peptide backbone conformation, and at 278 nm, is diagnostic of aromatic amino acid residues. In the particular case illustrated in Fig. 2b, the destabilizing agent was guanidinium hydrochloride (GuHCl). If P_N and P_D

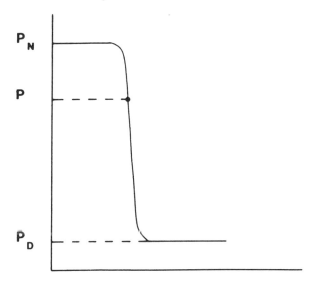

Fig. 2a. The effect of perturbant intensity on the N ⇌ D equilibrium; P is any physical or biological property that can be measured independently for the N and D states.

describe the experimental values adopted by P for the two states, then for any point on the curve:

$$K = f_U/f_N = (P - P_U)/(P_N - P) \qquad (2)$$

where f_U and f_D are the fractions of molecules in state U and D, respectively. In practice, experimental considerations limit the measurements to the range $0.1 < K < 9$, so that $+5 > \Delta G^0$ kJ/mol > -5.

The curve in Fig. 2a is highly idealized, because there is no reason why P_N or P_U should be independent of the perturbant intensity. Nevertheless, a comparison of the experimental data in Fig. 2b with the assumptions of the simple two-state model shows this to be a satisfactory representation of the unfolding of phosphoglycerate kinase. Such transition states as do exist (and they must of course exist) are not significantly populated.

The CD data need to be further processed to yield ΔG^0 extrapolated to zero GuHCl concentration. The available data in Fig. 2c graphically illustrate the limitation of most experimental techniques as regards the stability range that can be probed. Various extrapolation techniques have been suggested, all of a semiempirical nature (1).

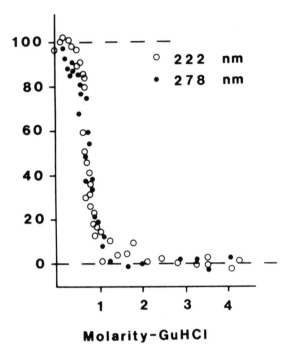

Fig. 2b. Reversible denaturation of yeast 3-phosphoglycerate kinase by GuHCl, monitored by CD.

One popular approach views denaturation as the equivalent of transferring the amino acid residues from the "native," aqueous solvent to the perturbing solvent. ΔG is then expressed in terms of the sum of the individual amino acid contributions, $\delta g_{tr,i}$ and a mean change in solvent exposure $\Delta \alpha_i$ to which residue i is subjected during denaturation:

$$\Delta G = \Delta G_0 + \sum_i \Delta \alpha_i \, \delta g_{tr,i}$$

Here $\delta g_{tr,i}$ is the free energy required for the transfer of residue i from the aqueous medium to the denaturing medium (aq. GuHCl in Fig. 2). Since $\delta g_{tr,i}$ is not a linear function of the GuHCl concentration, a plot of ΔG vs GuHCl concentration will not be linear (*see* Fig. 2c). The best fit is obtained by adjusting $\Delta \alpha_i$; the curve must also pass through the $\Delta G = 0$ point, as obtained by experiment (*see* Fig. 2b).

The extrapolated ΔG^0 represents the stability of the protein in the absence of the perturbant, i.e., the *intrinsic* stability. However, the extrapolation becomes the more uncertain the higher the con-

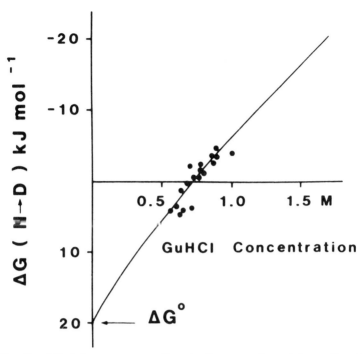

Fig. 2c. ΔG data, corresponding to the experimental results shown in Fig. 2b and illustrating the limited range over which the N/D equilibrium can be studied in practice (1).

centration of perturbant required for denaturation and the steeper the slope of the ΔG vs concentration curve. The denaturation of lysozyme by urea provides a good illustration: Experimental measurements are limited to the urea concentration range 4–7M, thus requiring a very long extrapolation to $c = 0$ for the estimation of ΔG^0 (50 kJ/mol).

2.1. Further Analysis of "All-or-None" Denaturation

For any meaningful comparison of the stabilities of different proteins, a knowledge of ΔG^0 and the derived thermodynamic quantities is essential. The latter can be obtained from temperature, pressure, or concentration dependences of ΔG^0. In certain favorable cases they can be obtained by direct measurement. Thus, ΔH^0, the heat of denaturation, can be obtained by the van't Hoff method from ln $K(T^{-1})$ or directly, by calorimetry (7). In fact, the agreement between the two ΔH^0 values is evidence that the two-state denaturation model is valid. Although the van't Hoff treatment of equilibrium data is popular, it is also subject to considerable experimental uncertainties, as shown in Table 1. The errors are particularly marked when the experimental temperature range is narrow, a frequent prerequisite for biochemical measurements. Direct calorimetric or volumetric measurements are much to be preferred for estimations of ΔH^0, ΔV^0, and their further temperature and/or pressure derivatives.

Table 1
Propagation of Errors (Standard Deviation)
in the van't Hoff Treatment of Spectroscopic Data (23)[a]

Parameter	Experimental design	
	20–30°C in 2-deg steps	5–50°C in 5-deg steps
log K	0.02	0.02
ΔH (J/mol)	4055	794
ΔS (J/mol/K)	14	1.4
ΔC_p (J/mol/K)	2784	117

[a]The accuracy that can be attained by the use of temperature-difference spectroscopy is usually lower and hence the errors will be greater than those shown in the table.

A further advantage of direct measurements is that they do not require the long extrapolation to zero perturbant concentration discussed above, because experimental determinations are possible on the N and D states independently, and not just in the narrow transition range. Temperature scanning calorimetry has been developed into a very sensitive experimental tool for the study of protein stability. The instrumental output measures specific heat as a function of temperature, as shown in Fig. 3 for lysozyme at different pH values (8). The following information can be extracted from the data:

1. $C_p(T)$ of the native protein below the transition temperature T^*
2. $C_p(T)$ of the denatured proein above the transition temperature T^*
3. The heat capacity change ΔC_p associated with the N → D transition

4. The enthalpy change ΔH^* associated with the transition
5. The temperature T^* of half-conversion, where $K = 1$ ($\Delta G^0 = 0$)
6. The van't Hoff enthalpy change $\Delta H_{vH} = 2(RT\Delta C_p)^{\frac{1}{2}}$

Indirectly $\Delta G(T)$ can also be obtained by the integration of the C_p data:

$$\Delta G(T) = \Delta H^*[(T^* - T)/T^*] - \int\Delta C_p dT + T \int(\Delta C_p/T)dT \quad (3)$$

The significance of Eq. (3) will be referred to in subsequent sections.

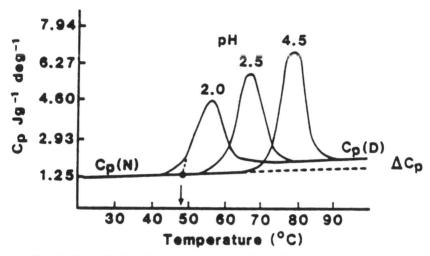

Fig. 3. Specific heat/temperature profiles of lysozyme as a function of pH. The area under $C_p(T)$ curve is a measure of ΔH (N $\rightleftharpoons$ D) and the change in the base line is equal to ΔC_p (N $\rightleftharpoons$ D) (8).

3. Thermal Denaturation

The effect of temperature as a means of destabilizing proteins has been studied in detail. High temperatures are of technological importance but irrelevant to the in vivo functioning of a protein. To study thermally induced denaturation, the treatment outlined in section 2 can be used. For enzymes, the activity profile is probably the most sensitive monitor. Otherwise any of the spectroscopic, optical rotation, or macroscopic (sedimentation, viscosity, calorimetry) methods find application.

Thermal denaturation is sensitive to the amino acid composition/sequence, as shown in Table 2 for different collagen types: T^* is seen to be a function of the proline (and hydroxyproline) content, which, in turn, is related to the species (9). The thermal denaturation temperature T^* is frequently sensitive to other forms of protein stabilization/destabilization treatments, such as pH, ionic strength, and so on, and is therefore used as monitor for the effectiveness of such treatments.

Table 2
Denaturation Temperatures (T^*) of Collagen of Different Imino Acid Contents in Aqueous Solution at pH 3.5 (9)

Collagen	Pro, HOPro/ 1000 residues	$T^*/°C$
Rat	226	40.8
Pike	199	30.6
Merlang	—	21.5
Cod	155	20.0

4. Low-Temperature Denaturation

The detailed shape of $\Delta G(T)$, as given by Eq. (3), depends on the magnitude of ΔC_p and its dependence on temperature (if any). In practice it is found that for all proteins studied in detail, $\Delta C_p >> 0$. This makes ΔG very sensitive to changes in temperature. In many cases $\Delta G(T)$ approximates to a parabola (10,11). There thus exist *two* denaturation temperatures and a temperature of maximum stability, T_{max}, which varies from protein to protein, but never appears to be at the physiological temperature of the organism. For tryptic enzymes, cytochorme c, and myoglobin, T_{max} = 12, 25, and 30°C, respectively.

The *low* denaturation temperature has not yet been extensively studied because it often falls below the equilibrium freezing point of the solution. However, the phenomenon of cold inactivation is well known. For instance, tubulin and TMV-coat protein depolymerize below 25°C (12,13), and many enzymes have been shown to become cold inactivated: Phosphofructokinase (PFK) dissociates from a tetramer into two dimers at +10°C (14). This is a fully reversible process, but occurs with loss of enzymatic activity. Cold inactivation and lability of proteins are important factors in physiological chill and freeze injury and resistance (11).

The extreme sensitivity of $\Delta G(T)$ to the amino acid composition is illustrated by the following two examples. The $\Delta G(T)$ curves for a number of T4 lysozyme mutants are compared in Fig. 4. Mutation 1 is arg-86 → his and mutation 2 is glu-128 → lys or ala. The other mutations have not been identified. All the mutations shown reduce T^* from its mesophilic value of 57°C, but also increase T_{max}, while at the same time reducing the degree of stability at T_{max}, in most cases quite substantially (15). The other example is a comparison of $\Delta G(T)$ and $\Delta H(T)$ for phosphoglycerate kinase (PGK) in 2.26M GuHCl (see also section 2), with enzyme taken from two different sources: yeast and a thermophilic microorganism (16). The experimental CD data were fitted to a simple polynomial for $\Delta G(T)$:

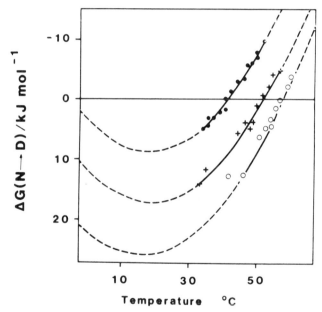

Fig. 4. Effect of mutations on the thermal stability of T4 lysozyme: mesophile (○), mutation 1 (+), and mutation 2 (●) (for details, see text).

$$\Delta G(T) = AT + BT^2 + CT^3 + D$$

from which $\Delta H(T)$ was obtained by differentiation. Figure 5a shows up the qualitative similarity, but quantitative difference, between the same enzymes isolated from two different organisms. The difference is further magnified in $\Delta H(T)$, shown in Fig. 5b. Whether the heat stability of the thermophilic PGK is caused by the constant, almost zero enthalpy is open to speculation. The second tempera-

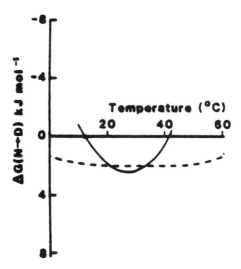

Fig. 5. (A) Free energy, and (B) enthalpy of denaturation of phosphoglycerate kinase from yeast (drawn-out-line) and a thermophilic bacterium (broken line) (16).

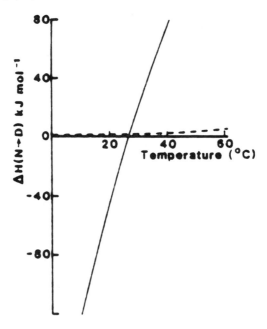

Fig. 5B

ture derivative of $\Delta G(T)$ yields ΔC_p, which is practically zero for the thermophilic PGK, but amounts to 7.3 kJ/(mol K) for yeast PGK, causing the latter only to be stable in the limited temperature range 10–40°C (in 2.26M GuHCl).

4.1. Freeze Denaturation

The main injurious effect of freezing is not the low temperature, but the concomitant concentration of all soluble species while ice separates from the mixture as a pure phase. Complex relationships exist between the initial protein concentration and the degree of freeze denaturation observed at different subzero temperatures; they are described in detail in ref. *10*.

Denaturation is especially marked at high subzero temperatures, damage being reduced as the temperature falls to below the eutectic temperature of the mixture, i.e., as the liquid phase volume approaches zero. Freezing also increases the concentrations of buffers and other additives present in the solution by orders of magnitude. This can lead to the precipitation of acids and/or salts and large changes in the perceived pH, which in themselves can cause damage to the protein. The physical chemistry of freeze-concentrated systems lies at the very basis of technical freeze drying, a fact that is not generally appreciated.

5. Pressure Denaturation

According to the van't Hoff equation

$$d \ln K/dP = \Delta V \tag{4}$$

where ΔV is volume difference between the D and N states. Conversely a knowledge of ΔV provides an indication of the effect of pressure on the N → D equilibrium. However, ΔV itself changes with pressure, reflecting the compressibilities of the N and D states. Table 3 summarizes some ΔV data and the "average residue" figures, based on the assumption that ΔV is the additive sum of the individual amino acid values. ΔV per residue is seen to vary from zero to values larger than can be accounted for in terms of simple α-helix → coil or β-sheet → coil transitions.

Limited information is available on the high-pressure denaturation of proteins (*17*). Figure 6 shows some pressure/temperature/pH

Table 3
ΔV for the Transfer of Proteins from the N State to the D State in 6M GuHCl

Protein	ΔV, cm³/mol Protein	"Average" residue
Ribonuclease	30 ± 30	0.24
Lysozyme	−30 ± 40	−0.23
Tubulin	0 ± 160	0.00
Chymotrypsinogen A	100 ± 50	0.40
α-Chymotrypsin	150 ± 50	0.62
BSA	750 ± 200	1.27
α-Lactalbumin	40 ± 30	0.31
Lactate dehydrogenase	70 ± 70	0.22
Catalase	240 ± 100	0.44
β-Lactoglobulin	400 ± 60	2.50

relationships for metmyoglobin (18). The curve in Fig. 6a represents the $P(T)$ locus at pH 2.07 for which $K = 1$, i.e., half-denaturation. It suggests that at low temperatures ($<10\,°C$) and pressures of 300 MPa, chymotrypsinogen is denatured. However, the protein refolds as the temperature is increased to $>10\,°C$, only to denature once again at $40\,°C$. This indicates that ΔV changes sign. The isobars for metmyoglobin in Fig. 6b, again corresponding to half-denaturation, also indicate the complex P–T–pH relationship.

6. Shear Denaturation

Closely related to pressure denaturation are the effects of shear on protein stability. Since protein purification procedures involve mixing, flow in tubes, ultrafiltration, passage through pumps, and so on, the effects of shear are of utmost importance. Similarly, immobilized proteins with fluid flowing past are also subject to shear degradation, e.g., a coating of anticoagulant or immobilized enzyme on a tube surface. The kinetics of enzyme reactions are also altered by shear. If a macromolecule is so oriented during flow that a critical bond becomes subject to the full shear effect, then under extreme conditions even covalent bonds can be broken.

The best correlation of the available data over a wide range of shear rates is

Conformational Stability

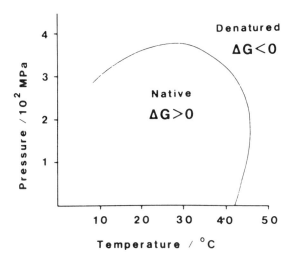

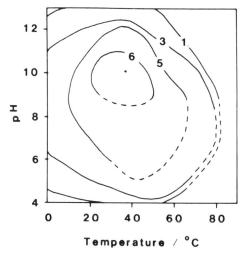

Fig. 6. Pressure denaturation (and renaturation) of metmyoglobin: (top panel) at pH 2.07, and (lower panel) at the indicated pressures (in 10^2 MPa).

$$\gamma t = a(x/x_0)^b$$

where γ is the shear rate, t the exposure time, and x/x_0 the degree of inactivation (19). The coefficients a and b depend on pH and temperature. Some representative data are shown in Fig. 7.

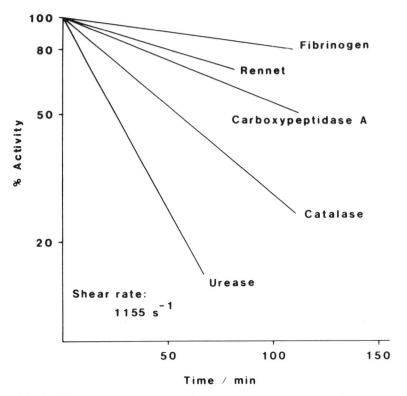

Fig. 7. The residual activities of representative proteins when exposed to constant shear conditions for varying periods.

Although the average shear under conditions of turbulent flow cannot be calculated, inactivation increases with increasing Reynolds number. Loss by pumping can be considerable: Fibrinogen solution pumped through a finger pump in tygon tubing loses up to 20% of its original activity after 3000 passes at a moderate flow rate ($\gamma = 145/s$) (20).

Shear inactivation data for proteins subjected to ultrafiltration, mixing, and passage through a chromatographic column have also been reported (21). In general, laminar flow causes less inactivation than turbulent flow, unless the latter disproportionately reduces the processing time. Recycling should be avoided.

7. Chemical Denaturation and Stabilization

Under this heading are included effects produced on the N–D equilibrium by additives (22). Such effects may rely on electrostatic interactions, or they may be caused by salting in/out phenomena, already described in Chapter 3. They may also be caused by specific binding phenomena or may be classified as general solvent effects, in that the additive so modifies the structure of water as to render it unable to maintain the properly folded N state. Conversely, some additives are able to enhance the stability of the N state, as evidenced by an increase in the thermal denaturation temperature T^*.

7.1. pH Denaturation

As expected, the stability of the N state is sensitive to the degree of protonation of the protein. The N → D transition can therefore be followed by means of a pH titration, as illustrated for lysozyme in Fig. 8 (23). The transition can be considerably sharpened by a measurement of ΔH, instead of ΔG. This is shown in Fig. 9 for carp

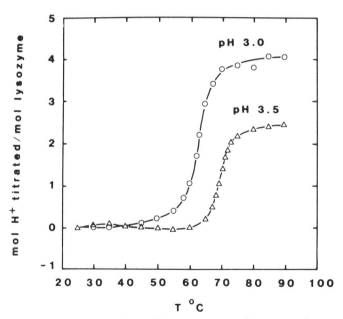

Fig. 8. The proton uptake of lysozyme as a function of temperature (23).

muscle parvalbumin (24). By combining thermal and pH denaturation data, it is possible to construct the ΔG-pH-T surface for any given protein and thereby to identify the pH and temperature regions where the native protein is stable. Figure 10 provides an example of such a surface for lysozyme (11,25). The $\Delta G = 0$ con-

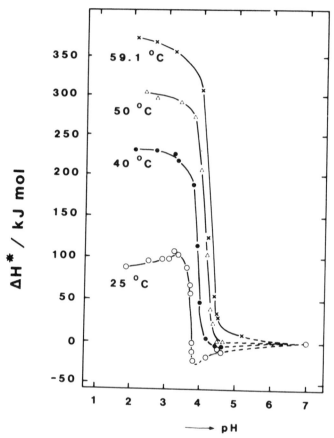

Fig. 9. Isothermal calorimetric pH titration of the N → D transition of carp muscle parvalbumin (24).

tour marks the limit of stability of the N state for this particular protein. Similar surfaces for other small globular proteins differ in detail, but in no case does the maximum stability of the N state exceed 70 kJ/mol. The stabilities of several proteins at moderate pH values are collected in Table 4, indicating also the method by which denaturation was achieved (1,26).

7.2. Salting In/Out and Protein Stability

A clear distinction must be made between salt effects at low concentration ($<0.15M$) and those produced by higher concentrations. In the former concentration range the effects on the protein

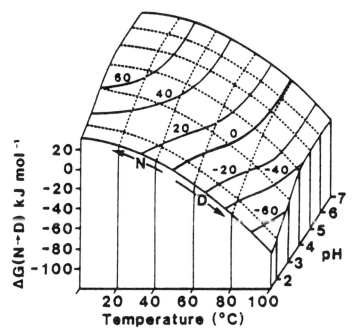

Fig. 10. Temperature-pH-stability profile for lysozyme (11,25). The $\Delta G = 0$ contour corresponds to the conditions for half-denaturation.

are of a nonspecific, electrostatic nature and can be expressed in terms of the ionic strength. (Note: It may be no coincidence that $I = 0.15$ corresponds to the isotonic salt concentration of many living organisms). At higher concentrations all the symptoms of the ion specific effects associated with the lyotropic series become apparent. This is graphically illustrated in Fig. 11 (27). Reference to the chapter 3 discussion on salting in/out shows that those salts that reduce the solubility of proteins also enhance the stability of the N state, whereas salting in electrolytes weaken the integrity of the N state and favor denaturation. The results for ribonuclease are typical of proteins in general, although the actual degree of stabilization or destabilization varies from one protein to another. The specific effects of ions are made much clearer when we compare the different salts of a common cation. This is done in Fig. 12 for GuH^+. Although GuH^+ salts are generally taken to be denaturing solvents and GuHCl is the most commonly used denaturing agent, such a generalization is seen to be mistaken, because $(GuH)_2SO_4$, like other sulfates, stabilizes proteins against thermal denaturation (28). This is also true for the phosphate (not shown in Fig. 12).

Table 4
Free Energies of Some Native Proteins
Relative to Their Unfolded States at 25°C[a]

Protein	$-\Delta G$ (kJ/mol)	Method
Hen egg-white lysozyme, pH 7	53	GuHCl equilibrium
Human lysozyme	117	GuHCl equilibrium
Ribonuclease, pH 6.6	62	GuHCl equilibrium
α-Chymotrypsin, pH 4	48	Heat
Penicillinase, pH 7	25	GuHCl equilibrium
Phosphoglycerate kinase	27	GuHCl equilibrium
Peptide (99-149) from staphylococcal nuclease	-21	Immunological
Bovine heart ferricytochrome c, pH 7	64	GuHCl equilibrium
	38	Heat
α-Lactalbumin, pH 6.5	27	GuHCl equilibrium
Ovalbumin, pH 7	25	GuHCl equilibrium
Sperm whale ferrimyoglobin-cyanide complex, pH 10	50	Heat

[a]All GuHCl equilibrium data have been extrapolated to zero GuHCl concentration.

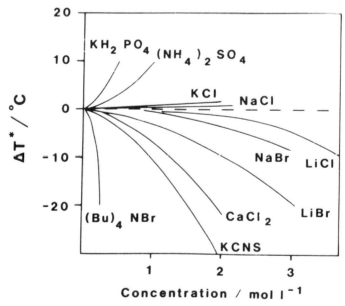

Fig. 11. The effects of salts on the denaturation temperature (T^*) of ribonuclease at pH 7 (28).

Conformational Stability 115

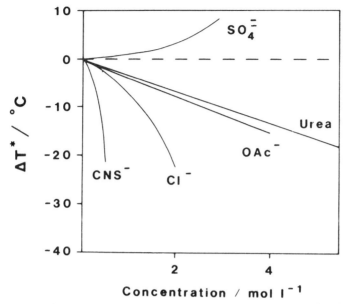

Fig. 12. The effects of urea and guanidinium salts on the denaturation temperature (T^*) of ribonuclease at pH 7 (28).

As indicated in chapter 3, nothing is known about the origin of the lyotropic series. In the protein literature one finds numerous interpretations relying on "binding" phenomena. However, such hypotheses are unrealistic, although they may be made to fit the data for individual experimental studies. Just as ion-specific effects determine the solubility and stability of proteins, so they also influence the solubility of argon gas (and other simple gases) in water (22). The origins of the lyotropic series must be closely related to the details of ion hydration and how such hydrated ions can interact with hydrated argon atoms or protein molecules. There are also indications that the lyotropic series is associated with the phenomenon of hydrophobic hydration (see chapter 5), but the details remain obscure (27).

Although nomenclature in terms of *salt* effects may be appropriate to discussions of the effects of ions on protein behavior, such nomenclature appears to be inapplicable to nonelectrolyte effects on protein stability and solubility. Nevertheless, the effects of some organic substances fit in well with those of the ions in the lyotropic series. Thus carbohydrates (sugars and sugar alcohols) tend to stabilize proteins against denaturing treatments, the magnitude of the protection being, to a first approximation, a function of the number of −OH groups (29). This is shown in Fig. 13, where the effects of various organic compounds are compared with those of ions.

The universal destabilizing action (chaotropism) of urea is well known (22). Urea enhances the solubility of nonpolar compounds

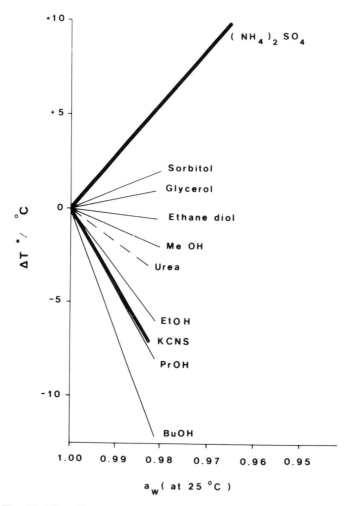

Fig. 13. The effect of organic reagents on the denaturation temperature (T^*) of ribonuclease at (nominal) pH 7. Additive concentrations are expressed in terms of water activity (a_w), mainly to demonstrate that a_w is not a reliable measure of enzyme stability.

Conformational Stability

in water (salting in?), reduces the tendency of surfactants to form micelles, and destabilizes phospholipid membrane structures. All these effects can be accounted for on a semiquantitative basis by the influence of urea on the intermolecular, hydrogen-bonded structure of liquid water.

The behavior of proteins in multicomponent mixtures should be of technological importance, but has not been widely studied. It is believed that in mixtures of salts the observed behavior is the additive sum of the various ionic contributions to the N–D equilibrium. However, this simple additivity is unlikely to apply in mixtures of urea and GuHCl, which are *worse* protein denaturants than would be expected from their individual effects, as seen in Fig. 14 (*30*).

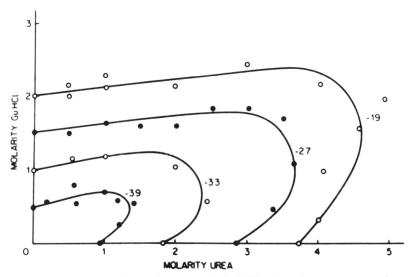

Fig. 14. Concentrations of urea and GuHCl plotted at a constant k_{app} of lysozyme unfolding. Lines represent the average contours of constant $\log_{10} k$ (*30*) [from B. Robson, in *Biophysics of Water* (1982) (F. Franks and S. F. Mathias, eds.) John Wiley, New York.

The effects of alcohols and their derivatives (alkoxyalcohols and haloalcohols) on protein stability is particularly complex. This is because of the general hydrophobic nature of these substances (*31*), the properties of which display complicated concentration and temperature dependences, as illustrated in Fig. 15 for the influence of ethanol concentration and temperature on the N-state stability of RNase (*32*). It is seen that at low temperatures and low concen-

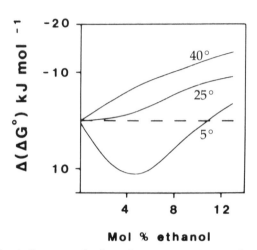

Fig. 15. The influence of ethanol concentration and temperature on the stability of ribonuclease, relative to its stability in aqueous solution at pH 7. $\Delta(\Delta G^0) > 0$ refers to stabilization of the native state (32).

trations ethanol acts as a typical salting-out compound that turns to salting-in behavior at higher temperatures and high concentrations. The destabilizing effect per mol additive increases with the carbon chain length of the additive, evidence for the involvement of hydrophobic interactions in the N → D process (22). The powerful destabilizing effect of butanol is shown in Fig. 13. Similar salting-in effects are observed for tetraalkylammonium halides, which do not behave as typical electrolytes, but as hydrophobic alkyl derivatives. The common description of such effects as additive "binding" are erroneous, since the hydrophobic interaction between protein and additive results from the *repulsion* of apolar residues or patches on the protein surface from water rather than from an inherent attraction (van der Waals type) between such domains (*see* chapter 5). The interactions between proteins and long-chain alkyl derivatives find application in SDS gel electrophoresis. Here the situation is further complicated by micelle formation of the surfactant additive. The exact mode of interaction between SDS and the protein is not yet clear.

7.3. Characterization of Protein D-States

The native state of a protein is unique and can be exactly characterized in great detail. For small globular proteins the N-state possesses a high degree of intramolecular order, which, in the case of more complex proteins, extends to the details of the intermolecular interactions and the architecture of the multisubunit structure. The D-state, on the other hand, is not as easy to define, since it approaches the random coil configuration more commonly found in synthetic macromolecules dissolved in "good" solvents. (The use of the word "good" in this context is in terms of the Flory-Huggins theory of polymer solutions and refers to solutions for which the χ-parameter does not exceed 0.5.)

The description of the D-state is made more difficult by the fact that it is definitely *not* a truly random coil. In fact, denaturation takes place in stages and, depending on the method used to denature a protein, several different D-states can be identified. For lysozyme and α-lactalbumin, four intermediates states have been identified, of increasing degree of unfolding, and depending on whether heat, LiCl, LiClO$_4$, GuHCl, or urea are employed as denaturants (33,34). Figure 16 illustrates the effects of chemical denaturing agents on the structural integrity of α-lactalbumin, as monitored by ORD. In each case the typical N–D curves are obtained (*see* Fig. 2a), but of different degrees of cooperativity (steepness) and with different limiting values. According to Fig.

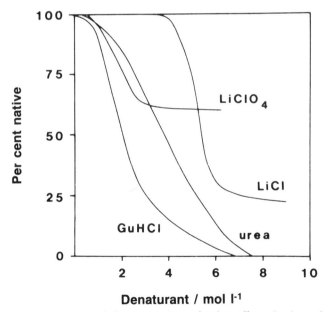

Fig. 16. The degree of denaturation of α-lactalbumin (monitored by ORD) induced by different chemical agents.

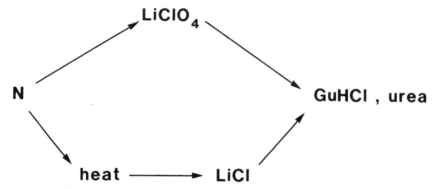

Fig. 17. Intermediate states in the denaturation of α-lactalbumin or lysozyme induced by different treatments. I, LiClO$_4$; II, heat; III, LiCl; IV, urea or GuHCl. For details, see text.

16, GuHCl and urea give rise to the same, maximally unfolded state, LiClO$_4$ has the least effect, with LiCl occupying an intermediate position. With the aid of sequential denaturation treatments it can be shown that state IV can be reached by two different routes, as shown in Fig. 17, but that conversions I → II or I → III do not take place. State II, which corresponds to the thermally denatured protein, has been observed for lysozyme and RNase but not for α-lactalbumin. All the evidence suggests that different reagents cause unfolding of different domains within the protein and that GuHCl and urea produce the most complete unfolding.

8. Surface Denaturation

Since proteins are amphiphilic polyelectrolytes, they exhibit some degree of surface activity, i.e., they adsorb to interfaces. Hence, proteins act as emulsifying/dispersing agents, as in the stabilization of fat in blood or in milk, or the stabilization of air bubbles in ice cream. When the surface forces are strong and coupled with a low N-state stability, sorption induces surface denaturation, as, for instance, in the precipitation of blood proteins on contact with some plastic materials.

Three types of properties are of importance in relation to technological applications of protein sorption:

1. The rate of sorption/denaturation
2. Sorption isotherms, i.e., the quantity of protein in the interfacial film
3. The mechanical properties of the sorbed film

Conformational Stability 121

These properties are discussed in more detail in chapter 19.

Quantitative, detailed information on the surface properties of proteins exist for β-casein, BSA, and lysozyme (35–37). The initial sorption of proteins to a clean air/water interface is rapid and diffusion-controlled. This is followed by a slower reorganization/ denaturation step that takes place with an apparent desorption, which is, however, not accompanied by a decrease in the surface pressure. The adsorbed protein therefore unfolds and occupies the surface more economically, i.e., fewer protein molecules in the surface layer will cover the available area. Eventually more protein will be sorbed to the already existing surface layer. The existence of the two processes is illustrated in Fig. 18. The slower process has a half-life of hours and depends on the protein concentration. Intermolecular effects are important (e.g., aggregation) and the process determines the rheological properties of the sorbed layer.

The initial fast process is associated with the irreversible formation of a monolayer of concentration 2–3 mg/m². Eventually the thickness increases to 5–6 nm. Although the film pressure is determined by the properties of the monolayer, films can build up to >10 nm thickness. The secondary sorption is reversible, indicating that the protein is in a native state.

The actual structure of the sorbed protein depends on the stability of the N-state. Lysozyme, which is particularly stable, can exist in the native state at an air/water interface, but at an oil/water interface even this stable molecule is denatured (compare the effects of alcohols on protein stability). The BSA is partially native, but no definite conformational transition can be observed at a particular packing density during sorption. β-Casein, which is a highly surface active molecule, is extended in the interfacial film; an increase in the concentration leads to densely packed films. The mechanical properties of interfacial protein films are of considerable importance to their role as stabilizers of disperse systems. The dilatational modulus, which measures the response of the film to a compression/expansion cycle, has been measured for β-casein (3–5), BSA (60–400), and lysozyme (200–600 mN/m²). A comparison of protein gels and interfacial films suggests that the sorbed layer has the properties of a thin gel layer.

9. Multistate Denaturation

In many cases the simple two-state model of protein stability, as discussed in section 2, is only a crude approximation to the ac-

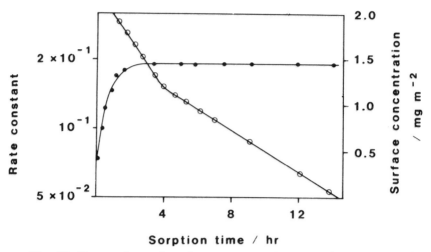

Fig. 18. Comparison of surface concentration-time dependence (●, right-hand ordinate) with rate of change of surface pressure (○, left-hand ordinate) for lysozyme. Note that the change in slope occurs when the surface concentration reaches its steady-state value (35,36).

tual course of denaturation, especially when intermediate states have in fact been identified, or even isolated. Various schemes can be written for multi-state denaturation, e.g.,

$$N \to I \begin{smallmatrix} \nearrow D_1 \\ \searrow D_2 \end{smallmatrix} \qquad \text{or } N \to D_1 \to D_2$$

One or the other of the steps might be fast or slow, reversible or irreversible (e.g., aggregation). A slow process I → D has been identified with a *cis–trans* rearrangement in proline and studied in detail for the unfolding of RNase (38). This protein contains four proline residues at positions 42, 93, 114, and 117, respectively. Of these, only pro-93 and pro-114 are involved in the slow process that accompanies refolding; the other two pro residues are in the *trans*-configuration both in the N and the D-state of the protein. On the other hand, RNase in the D-state contains 70% *cis*-proline-93 and 95% *cis*-proline-114. The refolding kinetics of the denatured protein depends on the *trans-cis* isomerization of these two residues, which are 100% in the *cis*-configuration in the N-state.

Stable intermediates are often produced by the stepwise dissociation of a multisubunit protein and have been implicated in the phenomenon of cold inactivation. In such cases the multimer reversibly dissociates into smaller multimers or even single subunits that, under physiological conditions, maintain their native conformations, but lose their biological activity (39,40). Examples of such behavior include phosphofructokinase, glucose-6-phosphate dehydrogenase, ATPase, pyruvate carboxylase, and carbamyl phosphate synthetase.

The various stages of denaturation have been studied in detail for β-lactoglobulin, which normally exists as an $\alpha\beta$-dimer at room temperature and $3 < \text{pH} < 6$. The dimer dissociates at $\text{pH} > 6$ and this is followed by a conformational transition that, in turn, leads to nonspecific, irreversible aggregation (41). The various processes can be summarized as follows:

pH	3–6	6–9	9–12
Temperature	20°C	0°C	
State	N_2	$N_2 \rightarrow 2R$	$2R \rightarrow S \rightarrow S_n$

where n is the native state ($\alpha\beta$-dimer), R is the dissociated monomer, S is the denatured monomer, and S_n the aggregated state.

Another way of studying the details of unfolding is by the removal of short peptide sequences. For instance, by removing the C-terminal residues 121–124 from RNase, the stability (thermodynamic) is reduced by 70%, but the enzyme activity, by 99.5%. If this modified protein is then denatured, the regain of activity is faster than the rate of refolding, suggesting the existence of intermediate states.

Urea gradient electrophoresis can be used to detect the existence of intermediate species, as shown schematically in Fig. 19, for various combinations of fast and slow reaction steps (42). Electrophoretic migration is from top to bottom with the urea concentration increasing horizontally from left to right. Kinetic transitions shown as slow are assumed to occur at all urea concentrations at negligible rates during the electrophoretic separations, whereas those shown as fast have half-times much less than the duration of the electrophoretic separation. Although other schemes are possible, all observed instances of slow unfolding and refolding correspond to the two patterns shown in the third scheme of Fig. 19.

10. Protein Engineering: The Means of Affecting Protein Stability

Since the assembly of amino acids into peptides and proteins is under the control of the genetic code, it follows that mutations can be programmed into the assembly process by genetic means, and this practice has become known as protein engineering. It has already been shown in previous chapters how the substitution of a single amino acid residue in lysozyme can substantially affect the temperature stability of the enzyme, without, however, affecting its function. It is not certain, on the other hand, how far similar stability changes would be observed for *any* random substitution along the peptide chain.

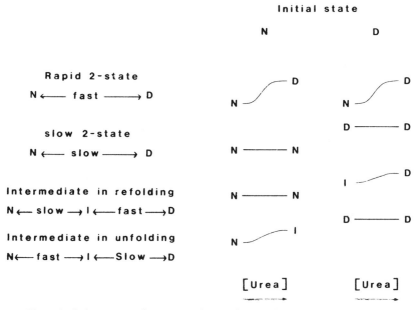

Fig. 19. Schematic of urea-gradient electrophoresis patterns for some simple protein unfolding transitions, showing four kinetic possibilities for fast and slow transitions with or without intermediates (42).

Similarly, enzymes from halophilic bacteria have been found to contain a higher proportion of acidic residues (glu, asp) than their mesophilic counterparts. The enhanced salt tolerance has been simply accounted for by the increased charge density on the protein surface.

The global stability of a native protein is highly marginal and hardly ever exceeds 60 kJ/mol. Purposeful protein engineering should therefore hardly require more than a few amino acid substitutions per peptide chain, provided that such substitutions are made at the correct locations. As long as we only have a sketchy

Conformational Stability

understanding of the detailed molecular origins of the stability/instability relationships, the substitution of the correct amino acid residues at the correct locations is likely to remain a haphazard process. Computer modeling (see next chapter) may help in pinpointing particular substitution-sensitive domains in tertiary and quaternary structures (43).

With a more profound understanding of the factors that affect resistance to changes in temperature, pressure, pH, salt concentration, proteolytic attack, and so on, it should become possible to design proteins to given performance specifications.

References

1. R. H. Pain, in *Characterization of Protein Conformation and Function* (F. Franks, ed.) Symposium Press, London (1979), p. 19.
2. R. Lumry, R. Biltonen, and J. F. Brandts, *Biopolymers* **4**, 917–944 (1966).
3. J. F. Brandts and R. Lumry, *J. Phys. Chem.* **67**, 1484–1494 (1963).
4. J. F. Brandts, *J. Am. Chem. Soc.* **86**, 4291–4301 (1964).
5. C. N. Pace, *Crit. Rev. Biochem.* **3**, 1–43 (1975).
6. C. Tanford, *Adv. Protein Chem.* **24**, 1–95 (1970).
7. W. Pfeil and P. L. Privalov, *Biophys. Chem.* **4**, 23–40 (1976).
8. P. L. Privalov and N. N. Kechinashvili, *J. Mol. Biol.* **86**, 665–684 (1974).
9. P. L. Privalov and E. I. Tiktopulo, *Biopolymer* **9**, 127–139 (1970).
10. J. G. Brandts, J. Fu, and J. H. Nordin, in *The Frozen Cell* (G. E. W. Wolstenholme and M. O'Connor, eds.) J & A Churchill, London (1970), pp. 189–208.
11. F. Franks, *Biophysics and Biochemistry at Low Temperatures*, Cambridge University Press, Cambridge (1985).
12. S. N. Timasheff, in *Physical Aspects of Protein Interactions* (N. Catsimpoolas, ed.) Elsevier North-Holland, New York (1978), pp. 219–273.
13. M. A. Lauffer, in *Physical Aspects of Protein Interactions* (N. Catsimpoolas, ed.) Elsevier North Holland, New York (1978), pp. 115–170.
14. W. L. Dixon, F. Franks, and T. apRees, *Phytochemistry* **20**, 969–972 (1981).
15. J. Schellman, in *Protein Folding* (R. Jaenicke, ed.) Elsevier North-Holland, Amsterdam (1980), p. 331.
16. H. Nojima, A. Ikai, and H. Noda, *J. Mol. Biol.* **116**, 429–442 (1977).
17. S. A. Hawley and R. M. Mitchell, *Biochemistry* **14**, 3257–3264 (1975).
18. A. Zipp and W. Kauzmann, *Biochemistry* **12**, 4217–4228 (1973).
19. S. E. Charm and B. L. Wong, *Biotech. Bioeng.* **12**, 1103–1109 (1970).
20. S. E. Charm and B. L. Wong, *Science* **170**, 466–468 (1970).
21. S. E. Charm and B. L. Wong, *Biorheology* **12**, 275–278 (1975).
22. F. Franks and D. Eagland, *Crit. Rev. Biochem.* **3**, 165–219 (1975).

23. W. Pfeil and P. L. Privalov, in *Biochemical Thermodynamics* (M. N. Jones, ed.) Elsevier, Amsterdam (1979), p. 75.
24. V. V. Filimonov, W. Pfeil, T. N. Tsalkova, and P. L. Privalov, *Biophys. Chem.* **8**, 117–122 (1978).
25. W. Pfeil and P. L. Privalov, *Biophys. Chem.* **4**, 41–50 (1976).
26. P. L. Privalov, *Adv. Protein. Chem.* **33**, 167–241 (1979).
27. P. H. von Hippel and A. Hamabata, *J. Mechanochem. Cell Motil.* **2**, 127–138 (1973).
28. P. H. von Hippel and K. Y. Wong, *J. Biol. Chem.* **240**, 3909–3923 (1965).
29. S. Y. Gerlsma, *J. Biol. Chem.* **243**, 957 (1968).
30. B. Robson, in *Water Biophysics* (F. Franks, S. F. Mathias, eds.) John Wiley & Sons, Chichester (1982), p. 62.
31. F. Franks and J. E. Desnoyers, *Water Sci. Rev.* **1**, 171–232 (1985).
32. J. F. Brandts and L. Hunt, *J. Am. Chem. Soc.* **89**, 4826–4838 (1967).
33. M. Kugimiya and C. C. Bigelow, *Can. J. Biochem.* **51**, 581–585 (1973).
34. R. N. Sharma and C. C. Bigelow, *J. Mol. Biol.* **88**, 247–257 (1974).
35. D. E. Graham and M. C. Phillips, *J. Colloid Interface Sci.* **70**, 403–414 (1979).
36. D. E. Graham and M. C. Phillips, *J. Colloid Interface Sci.* **70**, 415–439 (1979).
37. D. E. Graham and M. C. Phillips, *J. Colloid Interface Sci.* **76**, 227–250 (1980).
38. J. F. Brandts, H. R. Halvorson, and M. Brennan, *Biochemistry* **14**, 4953–4963 (1975).
39. P. E. Bock and C. Frieden, *Trends Biochem. Sci.* May 100–103 (1978).
40. P. E. Bock and C. Frieden, *J. Biol. Chem.* **251**, 5630–5643 (1976); *Trends. Biochem. Sci.* May 100–103 (1978).
41. P. Douzou, *Cryobiochemistry* Academic, London (1977).
42. T. E. Creighton, *J. Mol. Biol.* **129**, 235–264 (1979).
43. M. Karplus, S. Andrew, and M. C. Ammon, *Sci. Am.* **254**, 36 (1986).

Chapter 5

Protein Hydration

Felix Franks

1. Historical

In chapter 4 the marginal stability of native proteins was discussed with reference to the effects of environmental factors such as temperature, pressure, and additives. It must also be stressed that biological activity critically depends on a correctly folded state, and folding takes place on the ribosome in an aqueous medium. Even in vitro, crystalline proteins contain almost 40% water (*1*). Hydration is therefore likely to play a major role in the maintenance of the native state. The physical properties of water, too, are sensitive to the same factors that influence protein stability, so that some connection is likely. Nevertheless, the realization that hydration might be an important factor in biological and technological function only dates from the early 1970s. Figure 1 summarizes the chronological development of our present understanding of two aspects of enzyme function: specificity and catalytic activity (*2*). It is now accepted that protein hydration interactions are of crucial importance in the maintenance of higher order structures and in rendering proteins useful as technological macromolecules. As will presently be shown, however, there is as yet little understanding about the details of such interactions and their role in determining the functional attributes of proteins.

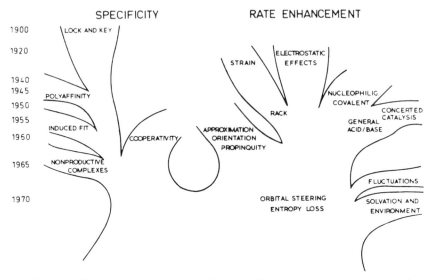

Fig. 1. Chronology of our understanding of the mechanisms that govern enzyme catalysis (ref. 2). Note that a consideration of solvation as a contributing factor dates only from 1970. [from J. A. Rupley, in *Characterization of Protein Conformation and Function* (1979) (F. Franks, ed.) Symposium Press, London].

2. Classification of Protein Hydration Interactions

In a discussion of protein hydration, three distinct types of solute–water interactions can be distinguished. They are ion hydration (side chains of asp, asn, glu, gln, lys), hydrogen bonding between polar groups and water, and so-called hydrophobic hydration (ala, val, leu, ile, phe), which describes the response of water to chemically inert residues. In principle, the second type should be characteristic of *all* amino acid residues because, quite apart from polar side chains (e.g., —OH in ser, thr, tyr), every peptide can act as proton donor and acceptor. Some amino acids behave, on balance, as typically hydrophobic molecules (3), however, whereas others more closely resemble polyhydroxy compounds in their hydration properties.

2.1. Ion Hydration

Conceptually the interaction of water with charged groups is the simplest of the three types of hydration effects. It is of an electrostatic nature (ion–dipole interaction) and has a long range (proportional to r^{-3}, where r is the distance of separation). It is therefore strong compared to most other types of intermolecular interactions. Phenomenologically, ion–water interactions in solution can be represented as shown in Fig. 2. Many quantitative models assign a spherical, monomolecular hydration sphere to the ion, beyond which water is then assumed to be unperturbed (4). A model of this type can be elaborated by including two or more layers of water molecules. Such subtle modifications can sometimes be detected by spectroscopic techniques (5).

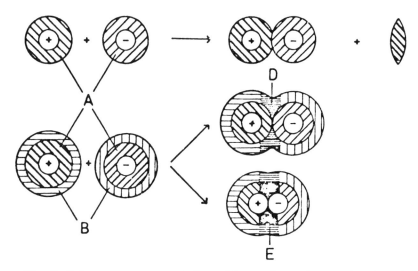

Fig. 2. Schematic representation of ion–solvent interaction in solution: (A) Primary hydration shells. (B) Secondary hydration shells. (D) Outer cosphere overlap (solvent-separated ion pair). (E) Primary cosphere overlap (contact ion pair) [from F. Franks, in *Water* (1983) Royal Society of Chemistry Paperbacks].

As regards the geometrical details of an ion hydration shell in solutions, neutron diffraction has provided the most detailed information, although mainly for monatomic ions (6), polyatomic ions of the type COO^- or NH_3^+ being too complex for the deconvolutions of the scattering intensity curves. Figure 3 illustrates the hydration geometry of the alkali metal ions (except Li⁺) and Ca²⁺ in solutions of their chlorides. In all cases the time-averaged primary hydration shell consists of six water molecules oriented with respect to the cation as shown. The organization of water molecules beyond the primary hydration sphere is still uncertain, but the spatial incompatibility of the primary hydration sphere with the tetrahedral disposition of water molecules in the bulk liquid demands a region of structural mismatch that is believed to have important implications for ion–ion and ion–molecule interactions in aqueous solutions (7). Figure 2 is therefore an oversimplification because it does not include this region of structural mismatch.

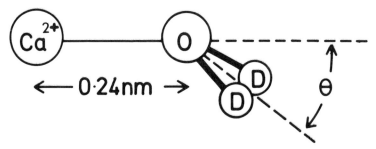

Fig. 3. Hydration geometry of alkali and alkaline earth metals (except Li⁺ and Mg²⁺), showing the tilt of the water molecules (six in the primary hydration shell). The angle θ decreases with decreasing concentration (*see* ref. 6).

The general hydration geometry of halide anions, as obtained from neutron diffraction, is shown in Fig. 4. Here again six water molecules form the primary hydration sphere, with one of the −OH bonds pointing toward the anion. Many thermodynamic models of ion hydration have been proposed; all but the most recent ones suffer from the limitation that the solvent is treated as a continuum, possessing only macroscopic physical properties (dielectric permittivity, viscosity) (8). Recent developments in the statistical thermodynamics of electrolyte solutions show promise of more realistic representations of ion–water interactions (4,9).

2.2. Hydrophilic Hydration

This term describes the *direct* interaction, by hydrogen bonding, between a functional group on a solute molecule and water, the water molecule being able to act either as proton donor or accep-

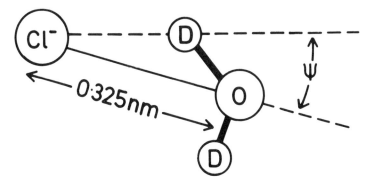

Fig. 4. Hydration geometry of halide anions, showing that the water molecules are not symmetrically disposed with respect to the Cl⁻—O axis. The angle of tilt ψ is a function of the concentration (see ref 6).

tor. This type of interaction is expected to figure largely in the hydration of the peptide bond, with both the >CO and the >NH groups able to participate in hydrogen bonding. The influence of this type of solvation on tertiary structure is graphically illustrated by polyproline, in which the solvent alone determines the respective stabilities of the polyproline-I and -II structures.

Among amino acid side chains that might hydrogen bond to water molecules, ser, thr, and tyr possess —OH groups, and his, arg, and trp can interact via the >NH group. The contributions of such side chain hydration interactions to protein conformational stability appear to be much less important than those made by main chain (peptide) hydration. Indeed, hydrophilic hydration may only play a minor role in the maintenance of higher-order structures of proteins, as distinct from carbohydrates, the solution configurational properties of which are determined largely by hydrophilic hydration (10,11).

2.3. Hydrophobic Hydration and Hydrophobic Interactions

In the vicinity of apolar groups, water molecules reorganize spatially and orientationally so as to be able to accommodate such groups with which they cannot interact directly (3,12). Slight distortions in the normal, predominantly tetrahedral geometry of the water molecules can give rise to a variety of differently shaped cavities, a simple example of which is shown in Fig. 5. Such structures can actually exist in the crystalline state, and they bear a

Fig. 5. Water cavity capable of accommodating an inert molecule. The particular structure shown consists of a hydrogen-bonded framework of 12 water molecules, eight of which (shaded circles) are nearest neighbors to the apolar molecule.

resemblance to ice, in the sense that each oxygen atom is hydrogen bonded to four other oxygen atoms, giving rise to three-dimensional water networks with cavities of various sizes and shapes, with the solute (guest) molecules occupying these cages (13).

As to the importance of the hydrophobic hydration phenomenon to protein function, the main point to note is that the *configurational* freedom of water molecules, i.e., the number of allowed mutual orientations, is reduced in the proximity of the apolar residue, because a water $-OH$ vector must not be directed toward the center of the cavity. This reduction in configurational freedom leads to the well-known result that when an apolar molecule or residue R is transferred from the gas phase or from an apolar solvent medium to water, this transfer occurs with a *loss* of entropy. In other words, $T \Delta S_t < 0$, where the subscript t describes the transfer

$$R \text{ (apolar solvent)} \rightarrow R \text{ (aqueous medium)} \quad (1)$$

It is also found that the transfer is thermodynamically *unfavorable*, i.e., $\Delta G_t > 0$. The above results imply that the reluctance of apolar molecules to mix with water derives not from a net repulsive energy of interaction (which would be expressed through $\Delta H_t > 0$), as is usually the case for low solubility. In other words, the low solubility of hydrocarbons in water is the direct result of the entropy lost by the water because of cage formation, rather than a lack of van der Waals attraction between hydrocarbon and water. Thus, for hydrophobic hydration, another important result is that $|\Delta H_t| < T|\Delta S_t|$, whatever the sign of ΔH_t (3). Yet another symptom of hydrophobic hydration is that $\Delta C_p >> 0$, consistent with the cage model, since a large specific heat change is indicative of the creation of structure or order in the liquid that is sensitive to thermal disruption.

The amino acid residues that are subject to hydrophobic hydration include ala, val, leu, ile, and phe; possibly sulfur-containing amino acids cys and met should also be included, as well as the imino acid pro. The above discussion demonstrates that apolar residue/water contacts are thermodynamically (configurationally) unfavorable and it is to be expected, therefore, that the apolar residues will be found predominantly in the interior of a folded protein structure, if the stability of the protein in its native environment is to be maximized. Indeed, it is believed that hydrophobic hydration (or rather, its converse, the hydrophobic interaction) provides much of the driving force for folding (14–17).

Since the transfer of an apolar residue R from hydrocarbon to water, as depicted in Eq. (1), is seen to be thermodynamically unfavorable, then the converse, i.e., the association of R residues in water, should be accompanied by a negative free energy change. At the simplest level, two hydrocarbon molecules or two alkyl residues, each with its hydration cage, would gain in stability by their association because this would "release" water molecules from the cages, which could then relax into their more stable, unperturbed configurational states. The process

$$2R(\text{hydrated}) \rightarrow R_2(\text{hydrated}) + \text{water} \qquad (2)$$

would therefore be expected to take place spontaneously. Once again, it is emphasized that the driving force for such an association does *not* derive from an attraction (e.g., by van der Waals forces) between alkyl groups, but from an extrusion of alkyl groups by water for configurational reasons. The process is said to be entropy-driven, in the sense that $T\Delta S_\phi > 0$ and $T|\Delta S_\phi| > |\Delta H_\phi|$, where the subscript ϕ describes the association in Eq. (2). Thus, what *appears* to be an attraction between alkyl residues (negative free energy) is actually the sum of two repulsions, and the term hydrophobic interaction, which is commonly used to describe the process, is really a misnomer.

It has been powerfully argued (18) that a lack of appreciation of the configurational origin of the hydrophobic interaction can lead to misleading conclusions and some confusion about protein structure. Thus, the free energy associated with processes Eq. (1) or (2) can easily be fitted by simple one-parameter models, e.g., the molecular surface area of the alkyl residues. ΔG_t is then expressed as the work done to create the cavity in water and described in terms of a surface tension. Such simplistic fitting devices may be able to model ΔG, but they fail to model ΔH and break down completely for the second T and P derivatives of ΔG (heat capacity and compressibility). Yet the literature abounds with such representations of hydrophobic effects. For up-to-date thinking and discussions of hydrophobic effects, the reader is referred to the 1982 Faraday Symposium (no. 17, The Hydrophobic Interaction) of the Royal Society of Chemistry.

Before leaving the subject of the "simple" hydrophobic interaction as represented by Eq. (2), we stress that the association of many R residues, according to the scheme

$$nR(aqueous) \rightarrow R_n(aqueous) + H_2O \qquad (3)$$

where $n >> 2$, as for instance in the formation of a surfactant micelle, cannot be correctly described by a stepwise aggregation starting from Eq. (2). The thermodynamics describing the association of pairs or small clusters of hydrated alkyl residues are *qualitatively* different from those describing multiple aggregation (19). This observation is of importance in calculations of protein stability, but is not even given a mention in most monographs dealing with the subject (18,20).

The simple model described by Eqs. (1) and (2) is in any case an oversimplification of the real processes that occur in water when it is perturbed by apolar residues. For instance, the cavity-forming effects produced by the introduction of the apolar group can hardly be confined to the primary hydration shell; the configurational perturbation is apparently felt by water molecules as far distant as 5 nm from the center of the primary cavity (21). This makes hydrophobic hydration a long-range interaction and calculations of hydrophobicity should take this into account, but rarely do so (19,20,22).

Equation (2) can be taken as the elementary unit step in calculations of protein folding and stability in the sense that in the D state most of the apolar residues are believed to be fully exposed to the solvent, but that the folding process reduces the solvent accessibility of these residues. As more apolar residues become involved in the folding process, so eventually the volume of apolar residues in the center of a globular protein increases and a considerable number of such residues will become completely inaccessible to the solvent. The interior of a protein (like that of a surfactant micelle) is often compared to a liquid hydrocarbon environment. This type of argument has led to the simple concept of hydrophobicity that is expressed as the free energy change ΔG_ϕ associated with the transfer of an apolar group from a hydrocarbon (typically octane) to an aqueous environment)

$$R(octane) \rightarrow R(aqueous) \qquad (4)$$

A problem arises when ΔG_ϕ is to be calculated for amino acids, peptides, and proteins because the contribution of the peptide group(s) has to be estimated. Different methods have been proposed and they have been discussed by Eisenberg et al. (23). Hydrophobicity scales, derived on the basis of different models, are summarized in Table 1. Here the values in the second column (24) refer to the transfer of the residue from ethanol to water, and the other columns all refer to other modifications of Eq. (4) in the sense that the transfer is not necessarily from liquid hydrocarbon.

Bigelow, who calculated the average hydrophobicities of 150 globular proteins, concluded that for a globular protein to be stable it must possess a certain minimum average ΔG_ϕ per residue (25). This is illustrated in Fig. 6. Of the 150 proteins examined, 50% were found to have ΔG_ψ (average) > 4.6 kJ/mol. The curve in Fig. 6 denotes the estimated lower limit of stability for a globular protein.

3. Medium Effect on Protein Stability

It is now possible to attempt an analysis of the various interactions that contribute to the stability of the N state relative to the D state and to speculate on the effects of temperature or additives on these various interactions. As discussed in chapter 3, $\Delta G(N \rightarrow D)$ for most single subunit proteins lies in the range of 50–100 kJ/mol. This is equivalent to only 3–5 hydrogen bonds in energy, whereas it is known from the secondary structures that the N state of a typical small globular protein (say of 100 residues) may contain approximately 150 hydrogen bonds. It must be concluded that $\Delta G(N \rightarrow D)$ is the sum of at least two contributions, one of which stabilizes the N state, and one that destabilizes it.

Table 2 is an attempt at a free energy balance sheet for a small globular protein (17). The first term is the experimentally determined $\Delta G(N \rightarrow D)$, which is assumed to be compounded of the contributions that follow. ΔG_{titr} expresses the transfer of ionized side chain groups from the protein interior to the bulk aqueous solvent during protein unfolding. An estimate can be obtained from the potentiometric titration curve. ΔG_{es} accounts for the net change in charge–charge interactions (repulsions) on the periphery of the protein. ΔG_ϕ is large, but its actual magnitude depends critically on the assumed range of the hydrophobic hydration effect. The model based on the simple water cage and Eq. (1) gives $\Delta G_\phi \approx 500$ kJ/mol, but by allowing for an increase in the range of the interaction, ΔG_ϕ can assume values of 2500 kJ/mol or larger.

Table 1
Order of Decreasing Hydrophobicity Estimated as the Free Energy
of Transfer (kJ/mole) for a Transfer From a Hydrophilic to a
Hydrophobic Phase

Residue	Transfer free energy		
	Water→ethanol	Protein folding	Water→vapor
Ile	21	2.9	9.0
Phe	21	2.1	−3.2
Val	12.5	2.5	8.4
Len	14.4	2.1	9.6
Trp	27.2	1.3	−24.7
Met	10.5	1.7	−6.3
Alg	4.2	1.3	8.0
Gly	0	1.3	10.1
Cys	0	3.8	−5.2
Tyr	18.8	−1.7	−25.6
Pro	6.3	−1.3	—
Thr	2.1	−0.8	−20.6
Ser	−2.1	−0.4	−21.3
His	4.2	−0.4	−43.3
Glu	—	−2.9	−43.0
Asn	−6.3	−2.1	−40.5
Gln	−4.2	−2.9	−39.4
Asp	—	−2.5	−45.8
Lys	—	−7.5	−40.0
Arg	—	−5.8	−83.6

^aData from ref. 23.

Table 2
Tentative Balance Sheet of Contributions to the Free Energy Change
Accompanying the Unfolding of Lysozyme and Ribonuclease (17)^a

	Term	Comments	Estimated magnitude, kJ/mol
$\Delta G_{N \to D}$	Free energy change on unfolding	Small ∼ 3–5 hydrogen bonds	50–100
ΔG_{titr}	pK_a changes on unfolding		−20?
ΔG_{es}	Changes in surface charge–charge interactions	Estimated at neutral pH	0?

(continued)

Protein Hydration

Table 2 (continued)
Tentative Balance Sheet of Contributions to the Free Energy Change
Accompanying the Unfolding of Lysozyme and Ribonuclease (17)[a]

ΔG_ϕ −	Hydrophobic free energy change	Surface exposure model, Allow for possible long range[b]	500 100–2500
ΔH_ψ	Several hydrogen bonding effects		
	Difference in strength of water–polar group and polar–polar group	Average of several quantum mechanical values	300
	Hydrogen bond distortion	Using water–water energy surface	−400 to −800
+	"Unsaturated" polar groups		−300
$n_c T\Delta S$ config. +	Configurational entropy change per residue	Value of between 8–20 J/K (residue mol)$^{-1}$ used	−300 to −800
$n_w T\Delta S$ release −	Water forms polar interactions with peptide groups	Assumes 25% of value for water freezing (underestimate?)	500
Other terms, e.g., ΔH_{VDW} −	Changes in van der Waals interactions	Protein denser than water, implying some stabilization? Possibly included in G_ϕ	?
$\Delta H_{\psi\text{-}\phi}$ −	Polar–apolar interactions within protein		?
$T\Delta S_{vibr}$	Vibrational entropy changes	Potentially large and solvent-related	?

[a] n_c and n_w are the respective nubmer of mols of residues and water involved in the unit process (17).
[b] Evidence is growing that the simple proportionality between buried accessible area and free energy change is probably not valid and that the hydrophobic interaction is of a long-range nature.

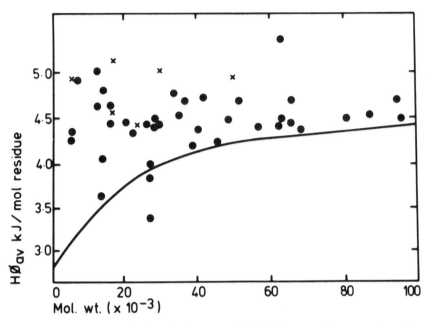

Fig. 6. Average hydrophobicities of globular proteins as a function of their molecular weights; crosses refer to associating proteins. The curve shows the lower limit of ΔG_ϕ compatible with a globular conformation. Approximately 50% of globular proteins have ΔG_ϕ = 4.6 kJ/mol (for details, see ref. 25).

The net polar interactions are treated as the sum of three contributions of which the last one arises from polar groups in the interior of the protein that, for steric reasons, are unable to participate in hydrogen bonds.

The configurational entropy term originates from the gain in configurational freedom of a residue once it is transposed to the solvent during unfolding. On the other hand, for each dissociated intrapeptide hydrogen bond, several water molecules become hydrogen bonded to >CO and >NH groups, so that the loss of the secondary structure contains a negative contribution to the entropy change that originates from this peptide hydration process.

There may well be other contributions to $\Delta G(N \to D)$, the magnitudes of which are not easy to estimate. Three possible effects are included in Table 2. The total van der Waals energies between amino acid residues in the compact folded protein are likely to differ from those between the residues and the surrounding water molecules in the D state, although it would at present be pointless to attempt to put a value on this contribution to the stability. The same applies to the interaction between polar and apolar groups in close proximity, although the indications are that such interactions are of a net repulsive nature (i.e., destabilizing), at least in aqueous solution (26,27).

The final contribution in Table 2 arises from changes in the vibrational and librational modes experienced by individual covalent bonds in the amino acid residues during unfolding. In principle (but hardly in practice), such information is obtainable from infrared and Raman spectra. At the present state of our knowledge it is only possible to speculate that such effect is certainly solvent-related and may be large.

Lest it be concluded that attempts at calculations of $\Delta G(N \rightarrow D)$ from a summation of the individual contributions have not been very successful, it must be stressed that other, empirical or semiempirical approaches, e.g., those based on so-called solvent accessibilities of individual residues in the folded and unfolded states of proteins, may have produced numbers that fit $\Delta G(N \rightarrow D)$ for individual proteins (28,29), but the theoretical basis for such calculations is weak. In addition, all such treatments are based on the simplistic (and in our view, incorrect) model of the hydrophobic interaction. Essentially, the crystal structure forms the starting point and the fraction of the total residue surface area of the protein that is "accessible" to water molecules (assumed to be spherical and closely packed) is estimated. A simple proportionality between the accessible area and $\Delta G(N \rightarrow D)$ is assumed. Clearly the results are extremely sensitive to the value chosen for the van der Waals radius (r^*) of the water molecule because cavities in the surface topology of a protein might be accessible to solvent molecules with $r^* = 0.14$ nm, but not to those with $r^* = 0.18$ nm (29).

4. Hydration Sites in Protein Crystals

High-resolution diffraction methods (X-ray and neutron) have made possible the assignment of spatial coordinates to water molecules in protein crystals (1). Such "structural" water molecules, which constitute only a small proportion of the total number of water molecules in the crystal, can be regarded as part of the secondary/tertiary peptide structure. Six typical locations for structural water molecules can be distinguished: (1) as metal ligand, e.g., in carboxypeptidase and carbonic anhydrase the Zn atom is coordinated to three his residues and one water molecule that is displaced by the substrate during the formation of the enzyme–substrate complex; (2) between residues at different locations within the same peptide chain. Such water bridges can consist of one or more molecules. A three-molecule water bridge linking >CO and —NH groups in papain is shown in Fig. 7 (30); (3) between ionogenic groups in close proximity. Thus in papain two water molecules are found in a hydrophilic cleft that contains two lys and one glu residue. The water molecules are linked to four more water molecules, which together constitute a chain to the protein peri-

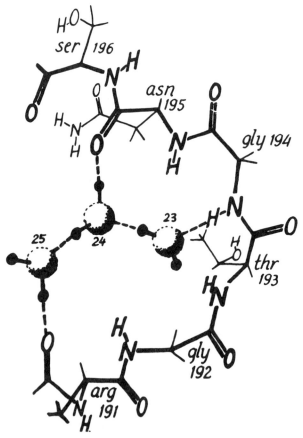

Fig. 7. Three-water-molecule bridge stabilizing the papain main chain by linking arg-191, gly-194, and asn-195 (*see* refs. *30* and *31*) [from J. H. C. Berendsen, in *Water — A Comprehensive Treatise* (1975) vol. 5 (F. Franks, ed.) Plenum, New York].

Protein Hydration

phery and the bulk solvent. Such a water chain provides an efficient means for proton transfer (31); (4) between main chain >CO and −NH groups and polar side chains, e.g., in papain the −OH groups of tyr-48 and tyr-82 are hydrogen bonded via water molecules to the >CO group of ala-104; (5) the active site of an enzyme is sometimes extensively hydrated. In carbonic anhydrase C, nine water molecules have been identified in the active site, four of which are linked to amino acid residues, but the remainder are only linked to other water molecules; water bridges are also operative in linking different peptide chains in a quaternary structure. Figure 8 shows one suggested hydration structure for collagen (32). It is by no means unique or generally accepted. Indeed, there is no agreement about the locations of water molecules in the collagen myofibril (33). Water can also form complex hydration bridges be-

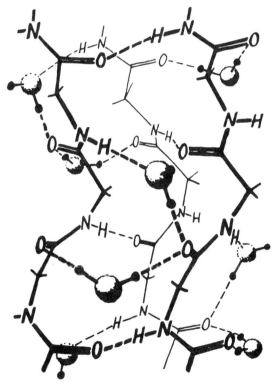

Fig. 8. Speculative stabilization of the three-stranded collagen structure by hydration interactions; water molecules are shown as spheres (from ref. 32) [from J. H. C. Berendsen, in *Water — A Comprehensive Treatise* (1975) vol. 5 (F. Franks, ed.) Plenum, New York].

tween different protein molecules in a crystal, as shown for α-chymotrypsin in Fig. 9. Here again, several of the water molecules participating in the bridge structure are not linked directly to the peptide, but only to other water molecules (34).

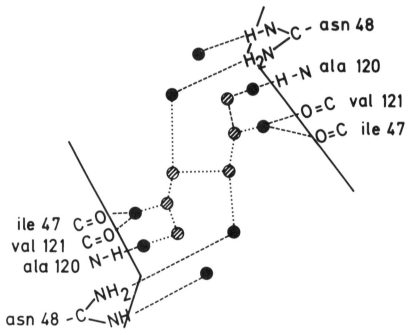

Fig. 9. Intersubunit hydration bridges in α-chymotrypsin (34). The symmetry relating the two protein molecules is retained by the water network. Water molecules marked in black constitute the primary hydration shell; those shaded are linked only to other water molecules (secondary hydration) [from J. L. Finney, in *Water — A Comprehensive Treatise* (1979) vol. 6 (F. Franks, ed.) Plenum, New York].

5. Dynamic Aspects of Water in Protein Solutions: Protein Hydration

Protein–water interactions are frequently described in terms of the effect of the macromolecule on the diffusional motions of the water molecules. Of all the techniques employed in studies of protein hydration, nuclear magnetic resonance (NMR) in its various modes has received most attention (35,36). In particular, nuclear magnetic relaxation rates of ^{1}H, ^{2}H, and ^{17}O, as well as line splittings, have been used to determine the rotational and translational diffusion of water. Various interpretations of the experimental data (decay of nuclear magnetization) in terms of diffusional parameters have been proposed, but there is as yet no universal agreement (37). The systems to be investigated are often heterogeneous and anisotropic because, for experimental reasons, high protein concentrations must be employed. The interpretation of experimental data is further complicated by water exchange between two or more sites; such exchange can be fast or slow. Proton exchange must also be a possibility, unless ^{17}O spectra are being considered. Furthermore, water can also perform anisotropic diffusion parallel to the protein surface.

Qualitatively, three types of motion that differ by orders of magnitude can generally be distinguished with time scales (τ). A small fraction of the molecules has $\tau > 10^{-7}$ s. Their motions are governed by those of the macromolecule, i.e., on the time scale of a molecular rotation of the protein such water molecules can be considered as "bound" to a specific site on the protein. The fraction of water molecules with $\tau < 10^{-11}$ s is almost unaffected by the protein. For intermediate diffusion times, with τ of the order of 10^{-9} s, the water molecules are believed to perform slow, anisotropic tumbling motions, characteristic of water in small pores or narrow capillaries or in the proximity of the protein surface. Table 3 provides a summary of typical water diffusion rates in proteinaceous systems.

Table 3
Typical Tumbling Times (τ) or Diffusion Times of Water in Proteinaceous Systems Under a Variety of Conditions. In All Cases Tumbling is Anisotropic

System	Experimental conditions	τ, s
Lysozyme	<0.3 g water/g protein	10^{-9}
	>0.3 g water/g protein	approaches 10^{-11}
Apotransferrin	10–70 mol water/mol protein	10^{-7}
	>70 mol water/mol protein	10^{-9}
Frog gastrocnemious muscle	250 K (unfrozen water)	2×10^{-9}
Collagen	0.2–0.5 g water/g protein	10^{-7}
Corneal stroma	0.8 g water/g protein, 145 K	10^{-9}
Striated muscle		2×10^{-8}

6. Protein Hydration and Bound Water: Some Popular Misconceptions

The protein literature contains numerous references to "bound" water (see, for instance, refs. 38,39), unfreezable water (40), and different "states" of water (41). In extreme cases it is even claimed that the water in living cells is structured in some peculiar way that differs from water in its normal state, and that this difference is intimately related to cellular functions such as the transport of ions across cell membranes and other life processes (42). It has also been argued that, taking into account the total surface areas of organelles, membranes, and cytoskeletal components, most of the cell water can be regarded as a sorbed layer, not many molecules thick, on the surface of the cell matrix components, most of which are proteinaceous. The water is thus said to be bound to the various protein components and to have properties that differ from those of "ordinary" or "free" water.

Water binding is seldom defined either in terms of specific sites of binding, energetics (presumably by hydrogen bonds) of binding, or lifetimes of the bound states. This is not the place to discuss the pros and cons of such views, but the reader should beware of hypotheses that offer simple explanations for very complex phenomena. Too frequently the evidence for such hypotheses is based on unwarranted assumptions and/or incorrect data handling or ambiguous interpretations of experimental results. This has been particularly true in the interpretations of NMR results.

The question is often asked: How many water molecules constitute the hydration shell of a protein? The literature contains many such estimates of protein hydration expressed as g water/g protein. However, the measured hydration number depends on the experimental technique used for the measurement. By and large, the degrees of hydration calculated from calorimetric and NMR measurements agree well, probably because both techniques really measure the unfrozen water. Differential scanning calorimetry (DSC) results are expressed as $C_p(T)$. At low water contents, e.g. <0.3 g water/g protein in the case of tropocollagen (43), no absorption of heat is apparent at 0°C and only at 0.465 g water/g protein is the normal melting behavior of ice observed. On the basis of a gly-pro-hypro model for tropocollagen, this corresponds to 2.4 mol water per residue, in good agreement with the value 2.7 obtained from NMR measurements (44). The same degree of correspondence is also achieved with several globular proteins and the results have been accounted for in terms of a protein model in which only polar groups on the exterior surface are hydrated (45).

Infrared spectroscopy is of particular value in the study of hydration at low moisture levels, but some results can also be utilized to describe the hydration in aqueous solution. McCabe and Fisher (46) were able to distinguish between the absorption caused by water excluded from the hydration sphere, the water affected by the peptide, and that caused by the protein itself. In this context there is general agreement that $-NH_2$ groups are more extensively hydrated than $-NH_3^+$ groups.

On a molecularly less detailed level, circular dichroism (CD) has been used to estimate the solvation contribution to the observed spectra of polypeptides (47). The assumption is made that the amide groups are the primary solvation sites, and the hydration estimate is based on the calculated molecular asymmetry introduced by placing water molecules at these sites.

Protein hydration can be expressed in terms of structure, energetics, and hydrodynamics, but no quantitative relationship exists between hydrodynamic and thermodynamic approaches to water binding. The former approach is very model-sensitive and difficulties arise in attempts to define the boundary of the hydrodynamic particle in the (continuum) solvent. Thermodynamic criteria such as water activity (a_w) are frequently used to distinguish (allegedly) between bound or free (or trapped) water. Their application becomes progressively more uncertain as the assumed pockets of trapped water decrease in size.

The assignment of notional hydration numbers, whether derived from spectroscopic or thermodynamic measurements, can sometimes lead to absurd results that illustrate the inherent weaknesses of such procedures. A striking example is provided by calculations of hydration numbers of a series of glycols, based on cryoscopic measurements (48). By postulating simple hydration equilibria and constructing a van't Hoff plot, the author was driven to the conclusion that the degree of water binding *increases* with increasing temperature, to the extent that at the boiling point of the solution the solute is quite extensively hydrated. This, and other, similar treatments show up the futility of approaches based on stoichiometric hydration equilibria.

In summary, it is unwise to express protein hydration in solution in terms of g water/g protein, as though there was a sharp cutoff between water of hydration and bulk water. (In technological applications, where protein preparations are dried to a given moisture content, such descriptions are admissible for purposes of quality control). The perturbation of water by the protein decays as a function of distance, the decay function depending on the particular hydration site on the polypeptide chain. Thus, for ionogenic side chains the interaction is strong and has a long range. For apolar sites, hydrophobic hydration is weak, but probably of long range (*12,21*). For polar sites ($-OH$, $>CO$, $-NH$) the interaction approximately equals in strength that between two water molecules and is likely to be short-ranged. The problem of expressing hydration in aqueous solution in a meaningful manner resembles that of describing a water sorption isotherm in which the sorbent is highly heterogeneous and contains an irregular array of different sorption sites with different sorption energies and is likely to orient the sorbed water molecules in different ways.

7. Proteins at Low Moisture Content: Sequential Hydration

If a protein in solution is carefully concentrated under conditions that do not disrupt the N state (i.e., moderate temperature), then in some cases water can be almost completely removed and an almost anhydrous protein obtained. If water vapor is then readmitted, the interactions in the system can be monitored by any one of several physical or biochemical techniques and the sites of interactions characterized as a function of the moisture content. Table 4 summarizes the techniques and the type of information that can be obtained (*49*).

A popular method for studying water-sensitive or -soluble polymers (and this includes proteins) is by the measurement of water vapor sorption isotherms. Figure 10 is typical of the behavior of fibrous proteins: the sorption and desorption isotherms do not coincide (*50*). Sorption hysteresis at low moisture levels is symptomatic of thermodynamic metastability. The hysteresis loops are experimentally reproducible and suggest conformational changes and/or slow water migration on the surface of the protein to sites of high sorption energy; *absorption* (plasticization) must also be considered a possibility.

Sorption isotherms of the type shown in Fig. 10 are often subjected to thermodynamic analyses based on the Langmuir, BET, or some other sorption theory. This makes possible the calculation of a notional monolayer coverage, the sorption energy, and other details of the sorption process. It should be remembered, however,

Protein Hydration

Table 4
Available Experimental Techniques for the Study of Protein–Water Interactions

	State of sample		Type of information			Information about	
	Solution	Solid	Time-average	Dynamic	Structural	Water	Protein
Diffraction		+	+		+	+	+
Spectroscopy	+	+	+	(+)	(+)	(+)	+
Thermodynamics	+		+			+	+
Sorption		+	+			+	+
Relaxation (nmr, esr)	+	+		+		+	+
Hydrodynamics	+			+			+
Kinetics (e.g., H⁺ exchange)	+	+		+			+

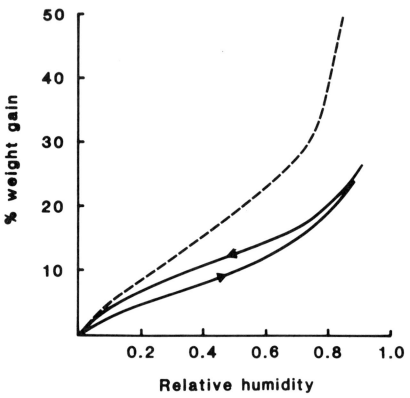

Fig. 10. Water sorption isotherms of collagen (broken lines) and human hair keratin (solid lines); note the marked sorption hysteresis (from ref. 50).

that the assumptions on which such *equilibrium* sorption theories are based cannot possibly apply to a nonequilibrium system that is subject to sorption hysteresis. Furthermore, under the experimental conditions employed, the measured relative humidity, p/p_o, is not to be equated to the water activity a_w. Indeed, a_w has no meaning where true thermodynamic sorption/desorption equilibrium does not exist, a fact that is still not widely appreciated, especially by food technologists who habitually express microbiological safety and textural quality in terms of a measured relative humidity that is mistakenly referred to as a_w.

Among thermodynamic properties, the heat capacity is the most sensitive indicator of overall hydration interactions, although various spectroscopic techniques provide information about interactions between water and specific chemical groups on the polypeptide. Figure 11 shows the changes in various experimental parameters as a function of the moisture content, expressed as g water/g protein, that occur when water vapor is admitted to previously dried

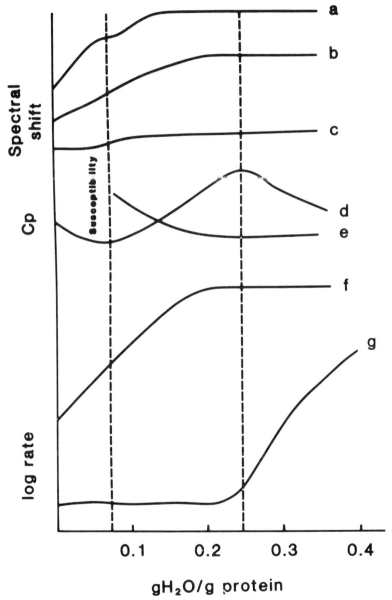

Fig. 11. Effect of sequential hydration on various experimental parameters of the lysozyme/substrate complex (for details, see text and ref. 49).

lysozyme (49). The curves labeled a, b, and c are, respectively, the infrared intensities of the carboxylate band (1580 cm^{-1}), the amide I band (1660 cm^{-1}), and the water OD stretching frequency (2570 cm^{-1}). Curve d is the apparent specific heat, e is the diamagnetic susceptibility, f is the log rate peptide proton exchange, and g is the log enzyme activity. Curve g coincides exactly with the rotational diffusion rate of the protein, as monitored by an electron spin probe.

From the moisture dependence of the various parameters in Fig. 11, a picture can be constructed that describes the events that take place on the surface of the protein as moisture is admitted. This is shown in Table 5. Where the various properties level off near 0.4 g water/g lysozyme, water is observed to take up its "normal" physical properties. The indications are that the first traces of moisture produce a "minor" conformational reordering of the protein resulting from the hydration and ionization of carboxylate groups. As the degree of hydration increases, so the acidic and basic groups become saturated and water molecules diffuse from there to other sites, specifically the peptide backbone $-$NH and $>$CO groups. At this stage a low level of enzyme activity is first observed that then increases with hydration, but only reaches its "normal" value at 9 g water/g protein. The decrease in C_p at hydration levels

Table 5
Events Associated with the Sequential Hydration (Water Vapor Uptake) of the Lysozyme/Substrate Complex, as Monitored by the Parameters Shown in Fig. 11

Degree of hydration gH$_2$O/g protein	Site of interaction	Event
0.1	Ionic side chains glu, asp, lys	Acids reach "normal" pK_a values
	Main chain $>$CO	Minor conformational transition
	Main chain $>$NH	Side chain mobility
0.2	Peptide groups "saturated"	$>$CO, $>$NH stretching frequencies attain "normal" values
	Side chain $-$OH	
0.3	Water clusters, bridges	τ H$_2$O assumes normal value ($\sim$ 3 ps). Detectable enzyme activity

above 0.25 is yet to be explained; it may be related to hydrophobic hydration phenomena or the water bridging of several hydration sites.

Controlled desiccation also enhances the thermal stability of globular proteins; the following water contents (per g protein) will stabilize the proteins against heating at 100°C: myoglobin, 8%; hemoglobin, 11%; chymotrypsinogen, 11%; collagen, 28%; lysozyme, 15%; and β-lactoglobulin, 22%. Similar effects of desiccation are observed in the inactivation of enzymes. Thus, the heat inactivation of glucose-6-phosphate dehydrogenase (also glucose dehydrogenase and leucine dehydrogenase) for which the inactivation rate constant at 50°C (in dilute solution) is 0.12/min, is reduced to 0.015/min at 145°C and $a_w = 0.11$.

8. Computer Modeling of Protein-Water Interactions

The in vivo generation of proteins on the ribosome takes place in an aqueous medium, and the stability and activity of any given protein in vitro are critically affected by minor changes in the physical and chemical conditions of the medium. It follows, therefore, that the hydrogen bonding energetics and topology in liquid water and between water and the peptide chain(s) are important determinants of folding, stability, and function.

The analysis of hydration processes presented earlier in this chapter illustrates the problems associated with experimental or theoretical estimations of water–water and water–peptide interactions. Recent years have witnessed the first attempts to "calculate" native protein conformations by computer simulation techniques that had earlier been used with reasonable success to describe liquid water itself (51) and dilute aqueous solutions of simple molecules (52,53).

In principle, the forces between all pairs of protein and water atoms are used to derive the global potential energy of the system. For example:

$$U = \sum_{\text{bonds}} K_b(b - b^0)^2 + \sum_{\text{angles}} K_\theta(\theta - \theta^0) + \sum_{\text{torsions}} K_\phi[1 - \cos(n\phi + \delta)]$$
$$+ \sum_{\substack{\text{nonbonded}\\\text{distances}}} \epsilon[(r^0/r)^{12} - 2(r^0 - r)^6] + \sum_{\substack{\text{partial}\\\text{charges}}} q_i q_j/r + \sum_{\substack{\text{hydrogen}\\\text{bonds}}} F(\theta, r) + \ldots \quad (5)$$

The variables are the bond lengths, b, the bond angles, θ, the torsion angles, ϕ, and the interatomic distances, r, which all depend on the set of all atomic coordinates that define the protein–water system. The energy parameters (K_b, b^0, K_θ, θ^0, K_ϕ, n, δ, ϵ, r^0, q, and so on) define U quantitatively.

The atomic positions can now be changed and U recalculated. If the new value of U is less than the previous one, the alteration is accepted. U is minimized by iteration, it being assumed that the minimum value of U corresponds to the native protein state. This type of computation (Monte Carlo simulation) can provide estimates of the distribution of water molecules in the neighborhood of the protein molecule and also the equilibrium properties of the system. A major weakness of the procedure lies in the implicit assumption that the calculated *potential* energy U can replace the *free* energy G in defining the conformational stability. The computer model therefore does not allow for the role of configurational entropy or hydrophobic effects in determining the stability of the native state. A further assumption is that the configuration with the minimum potential energy corresponds to the native state. In practice, the starting point for the simulation is usually the known crystal structure (54).

A potentially even more powerful technique, known as molecular dynamics, can provide additional information about the dynamic properties of the protein–water system. Once again the atomic positions are altered, but the atoms are moved along the trajectories that are defined by solving Newton's equations of motion at given time intervals. The total kinetic energy of the system provides a measure of the effective temperature, which can be altered simply by changing the velocities of the molecules. Molecular dynamics simulations yield not only the equilibrium (thermodynamic) properties of the system, but also properties such as bond vibration and libration frequencies and diffusion constants.

The modeling of systems containing macromolecules and many water molecules requires large capacity computers. Most useful computer studies so far have been limited to small proteins, e.g., the 58-residue pancreatic trypsin inhibitor (55). Approximations are also usually required in the manner in which sufficient water molecules can be incorporated into the computations. Finally, *all* computer simulations of real molecular systems depend critically on the quality of the various potential functions employed in Eq. (5) and the energy minimizations.

References

1. J. L. Finney, in *Water — A Comprehensive Treatise* vol. 6 (F. Franks, ed.) Plenum, New York (1979), pp. 47–122.
2. J. A. Rupley, in *Characterisation of Protein Conformation and Function* (F. Franks, ed.) Symposium Press, London (1979), pp. 54–61.
3. F. Franks, in ref. 1, **4**, 1–94 (1975).
4. H. L. Friedman and C. V. Krishnan, in ref. 1, **3**, 1–118 (1973).
5. T. H. Lilley, in ref. 3, **3**, 265–300 (1973).
6. J. E. Enderby and G. W. Neilson, in ref. 3, **6**, 1–45 (1979).
7. H. S. Frank and W. Y. Wen, *Disc. Faraday Soc.* **24**, 133 (1957).
8. B. E. Conway, *Ionic Hydration in Chemistry and Biophysics* Elsevier, Amsterdam (1981).
9. H. L. Friedman, *Faraday Disc. Chem. Soc.* **64**, 1(1978).
10. A. Suggett, *J. Solution Chem.* **5**, 33 (1976).
11. F. Franks, in *Polysaccharides in Foods* (J. M. V. Blanshard and J. R. Mitchell, eds.) Butterworths, London (1979), pp. 33–50.
12. F. Franks and J. E. Desnoyers, *Water Sci. Rev.* **1**, 171–232 (1985).
13. D. W. Davidson, in ref. **3**, 115–234 (1973).
14. D. Eagland, in ref. 1, **4**, 305–518 (1975).
15. F. Franks and D. Eagland, *Crit. Rev. Biochem.* **3**, 165 (1975).
16. R. H. Pain, in ref. 2, 19–36.
17. J. L. Finney, in *Biophysics of Water* (F. Franks and S. F. Mathias, eds.) John Wiley & Son, Chichester (1982), pp. 55–57.
18. D. Y. Chan, D. J. Mitchell, B. W. Ninham, and B. A. Pailthorpe, in ref. 1, **6** (1979).
19. C. Tanford, *The Hydrophobic Effect* John Wiley & Sons, New York (1973).
20. C. Chothia, *Nature* **248**, 338 (1974).
21. S. Okazaki, K. Nakanishi, and H. Touhara, *J. Chem. Phys.* **78**, 454 (1983).
22. S. F. Dec and S. J. Gill, *J. Solution Chem.* **14**, 417 (1985).
23. D. Eisenberg, R. M. Weiss, T. C. Terwilliger, and W. Wilcox, *Faraday Symp. Chem. Soc.* **17**, 109 (1982).
24. Y. Nozaki and C. Tanford, *J. Biol. Chem.* **246**, 2211 (1971).
25. C. C. Bigelow, *J. Theor. Biol.* **16**, 187 (1967).
26. L. Nord, E. E. Tucker, and S. D. Christian, *J. Solution Chem.* **12**, 889 (1983).
27. F. Franks and M. D. Pedley, *J. Chem. Soc. Faraday Trans I* **79**, 2249 (1983).
28. B. Lee and F. M. Richards, *J. Mol. Biol.* **55**, 379 (1971).
29. A. Shrake and J. A. Rupley, *J. Mol. Biol.* **79**, 351 (1973).
30. J. Drenth, J. N. Jansonius, R. Koekoek, and B. G. Wolthers, *Adv. Protein Chem.* **25**, 79 (1971).

31. H. J. C. Berendsen, in ref. 3, **5**, 293–330 (1975).
32. G. N. Ramachandran and R. Chandrasekharan, *Biopolymers* **6**, 1649 (1968).
33. G. E. Chapman, S. S. Danyluk, and K. A. McLauchlan, *Proc. Roy. Soc.* **B 178**, 465 (1971).
34. J. J. Birktoft and D. M. Blow, *J. Mol. Biol.* **68**, 187 (1972).
35. K. W. Packer, *Phil. Trans. Roy. Soc. Ser.B*, **278**, 59 (1977).
36. W. Derbyshire, in ref. 3, **7**, 339–430 (1982).
37. R. G. Bryant and B. Halle, in ref. 17, 389–393.
38. I. D. Kuntz and W. Kauzmann, *Adv. Protein Sci.* **28**, 239 (1974).
39. R. Cooke and I. D. Kuntz, *Ann. Rev. Biophys. Bioeng.* **3**, 95 (1974).
40. F. Franks, *Biophysics and Biochemistry at Low Temperatures*, Cambridge University Press, Cambridge, UK (1985).
41. J. S. Clegg, in ref. 17, 365–383.
42. G. N. Ling and W. Negendank, *Persp. Biol. Med.* **24**, 215 (1980).
43. P. L. Privalov and G. M. Mrevlishvili, *Biofizika* **12**, 22 (1967).
44. I. D. Kuntz, *J. Amer. Chem. Soc.* **93**, 516 (1971).
45. I. D. Kuntz, *J. Amer. Chem. Soc.* **93**, 514 (1971).
46. V. C. McCabe and H. F. Fisher, *J. Phys. Chem.* **74**, 2990 (1970).
47. E. S. Pysh, *Biopolymers* **13**, 1557 (1974).
48. H. K. Ross, *Ind. Eng. Chem.* **46**, 601 (1954).
49. J. A. Rupley, E. Gratton, and G. Careri, *Trends Biochem. Sci.* p. 18 (1983).
50. M. Escoubes and M. Pineri, in *Water in Polymers* (S. P. Rowland, ed.) ACS Symp. Ser., Washington, DC, 127 (1980), pp. 235–250.
51. D. W. Wood, in *Water — A Comparative Treatise* vol. 6 (F. Franks, ed.) Plenum Press, New York (1979) p. 279.
52. G. Ravishanker, M. Mezei and D. L. Beveridge, *Faraday Symp. Chem. Soc.* **17**, 79 (1982).
53. H. Tanaka, K. Nakanishi and H. Touhara, *J. Chem. Phys.* **81**, 4065 (1984).
54. B. Gelin and M. Karplus, *Proc. Nat. Acad. Sci. USA* **72**, 2002 (1975).
55. M. Levitt, in *Protein Folding* (R. Jaenicke, ed.) Elsevier, Amsterdam (1980) p. 17.

Chapter 6
Characteristics of Proteins and Peptides *In Situ*

An Overview

P. J. Thomas

1. Proteins and Peptides as Functional Units

In my contributions to this book I shall consider proteins and peptides as functional biological units, rather than as chemical compounds. I shall try to show that, from the point of view of a biologist, proteins should not be considered as individual entities, but rather as components of complex interacting cyclical control systems that involve not only multitudes of other proteins, but also nonprotein-components of cells and of the extracellular environment. I have in mind, for example, the phospholipids of the cell membrane, nonproteinacious chemical messengers (steroids and neurotransmitters), and drugs and toxins of low molecular weight.

Clearly, because of the numbers of proteins that have been described, I have to be highly selective in my treatment of the subject, and have concentrated, in this and succeeding chapters, on proteins as parts of control systems, digressing to other classes of proteins when they affect such systems either functionally or experimentally.

For convenience, I have arbitrarily divided proteins into 11 groups, and describe these before showing how they can interact at different spacial and temporal levels. These divisions are depicted schematically in Fig. 1, and will be discussed in more detail below.

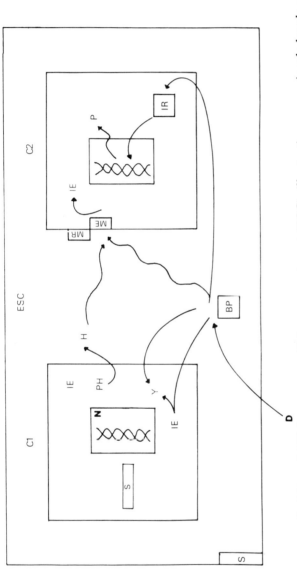

Fig. 1. Two cells (C1 and C2) in an extracellular compartment (ESC), illustrating some simple levels of communication between them. C1 synthesizes a prohormone (PH) that is degraded by intracellular enzymes (IE) to yield a protein hormone (H) that binds to membrane receptors (MR) on C2. This may activate membrane enzymes (ME) leading to a cascade of activation of intracellular enzymes (IE) and thence cellular responses. IEs (see C1) also induce the synthesis of nonprotein messengers (Y) that may act within the cell or be externalized. Y may effect C2 either via MR, as before (though binding may initiate a different class of event, for example, opening ion channels in the membrane), be taken back into C1, or bind to intracellular receptors (IR) in C2. Activation of IR leads to modifications of genetic expression, leading to the syntheses of new proteins (P). The free concentrations of Y [or indeed of foreign drugs and toxins (D)] may be modified by extracellular binding proteins (BP). Also indicated in the figure are intracellular and extracellular structural and contractile proteins (S) that modulate the distribution of the substances already mentioned.

1.1. Carrier Proteins

Carrier proteins bind to other molecules in accordance with the law of mass action, albeit that, because of ligand or concentration-induced changes in their conformation, they frequently exhibit positive (allosteric) or negative cooperation in their binding parameters. They bind with varying degrees of specifity and affinity, and alter the effective concentrations and biological half lives of their ligands. They may be vascular (for example, hemoglobin, serum albumin, alpha-fetoproteins, and sex steroid-binding globulin) or intracellular [for example, calmodulin (1) and cyclic AMP-binding proteins (2)]. In the case of some intracellular binding proteins, particularly those associated with neurotransmitters and neuropeptides within vesicles (3,4), it may be debatable whether they should properly be considered as carrier or storage proteins.

It has been suggested that in certain cases extracellular carrier proteins can enter target cells together with their ligands (5), thus conferring a strong degree of anatomical specificity. When this is not the case, they will merely sequester their ligands at regions of high concentration and release them along a concentration gradient, where they will affect responsive cells (Fig. 2).

Extracellular carrier proteins are of particular importance in therapeutics. Dosage and timing of drug administration is to a large extent dependent upon the relative proportion that is bound in the plasma at a given time, and this may be affected by both cyclical physiological changes and pathological conditions (Fig. 2). For example, the concentration of sex steroid binding globulins varies during the female sexual cycle (5), and serum albumin concentrations are dependent upon renal and hepatic function.

In summary, the bioavailability of many endogenous and exogenous biologically active compounds is dependent upon their interaction with carrier proteins.

1.2. Hormones and Prohormones

Hormones are generally defined as being chemical messengers in the body that act at a distance from their sites of origin (Fig. 3). In the case of protein and peptide hormones, this definition may

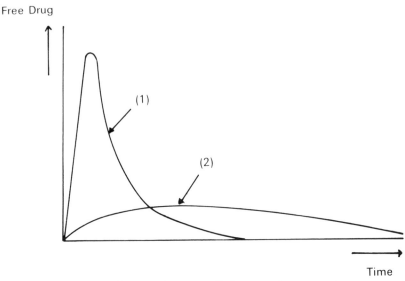

Fig. 2. The effects of extracellular carrier proteins upon the bioavailability of a drug, in (1) its absence, and (2) its presence.

need some modification, since the discovery (6,7) that many proteins and peptides that were originally defined as hormones are also found in nerve terminals in the central and peripheral nervous systems, where they coexist with, and are released together with, classical neurotransmitters (chemical messengers that act locally). Here they perform a neuromodulatory function that is distinct from their classical hormonal function (Fig. 4). However, although I am at present engaged in the business of definition, I regard the above problem of redefinition as trivial.

Not all hormones are proteins or peptides, steroid and thyroid hormones being obvious exceptions. However, as we shall see later (chapter 8), the control and activity of all of them involve the mediation of proteins.

I shall now introduce protein and peptide hormones, but before doing so must draw the readers attention to the fact that individual examples will be described in more detail in chapter 9.

Protein and peptide hormones are secreted principally by the anterior and posterior parts of the pituitary gland (the adenohypophysis and the neurohypophysis, respectively) and by the pancreas (8,9).

Peptides and Proteins

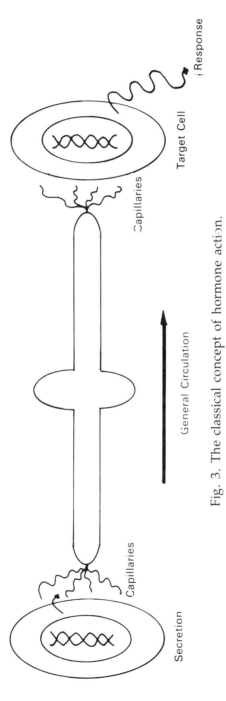

Fig. 3. The classical concept of hormone action.

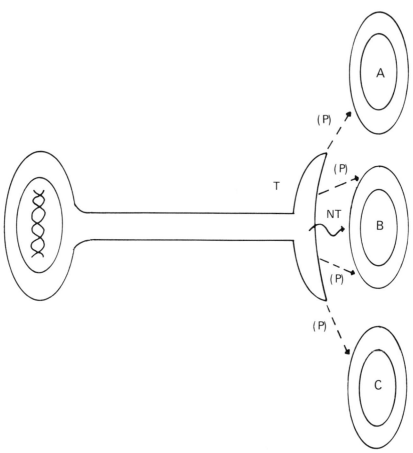

Fig. 4. Cotransmitters. Neurotransmitters (NT) are released from nerve terminals (T), and because of their rapid metabolism or reuptake, only impinge on cells in the immediate vicinity (B). Cotransmitter peptides (P) are less rapidly destroyed, so may affect not only cells in the immediate vicinity, but also potential target cells somewhat further away (A and C). They will not reach the general circulation in sufficient concentration to influence potential target cells further afield. Furthermore, neurotransmitter levels in nerve terminals can be rapidly replenished, whereas resynthesis of cotransmitters is relatively slow, since it involves *de novo* protein synthesis. Thus there are temporal and spatial, as well as chemical, differences between the two classes of compounds.

The anterior pituitary, or adenohypophysis, secretes several protein and glycoprotein hormones (Fig. 5) that are described in some detail in chapter 9. Their functions are trophic. That is to say that after being released into the circulation they impinge upon their target endocrine organs to modify the secretion of peripheral hormones, which in turn exert their effects upon cells throughout the body. The anterior pituitary is situated in close juxtaposition to a region of the brain known as the hypothalamus (Fig. 5), which, among many other things, secretes short-chain oligopeptides into a local network of blood vessels that bathe the anterior pituitary and that either enhance or inhibit the synthesis and release of individual pituitary hormones. These peptides are known as release (or release-inhibiting) factors or hormones (10).

This leads us naturally to our first example of a control system, one that involves protein, peptide, and nonprotein hormones. The hypothalamic secretion of release and release-inhibiting factors, followed by the secretion by the pituitary of trophic hormones and the consequent modulation of the secretion of peripheral hormones, may be considered as steps in a one-way (downward) control hierarchy (Fig. 6). This, of course, would be inherently unstable, and is modified by so called "feedback" controls, whereby peripheral hormones alter the secretion of their appropriate hypothalamic and pituitary regulators. These feedback controls may be positive or negative, and an example, the neural and pituitary control of ovarian steroids, is illustrated in Fig. 6. It should be noted in particular that signals from the external environment, transcribed by hypothalamic and extrahypothalamic areas of the brain, can influence these feedback loops (*see* ref. 9), and that the control system may be further influenced by variations in the concentrations of circulating sex steroid binding proteins (*see* ref. 5).

The control of the secretion of the peptide hormones of the posterior pituitary and the protein hormones of the pancreas is different in detail, but analogous in principle to that described above. It will be considered in chapter 9.

1.2.1. Prohormones

Most protein and peptide hormones are synthesized initially as longer chains of amino acids (*see* ref. 8), and after packaging into the Golgi apparatus (10) are broken down enzymatically into active (hormonal) and inactive components before secretion (Fig. 7). The precursors are called prohormones. An example, the relationship between insulin and proinsulin, is shown in Fig. 8. In some cases the breakdown may be sequential, leading to the formation of more than one active component and to the inelegant word, preprohormone.

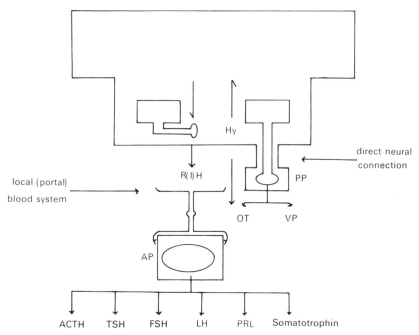

Fig. 5. The hypothalamus (Hy) is a relatively primitive area of the brain that has two-way communication with other brain areas. It contains many neurons, the axons of some of which terminate in the posterior pituitary and secrete the polypeptide hormones oxytocin (OT) and vasopressin (VP) into the blood stream. Other neurons terminate in the median eminence (on the surface of the brain), whence they discharge into a discrete part of the circulation (the pituitary portal system), which carries them to the anterior pituitary where they cause the secretion and release of much larger proteins and peptides; the trophic hormones of the pituitary. These are ACTH (adrenocorticotrophic hormone), TSH (thyroid stimulating hormone), FSH (follicle stimulating hormone), LH (luteinizing hormone), prolactin (PRL), and S (somatotropin). The short chain peptides of the hypothalamus that control anterior pituitary function are known as release (or release-inhibiting) hormones [R(I)H], and are relatively specific to the different trophic hormones. Their full biological names refer to their functions, for example, thyroid stimulating hormone releasing hormone (TRH: pyroglutamyl-histidyl-proline amide). There can be different conventions in nomenclature: for example TRH is also referred to as thyrotrophin releasing hormone, but these can be easily understood. The only known exception to the general rule that the hypothalamic factors are peptides is prolactin release-inhibiting factor that appears to be a neurotransmitter, dopamine.

Peptides and Proteins

In summary, the control of protein hormone secretion is cyclical and involves nonproteinaceous compounds. Active hormones are first synthesized as longer chain precursors.

1.3. Neuropeptides and Cotransmitters

Before defining neuropeptides and cotransmitters, I must say a few words about neurotransmission. Neurons affect other cells (whether they are other neurons or not) that are in close anatomical proximity to their efferent processes (axons). A neuronal impulse is transmitted from the main part of the cell body (the perikarion) along the axons in the form of waves of depolarization. When these waves reach the nerve terminal, they cause the release of neurotransmitters (which are small molecules, such as acetycholine, monoamines, glutamine, and gamma-aminobutyric acid) are then released into the synaptic cleft (the minute space between the nerve terminal and the target cell; Fig. 9); they bind to receptors and initiate a chain of events that ends in a response: in the case of neurons, a further wave of depolarization. Their action is limited either by reuptake into the nerve terminal whence they came, by enzymatic breakdown in the region of the synaptic cleft, or within the nerve terminal (Fig. 10) (11–15): The special case of their incorporation into target cells will be dealt with later under the heading of internalization. They are thus distinct from hormones not only in their biochemical characteristics, but by the fact that they act locally and their activity is rapidly terminated. Until relatively recently they were considered to be the only chemical messengers secreted by neurons, and that one nerve terminal secreted only one class of chemical messenger.

It has now been shown that many peptides, including a great number that had originally been classified solely as hormones, coexist with neurotransmitters within nerve terminals in the central and peripheral nervous systems and are released together with classical neurotransmitters after neuronal stimulation (6,7,16).

In functional terms they differ importantly from classical neurotransmitters. As stated before, the action of neurotransmitters is rapidly terminated by local uptake and/or degradation. Insofar as I know, no such precise and rapid mechanism for terminating the actions of neuropeptides has been described. Thus, if responsive cells are nearby, they have the potential to act at a greater distance than do neurotransmitters. However, because neuropeptides, in

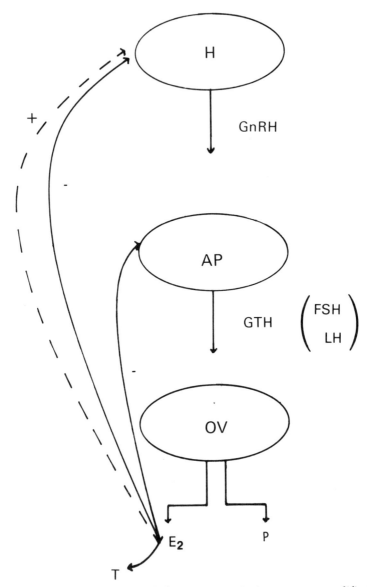

Fig. 6. Feedback control of the anterior pituitary as exemplified by the cyclical secretion of the ovarian steroid, 17β-estradiol. The hypothalamus (H) secretes gonadotrophin releasing hormone (GnRH), which stimulates the anterior pituitary (AP) to secrete and release two gondotrophic hormones (GTH). These are follicle stimulating hormone (FSH) and luteinizing hormone, which act sequentially upon the ovary (OV) to cause secretion of the steroid hormones 17β-estradiol (E_2) and pro-

their capacity as neuromodulators rather than in their capacity as hormones, are secreted in relatively low concentration (neural tissue is much more heterogeneous and therefore effectively more diffuse than endocrine tissue), they cannot act at as a great a distance as they do in the latter capacity.

Neuropeptides exhibit another important functional difference from neurotransmitters. Because of reuptake and rapid enzymatic synthesis within the nerve terminal, neurotransmitters are difficult to deplete. Neuropeptides, on the other hand, are synthesized as precursors within the perikaryon, and the active principles are formed by proteolytic cleavage during transport to the nerve terminal (*see* section 1.2). After depletion, therefore, recovery takes longer.

In summary, peptides and proteins that function as hormones elsewhere in the body also coexist with neurotransmitters in nerve terminals. Because their rates of synthesis and degradation differ from those of classical neurotransmitters, their effects are different and they are considered to be neuromodulators rather than neurotransmitters. This concept is illustrated in Fig. 4.

1.4. Receptors

Before a chemical can affect a living organism, tissue, or cell, it must (a) come into close contact with it, and (b) associate chemically with some component or components thereof. It has long been recognized that most tissues respond to chemical messengers not only in a saturable manner, but often with a fairly high degree of specificity, the effects of a given compound being inhibited, in a dose-dependent manner, by the presence of chemically related compounds that happen not to elicit the same biological response. From such observations the concept of receptors was developed by pharmacologists beginning at around the turn of the century (*17*), long before sophisticated modern methods of analysis were available

---

← Fig. 6 (*continued*)

gesterone (P). These act upon the pituitary and hypothalamus. In general, E_2 exerts a negative feedback upon both tissues, decreasing the secretion of GnRH and GTH and thus its own secretion (homoeostasis). However, a cyclical secretion of estradiol is necessary for ovulation and, under certain conditions that are dependent upon the sequential bombardment of the female hypothalamus by E_2 and P, it can respond to high and increasing levels of E_2 by increasing its output of GnRH. This positive feedback is, as far as I know, unique to the female brain.

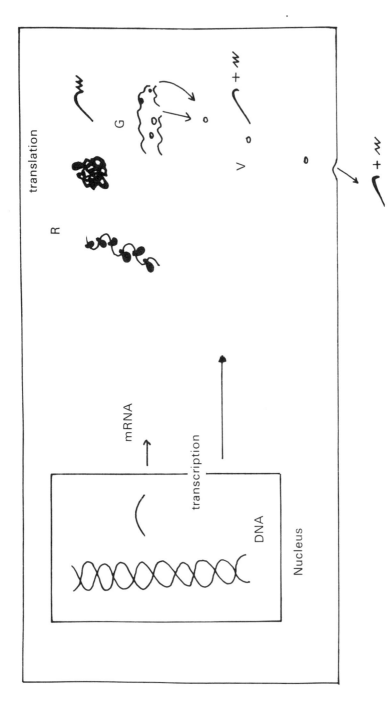

Fig. 7. The relationship between hormones and prohormones. Messenger RNA (mRNA) production leads to the formation of long chain peptides by the ribosomes (R). These are packed into vesicles (V) by the Golgi apparatus and are broken down into their hormonal and nonhormonal constituents during their transport to the cell surface, but before their extrusion.

Peptides and Proteins

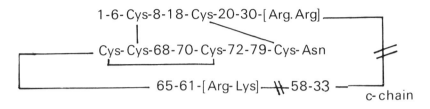

[ ] = excision   Remainder = Insulin

Proglucagon

N terminal   His- Ser- Glu- Gly- Thre-
Phe- Thre- Ser- Asp- Tyr- Ser- Lys-
Tyr- Leu- Asp- Ser-(Arg)2- Ala- Glu-
Asp- Phe- Val- Glu- Trp- Leu- Met-
Asp- Thre ─────────────────────┐
                               │← CLEAVAGE
Lys- Arg- Asn- Asn- Lys- Asn- Ileu-
Ala

C terminal

1-29 = Glucagon

Fig. 8. Two examples of the relationship between prohormones. (Top panel) Proinsulin and insulin, and (lower panel) proglucagon and glucagon.

(I stress the latter point, despite its obvious nature, because it indicates the fact that high technology is but a tool, albeit a very useful one).

These observations of dose-dependent competition and saturability led to the concept that target cells are characterized by the possession of a finite number of recognition sites for biologically active compounds, and that these recognition sites are linked to effector mechanisms (and also to a minor problem of definition:

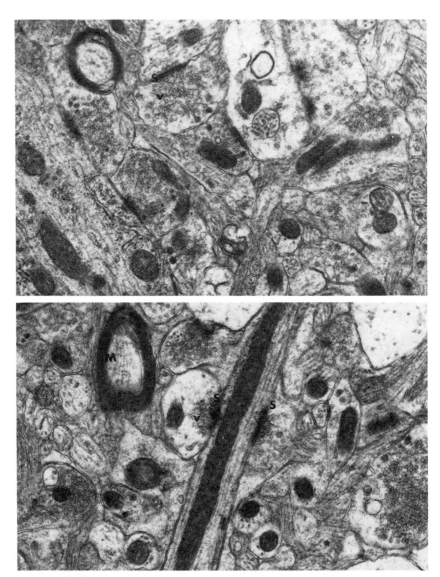

Fig. 9. Electron microscopic appearance of some synapses. Two electronic micrographs of the portions of the neuropil of the suprachiasmatic nucleus of adult male rats (×30,000). (A) A synapse (S) with its associated vesicles (v) is shown. There is a slight thickening on the presynaptic (vesicular) side that is separated from the postsynaptic cleft by a space—the synaptic cleft. (B) The synaptic clefts are not seen. This is because of the angle of the section. Also shown is the amylenated nerve fiber (M). Numerous mitochondria are seen in both sections.

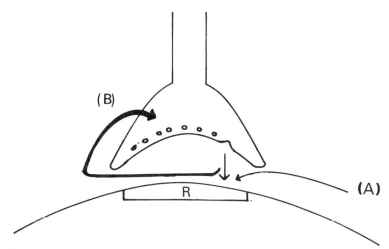

Fig. 10. The termination of neurotransmitter actions. Neurotransmitter action is rapid. After release, any that is not bound to the recognition site (R) is either (A) broken down by extracellular enzymes or is (B) taken back (reuptake) into the axon, where it may be either repackaged into vesicles, or broken down by intracellular enzymes.

It is not always clear whether the term "receptor" refers to the recognition site, as it does, by default, in radioligand competition studies, or to the active complex *in toto*).

In my contributions to this book I use the word "receptor" to mean the recognition site. I should also make it clear that although most, if not all, protein and chemical messengers initiate their effects by binding to receptors, as do the majority of other drugs and hormones, this is not necessarily a universal truth. For example, some drugs may work by differential solubility in membranes (general anesthetics, ethanol) and some toxins by binding to carrier proteins and enzymes that are not generally recognized as receptors.

Thus the term "receptor" implies that the site is present in only a specific group of target cells, even if these may make up the greater part of an organ.

Receptors are basically proteins. They are divided into two classes: membrane and intracellular (Fig. 11). Receptors for most chemical messengers are membrane bound: The recognition sites are exposed on the outer surface of the cell membrane and their occupation by a ligand is the first step in the initiation of a response. The above statement requires some qualification since some membrane receptors may be "internalized" (that is to say that the recep-

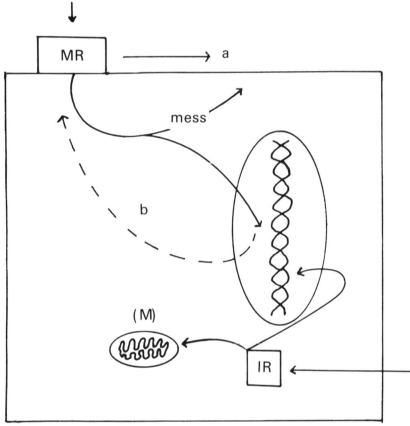

Fig. 11. A distinction between membrane (MR) and intracellular receptors (IR). Substances that activate MR initiate chains of responses that may affect either (a) the cell membrane, or (b) the intracellular components, including the genome. Substances that act upon IR affect, in the first instance, either mitochondrial (M) activity (thyroid hormones) or that of the genome (thyroid, steroid hormones, and vitamin D). Proteins synthesized as a result of genetic activation (b) can affect membrane receptors. Thus there are many levels at which these two types of pathway can interact.

tor, its associated enzymes, and adjacent parts of the cell membrane may invaginate and move into the cell shortly after the initial binding reaction). This concept is illustrated in Fig. 12. The biological importance of internalization has yet to be worked out in detail.

In contrast, receptors for a few groups of biologically active compounds appear to be intracellular at the time when they recognize

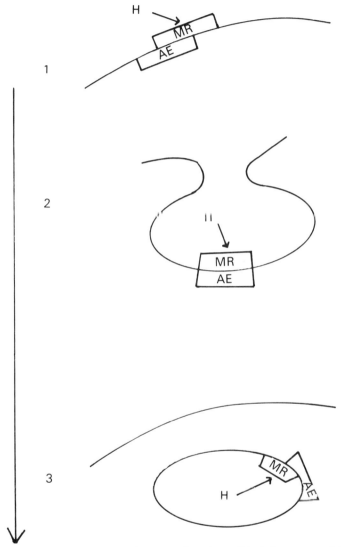

Fig. 12. The concept of internalization. (1) A hormone binds to a member receptor (MR). (2) The complex, together with the adjacent membrane and associated enzymes (AE) is shown. This complex moves into the cell (3) in the form of inverted vesicles.

their specific ligand. Examples of these intracellular receptors are those for steroid (18) and thyroid hormones (19) and for vitamin D (20). In essence, those chemical messengers that bind to intracellular receptors are believed to pass through the cell membrane without being recognized by any component thereof. Once within the cell, they come into contact with intracellular binding sites with which they associate specifically and with high affinity. This binding alters the configuration of these intracellular receptors, and the hormone–receptor complex so formed moves to the nucleus where it interacts with proteins associated with the genetic material, leading to modification of genetic expression and thence (via RNA) to alterations in protein biosynthesis. This concept is depicted in Fig. 13, in which both classical and recent models of the action of a particular steroid hormone, 17β-estradiol, are summarized. It should be emphasized that the classical model (the steroid binding to a cytosolic receptor that moves to the nucleus) and the modern model (21,22), in which the initial receptor is believed to be loosely associated with the nuclear membrane, do not differ in substance. The hypothesis is still two-stage: The conflict is only concerned with whether the initial receptor is torn away from the nuclear membrane during the process of cellular homogenization.

It should be made clear that this model only applies to steroid hormones and vitamin D: thyroid hormones may act on the mitochondrion as well as on the nucleus (19).

In summary, receptors are sites that are specific to individual cell types. They are proteins and recognize individual chemical messengers. The bulk of them reside on the cell surface and are termed membrane receptors: Others lie within the cell and are termed intracellular receptors. We shall see later that intracellular and extracellular messengers can induce or inhibit common intracellular messengers (cyclic nucleotides) and so have the capacity to interact (*see* chapter 8).

1.5. Enzymes

Enzymes are proteins that recognize specific chemicals, but that, unlike binding proteins and receptors, do not merely sequester them, or because of binding, alter in configuration and thus in ability to interact with other proteins. They catalyze the conversion of the initial messengers into other compounds. Substrates and products are many and varied, and I must be selective in my treatment. I shall only consider those enzymes that are involved in the transfer of messages from cell to cell, and in particular with enzymes that are associated with membrane and intracellular receptors.

Peptides and Proteins

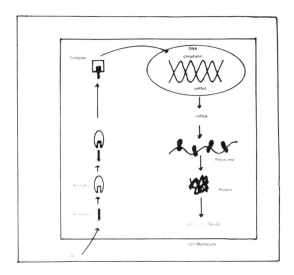

Intracellular cytoplasmic receptors (proteins)

$$8s \underset{}{\overset{KCl}{\rightleftarrows}} 4s$$

$$E_2 + 4s \longrightarrow 5s$$

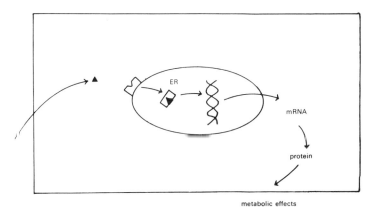

Fig. 13. (Upper panel) The classical model of steroid hormone action; (lower panel) A new possibility. The initial receptor does not necessarily sit in the cytoplasm, but may be bound loosely to the nuclear membrane and dissociated during homogenization. The difference between the two models is merely one of experimental technique.

Membrane receptors open many pathways (Fig. 14). When they are activated their structure is modified, and depending upon their nature, they can either modulate ion channels (23) or activate or inhibit intracellular enzymes. Among the intracellular enzymes they modulate are adenylate cyclase and guanylate cyclase, enzymes that catalyze the formation of cyclic 3'5'-adenosine monophosphate (cAMP) from adenosine triphosphate (ATP) and cyclic 3'5'-guanosine monophosphate (cGMP) from guanosine triphosphate (GTP).

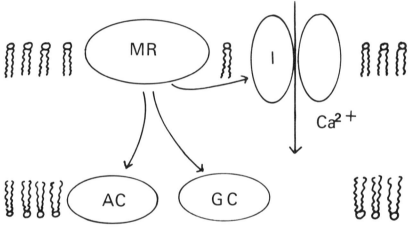

Fig. 14. Examples of pathways opened by membrane receptors. (An ideal pathway is depicted here; it does not necessarily follow that any individual receptor may affect all these pathways.) Activation of membrane receptors (MR) can lead to the opening of proteinaceous ion channels (I), leading, for example, to the influx of Ca^{2+} into the cell. It can also modulate the activities of enzymes such as adenylate cyclase and guanylate cyclase. (AC and GC).

cAMP is the classic intracellular messenger (24). Its concentrations within the cell determine the extent and duration of a vast number of intracellular responses. Its effects are in both the short and long term (25–29). Its short-term effects are, for example, upon membrane function, cellular secretion, and tissue contraction. Its long-term effects are upon cellular differentiation: It promotes the structural and functional development of cells (in culture, at least) at the expense of their growth. cGMP has, in general, opposing effects to those of cAMP (29), in both the short and long term, and its concentration tends to rise when that of cAMP falls, and vice versa. In developing cells it also tends to act in opposition to cAMP, promoting growth at the expense of differentiation (30–32). This may have importance to the development of cancers.

Neither adenylate cyclase nor guanylate cyclase are directly linked to the receptors that activate or inhibit them. Theories about the linkage mechanisms will be dealt with later.

1.5.1. Mode of Action and the Termination of Action of cAMP and cGMP

The cyclic nucleotides are broken down by enzymes termed phosphodiesterases (33). These are divided into several classes and may be specific to either cAMP or cGMP, or may effect both substances. They will be dealt with in detail in chapter 8; at this stage I only draw attention to their presence and point out that their activity, as well as that of the cyclases, is influenced by local concentrations of calcium (which are in turn influenced by calmodulin); at least one hormone (insulin) has been reported to have a direct effect upon their activity.

In general, it may be asserted that the cyclic nucleotides exert their effects via protein phosphorylation (34). They associate with and activate protein kinases, allowing them to catalyze the reactions whereby inorganic phosphate becomes protein bound, thus affecting specific activity. Comments on the regulatory and catalytic subunit structure of these protein kinases will be made in chapter 8, but it should be emphasized here that (a) not all protein kinases are cyclic nucleotide dependent, and (b) cyclic nucleotide-dependent protein phosphorylation is not restricted to parts of the cell adjacent to hormone recognition sites at the membrane. Proteins as far away as the cell nucleus may be affected. Thus, both membrane-active hormones and those that act via intracellular receptors have the potential to affect genetic expression as well as to initiate short term responses (Fig. 15).

In summary, activation of membrane receptors may lead to immediate (ion channel), short term (membrane protein phosphorylation), and possibly long-term (genetic expression) effects. In the latter case they may indirectly interact with those hormones that act via intracellular receptors.

1.6. Proteins Associated with Genetic Material

Every cell in the body (apart from the germ cells) has the same complement of DNA, and this DNA eventually determines the composition of the proteins that in turn determine its structure and function. This poses a problem since cellular structure and function is diverse. The problem could, however, be solved, in princi-

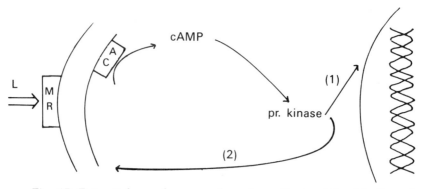

Fig. 15. Potential membrane and nuclear effects of the binding of a ligand to a membrane receptor (MR). Activation of adenylate (AC) (or guanylate) cyclase leads to the activation of protein kinases. This leads to the phosphorylation of some proteins and thus alters their functions. Their effects may be long-term (upon the genome) (1) or short-term (2). Intermediate temporal effects may also be involved.

ple if not in detail, if one considers the proteins associated with the genetic material and if one takes the view that these proteins modify certain aspects of genetic expression—a plausible view when one considers the binding of steroid receptors to nuclear proteins and subsequent modification of mRNA and protein biosynthesis. However, we are still faced with a problem since the bulk of the proteins associated with the genetic material are histones that are highly conserved (that is, their composition differs little from species to species or from tissue to tissue) and that the nonhistone proteins, which are less highly conserved, are relatively few in number (35). We are faced with another problem: the DNA is (by dogma) identical in all somatic cells. I see no reason, even if the nuclear proteins were polydisperse, that they should bind to different regions of the DNA in different cells. Nevertheless there must be differences. Steroid–receptor complexes bind to the chromatin and initiate responses in target cells, but not elsewhere. Thus, nuclear proteins are likely to be a key to the problem of morphogenisis and to an understanding of the long-term effects of hormones.

In summary, nuclear proteins may modulate the expression of DNA. They can also be affected by both steroid and (indirectly, via cyclic nucleotides) cell-membrane-acting hormones (Figs. 15 and 16). This should be considered in conjunction with the phosphorylation of serine residues by protein kinases and the methylation of DNA. These observations may provide an insight into the problems of understanding morphogenesis.

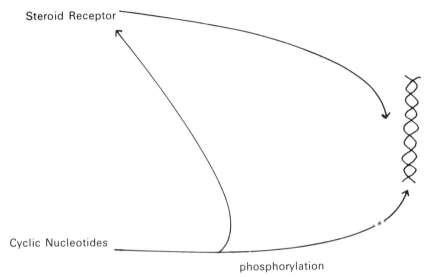

Fig. 16. Two hormonal means of affecting genetic expression.

1.7. Intracellular Structural Proteins

We have so far considered a few aspects of inter- and intracellular communication, but have made no comment upon gross or microscopic anatomy. These are important omissions. The former is beyond the scope of this work, except to say that the distance between secretory cells and their targets (for example, the distance between the anterior pituitary and, among other things, the thyroid and the gonads) or their juxtaposition (for example, the close anatomical relationship between the processes of different neurons within the brain) is critical; as a general rule chemical messengers that act at a distance have a high affinity and specificity for their receptors: those that act locally are less specific and bind with a lower affinity. This is probably a reflection of the relative likelihood of messenger and receptor coming close together. Gross anatomy is obviously dependent upon, among other things, extracellular structural proteins (for example, those involved in the deposition of bone, cartilage, and ligaments), and at a finer level, proteins that guide, by chemotaxis, cells and cell processes to their ultimate location in the developing animal. These aspects will not be commented upon here and the interested reader is referred to any up-to-date text book on anatomy, physiology, or embryology for information. We shall restrict our comments to intracellular and contractile proteins, and shall introduce them by describing an apparent paradox arising from our earlier treatment of intracellular control systems.

There is an abundance of intercellular messengers (neuropeptides, neurotransmitters, hormones, and so on). When the full count is known, it will probably run into hundreds. The number of their recognition sites is probably even greater (subclasses of many of their receptors have been described) (*see* chapter 8). Their ultimate effects are many and varied, but the initial intracellular pathways through which they are known to work are few, numbering, perhaps in tens. These pathways include the opening of ion channels in cell membranes and the modulation of the activities of the enzymes controlling cyclic nucleotide concentrations. Thus, multiple causes work through a few intermediates to produce multiple effects. This apparent paradox is illustrated in Fig. 17.

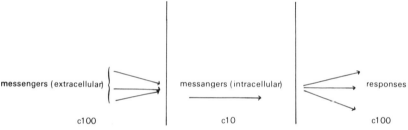

Fig. 17. Many different extracellular chemicals affect only a few intracellular messengers, to produce a multiplicity of responses. This apparent paradox suggests that the final effects of the intracellular messengers might depend upon the microscopic anatomy of the cell.

This would be a problem if cells were homogeneous, but they are not. They contain not only nuclei and other subcellular particles such as mitochondria and microsomes, but also filamentous and tubular structures.

Let us first consider the microtubules. These are composed of aggregates of tubulin, with apparently empty lumens (36). The microtubule has the ability to transport substances within the cell in both centrifugal (37) and centripetal direction. Therefore the fine anatomy of microtubules and their functional state (Ca^{2+}, influencing their aggregation and disaggregation) probably influences the ultimate subcellular location of chemical messengers, and thus their ultimate distribution and possible exocytosis. We can see that more knowledge of subcellular architecture may enable us to resolve the problem of how a multiplicity of extracellular messengers causes such diverse effects with so few common intermediates. I do not know how this should be done, or whether the technology is presently available, but offer my next figure as a possible guide to future workers (Fig. 17).

1.8. Storage Proteins

Storage (or synaptic) vesicles are currently believed to contain small molecules that may be exported from the cell. There is considerable evidence suggesting that they also contain proteins. For example, the posterior pituitary secretes the octapeptides, oxytocin and vasopressin, that in turn affect the uterus and mammary gland (oxytocin) and the kidneys and blood pressure (vasopressin). In their respective vesicles they are found to be associated with proteins that are called neurophysins (37) and are probably the nonhormonal remnants of their precursor molecules (see Fig. 8). Oxytocin and vasopressin bind to these neurophysins within the vesicles and are expelled from the cells at the same time. I do not know whether the neurophysins have an endocrine function in their own right, but they may be useful (within the cell) to preserve osmotic balance.

There are other proteins within synaptic vesicles, e.g., serotonin-binding protein (34). The action of neurotransmitters can be stopped in several ways (see section 1.3). Vesicles that contain serotonin (5-hydroxytryptamine or 5-HT) also contain a protein that binds the neurotransmitter with an avidity comparable to that of membrane receptors, and this means that the storage capacity of the vesicles is saturable. It has an unusual property in that the dissociation of the complex appears to be very slow: some binding is retained after subjection to polyacrylamide gel electrophoresis.

If this phenomenon proves to be generalizable to other neurotransmitters, it suggests that the enzymic "pumps" (13,14) that carry them into the nerve terminals are merely the first link in a chain of reactions.

In summary, many small molecules are stored within cells in association with proteins and this, among other things, maintains their subcellular location.

1.9. Toxins

So far the only proteins that have been mentioned are those that have a physiological role. Other proteins that are foreign to the body may be poisons. Some of the latter are useful scientific tools: for example the toxins purified from the venoms of various snakes bind strongly, but reversibly, to the acetylcholine receptor, preventing the physiological actions of the neurotransmitter itself and playing a critical role in its assay, isolation, and purification (38).

Other toxins may act by misdirecting the immune system, leading, in the long term, to problems such as anaphylaxis or autoimmune disease.

2. Multiple Functions of Individual Proteins

There is something arbitrary about my system of protein classification and probably about any such system. I have ignored the fact that proteins may have more than one function and the active sites involved may be different. I shall use hormones as examples.

Hormones were originally discovered because in their absence major bodily disturbances occurred. Their main actions were determined and they were named accordingly. Later they were purified and in many cases their structures have been determined. Purification led, in turn, to the raising of antibodies and to a more subtle analysis of their anatomical location (or at least of the anatomical location of substances with similar antigenic determinants). It was then found that many proteins and peptides that were originally described as being of peripheral in origin are also found in the brain.

Insulin provides an example. It was described originally as an antidiabetogenic hormone that is secreted by the β cells of the islets of Langerhans in the pancreas. Recently insulin itself or at least an insulin-like immunoactive material has been found in the brain, as indeed have binding sites that appear to be insulin receptors (39–41). Insulin cannot pass the blood–brain barrier: In other words its molecular properties are such that it cannot pass from the blood into most parts of the brain. Furthermore, the concentrations of "insulin" in the brain do not fluctuate with those of the bloodborne hormone. It seems likely therefore that it is synthesized and secreted by the brain, as well as by the pancreas, and that since its concentration in the brain, unlike in the periphery, does not depend upon that of blood glucose, its functions are distinct from those of peripheral insulin.

Adrenocorticotropic hormone (ACTH) comes from the anterior pituitary and modulates the secretion of glucocorticoids by the cortex of the adrenal glands. It also has effects (essentially those of enhancement) upon memory. It has been shown by proteolytic cleavage (and synthesis) that the parts of the molecule necessary to its original hormonal action and its effects upon memory are distinct; indeed, the latter effect can take place only when that part of the peptide chain represented by the amino acid residues 4–10 are present. This portion of the molecule has no classical hormonal activity (42,43).

Another example of a hormone with distinct peripheral and central nervous activities is vasopressin (44). This octapeptide hormone is secreted by the posterior pituitary and classically has both pressor and antidiuretic effects. It also has memory-enhancing effects, in general similar to, but in detail distinct from, those of ACTH. Once again the two activities of the hormone can be distinguished at a molecular level. des-Glycinamide vasopressin has no classical peripheral hormonal effects, but retains its effects upon the brain.

If we also consider the multiplicity of hormones that also appear to function as cotransmitters (*see* section 1.3), we see that, in summary, individual proteins and peptides may have more than one role to play in the body.

3. Cyclical Interactions

3.1. The Multicellular Level

I have already illustrated the hypothalmo-pituitary-gonadal axis and now restrict myself to stating that this is but one of many cyclical control systems involving both proteins and nonproteinaceous chemical messengers that may extend over much of the height of the body and has a relatively short duration (e.g., the entire sexual cycle in women takes about a lunar month). The example given is peculiar in that it involves both positive and negative feedback systems, but many others exist that involve only negative feedback systems. All of these cyclical control systems can be modulated by information fed in from outside the cycle that is not influenced by it. For example, the sexual cycle in females is influenced by photoperiodicity: Information from the retina is transferred via nerve impulses in the retinohypothalmic tract to the hypothalamus (*see* Fig. 18 and section 1.2), where it leads to the modification of releasing hormones, but positive feedback appears to be restricted to the control of gonadotrophins in females.

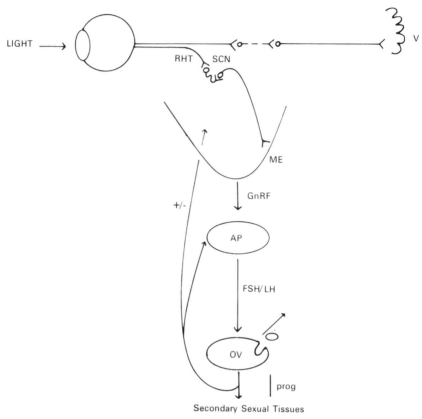

Fig. 18. An example of the modulation of multicellular cycles by external neural stimuli. The cyclical control of gonadotrophin (FSH/LH; see Fig. 6 for definition) from the anterior pituitary (AP). The hormones act synergistically to stimulate the ovaries to produce ova and the steroid hormones estradiol and progesterone (prog). These have negative feedback effects upon the hypothalamus. Light is perceived by the eye and eventually converted into visual images in the visual cortex (V). However, some nerve fibers from the eye go to the suprachiasmatic nucleus of the hypothalamus (SCN), via the retino hypothalamic tract (RHT), and impose an additional control upon the system: The timing of the sequence of daylight and darkness influences ovulation.

3.2. The Intracellular Level

Proteins and other chemical messengers also interact at an intracellular level. This is illustrated in Fig. 19, in which I indicate that this can happen in a manner that involves many different interlinking cycles. It must be stressed that the information given in this figure is merely illustrative and not intended to exhaustive, but is placed here as summary of the foregoing discussion. In essence, membrane receptors activate intracellular chemical messengers: ions and cyclic nucleotides, and do this through the mediation of membrane proteins. The activity of some of these proteins is influenced by calmodulin, a protein that itself is dependent upon calcium ingress for its regulatory activity and appears to be widely distributed in different cells and species, if it is not ubiquitous.

3.3. Long-Term Interactions

We must also consider long-term interactions, such as the sexual differentiation of the brain and the selective stabilization of synapses, which may last for months to years. Let us consider the former. The brains of males and females are different (45), and these differences are not solely under genetic control, but are brought about by sex hormones acting during infancy. Some of these effects of sex steroids are mediated via estrogen-induced changes in adenylate and guanylate cyclase activity (9,46). These changes are brought about by the synthesis of new proteins that inhibit adenylate, but enhance guanylate cyclase activity. Sexual differentiation involves modulation of the "wiring pattern" of the brain (47) and is analogous to the selective stabilization of synapse models proposed by Changeux and Danchin as long ago as 1976 (48), when it was suggested that use stabilizes some interneuronal connections at the expense of others. They proposed (48) that activation of a neuron has long- as well as short-term effects and suggested that intracellular coupling factors were involved. We have suggested (9,46) that these internal coupling factors are cyclic nucleotides. Our reasons for this assertion are based on the effects of steroid hormones upon brain differentiation and cyclic nucleotide activities.

In brief, estrogens cause the sexual differentiation of the brain. In doing so they reduce, but specify more precisely, the number of synapses available (47). They simultaneously "up-regulate" the activity of guanylate cyclase and "down regulate" that of adenylate cyclase. Since cyclic AMP promotes different differentiation at the expense of growth, and cyclic GMP does the reverse, we argued that they were the intracellular intermediates. This concept is illustrated in Fig. 20.

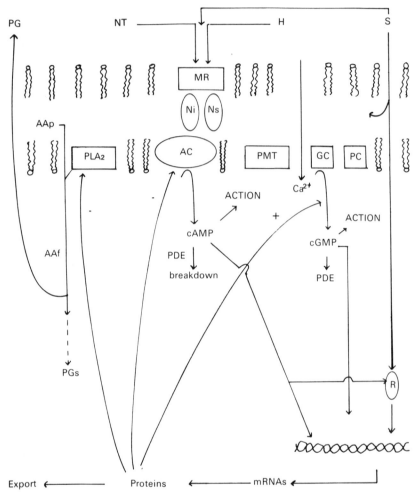

Fig. 19. Some potential interactions between different compounds at the cellular level. Not all these pathways necessarily apply to any individual cell. Neurotransmitters (NT) and peptide hormones (H) act upon membrane receptors (MR) to influence adenylate cyclase (AC) via the guanyl binding proteins Ni and Ns. Binding may also open ion channels in the membrane (e.g., C^{2+}) and modulate the activity of enzymes controlling phospholipid metabolism [e.g., phosphomethyl transferases 1 and 2 (PMT) (*see* ref. 49) and phospholipase C (PC) (*see* ref. 50). These latter effects may lead to the modulation of guanylate cyclase (GC). The cyclic nucleotides (cAMP and cGMP) formed as a result are broken down by phosphodiesterases (PDE), but before this activates protein kinases (not shown in diagram), which phosphorylate proteins in many places, in-

Peptides and Proteins

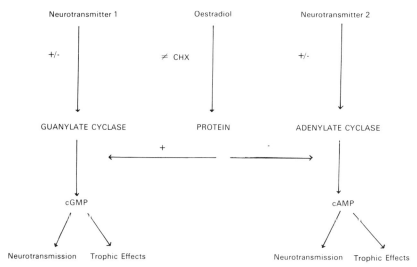

Fig. 20. Neurotransmitters affect the activities of adenylate and guanylate cyclase. The products of the reactions (cAMP and cGMP) not only have immediate effects, but also influence cell development, the former promoting differentiation at the expense of growth and the latter promoting the reverse. Use stabilizes synapses at the expense of others. Some experimental observations indicate that estradiol may inhibit this stabilization (45). This observation may be pertinent to the role of steroids in hormone-dependent tumors.

Fig. 19 (continued)

cluding the nucleus and the membrane itself. Steroid hormones (S) enter the cell and act via intracellular receptors (R), which are themselves modified by phosphorylation. Steroid hormones can also affect membranes, both directly and indirectly. For example, some of the metabolites of estrogens can compete with catecholamines for their membrane receptors (51). At the indirect level, estrogens can induce the synthesis of proteins that modulate AC and GC activities (46) and glucorticoids of a protein, which decreases the activity of phospholipase PLA_2, thus decreasing the transformation of arachidonic in phospholipid form (AAp) to the free acid (AAf), which is a precursor of those ubiquitous compounds, the prostaglandins (PG) (ref. 52).

4. Conclusions

We have seen, from the point of view of the biologist, that proteins should not only be considered as individual entities, but that their relationship with other proteins and with nonprotenaceous compounds is important. We have discussed many cycles. These, however, could be likened to the tip of an iceberg. It may or may not be that in the future there will be a simplifying hypothesis.

References

1. W. Y. Cheung, *Science* **207**, 19–27 (1980).
2. J. Kallos, *Nature* **265**, 705–710 (1977).
3. H. Tamir, A. Klein, and M. M. Rapport, *J. Neurochem.* **26**, 871–878 (1976).
4. M. D. Gershon, K. P. Liu, S. E. Karpiak, and H. Tamir, *J. Neuroscience* **3**, 1901–1911 (1983).
5. P. Englebienne, *Molec. Asp. med.* **7**, 313–396 (1984).
6. L. L. Iverson, *Trends Neurosci.* **6**, 293–294 (1983).
7. L. W. Swanson, *Trends Neurosci.* **6**, 294–295 (1983).
8. M. E. Hadley, *Endocrinology* Prenctice Hall, New Jersey (1984).
9. P. J. Thomas, *Molec. Asp. Med.* **5**, 1–61 (1982).
10. A. Peters, S. L. Palay, and H. de F. Webster, *The Fine Structure of the Nervous System* W. B. Saunders, Philadelphia, London, Toronto (1976), p. 36.
11. A. Goth, *Medical Pharmacology* Mosby, St. Louis, Toronto, London, p. 155–122 (1981).
12. P. Greengard, *Nature* **260**, 101–108 (1976).
13. L. L. Iverson, *Biochem. Pharmacol.* **23**, 1927–1233 (1974).
14. E. E. Muller, G. Nistico, and U. Scapagnini, *Neurotransmitter and Anterior Pituitary Function* Academic, London (1977), pp. 102–115.
15. K. Krvjevic, *Physiol. Rev.* **54**, 418–540 (1974).
16. J. M. Lundberg and T. Hokfelt, *Trends Neurosci.* **6**, 325–332 (1983).
17. J. Parascandol, in *Towards Understanding Receptors* (J. W. Lamble, ed.) Elsevier/North Holland (1981), pp. 1–8.
18. P. J. Thomas, *J. Endocrinal.* **57**, 333–359 (1973).
19. K. Sterling, *N. Eng. J. Med.* **300**, 117–123; 173–177 (1979).
20. M. R. Haussler, M. R. Hughes, T. McCain, and S. A. Pike, in *Molecular Endocrinology* (I. MacIntyre and M. Szelke, eds.) Elsevier/North Holland (1977), pp. 101–116.
21. W. J. King and G. L. Green, *Nature* **307**, 745–747 (1984).
22. W. V. Welshons, M. E. Lieberman, and J. Gorski, *Nature* **307**, 747–749 (1984).

23. D. Colquhoun, in *Towards Understanding Receptors* (J. W. Lamble, ed.) Elsevier/North Holland (1981), pp. 17–27.
24. E. W. Sutherland, T. W. Rall, and T. Menon, *J. Biol. Chem.* **237**, 1220–1222 (1962).
25. R. Lim, K. Mitsunobu, and W. K. P. Li, *Exp. Cell Res.* **79**, 243–246 (1973).
26. D. L. Schapiro, *Nature* **241**, 203–204 (1973).
27. Y. H. Erlich, E. G. Brunngraber, P. K. Sinha, and K. N. Prasad, *Nature* **265**, 238–240 (1977).
28. R. Siete, J. Luciano-Vuillet Leonetti, and M. Vio, *Brain Res.* **124**, 41–51 (1977).
29. J. Zwiller, C. Goridis, J. Ciesielski-Treska, and P. Maridel, *J. Neurochem.* **29**, 273–278 (1977).
30. T. W. Stone, D. A. Taylor, and F. E. Bloom, *Science* **187**, 845–846 (1975).
31. R. J. Coffey, E. M. Hadden, C. Lopez, and J. W. Hadden, *Adv. Cyclic Nucleotide Res.* **9**, 661–676 (1978).
32. W. G. George, G. M. Rodgers, and L. A. White, *Adv. Cyclic Nucleotide Res.* **9**, 517–523 (1978).
33. W. J. Thomson and W. W. Appleman, *Biochemistry* **10**, 311–316, (1971).
34. M. Williams and R. Rodnight, *Prog. Neurobiol.* **8**, 183–240.
35. A. L. Lehninger, *Biochemistry* Worth Publishers, New York (1975).
36. A. Peters, S. L. Palay, and H. de F. Webster, *The Fine Structure of the Nervous System* Saunders, Philadelphia, London, Toronto (1976), pp. 107–112.
37. B. Pickering, *Essays in Biochemistry* **14**, 45–81 (1978).
38. J-L. Popot and J-P. Changeux, *Physiol. Rev.* **64**, 1162–1238 (1984).
39. M. L. Barbaccia, D. M. Chung, and E. Costa, in *Regulatory Peptides; From Molecular Biology to Function* (M. Costa and E. Trabuchi, eds.) Raven, New York (1982), pp. 511–518.
40. J. Havrankova, M. Brownstein, and J. Roth, *diabetologica* **20**, 268–273 (1981).
41. J. Havrankovaa, D. Schnechel, J. Roth, and M. Brownstein, *Proc. Natl. Acad. Sci USA* **73**, 5737–5741 (1978).
42. B. Bohus, *Pharmacology* **18**, 113–122 (1979).
43. D. De Weid, *Trends Neurosci.* **2**, 79–82 (1979).
44. J. M. Moeglan, A. Audibert, M. Timisit-Berthier, J. C. Oliveros, and M. K. Jandali, in *Neuroendocrinology; Biological and Clinical Aspects* (A. Pollen and R. M. MacLeod, eds.) Academic, London (1979), pp. 47–57.
45. B. S. McEwen, *Trends Neurosci.* **6**, 22–26 (1985).
46. O. A. Amechi, P. J. Butterworth, and P. J. Thomas, *Brain Res.* **342**, 158–161 (1985).
47. C. Le Blond, R. Powell, G. Karkiulakis, S. Morris, and P. J. Thomas, *J. Endocrinol.* **95**, 137–145 (1982).

48. J-P Changeux and A. Danchin, *Nature* **264**, 705–710 (1980).
49. M. Hirata and J. Axelrod, *J. Science* **209**, 1082–1087 (1980).
50. M. J. Berridge, in *Towards Understanding Receptors* (J. W. Lamble, ed.) **1**, 122–131 (1981).
51. A. Biegon and B. S. McEwen, *J. Neuroscience* **2**, 199–202 (1982).
52. G. J. Blackwell, R. Carnuccio, M. Di Rosa, R. J. Flower, L. Parente, and P. Perisco, *Nature* **287**, 147–149 (1980).

Chapter 7
Methods of Measuring Binding, Some Extracellular Carrier Proteins, and Intracellular Receptor Proteins

A Selective Introduction

P. J. Thomas

1. Introduction

Carrier proteins and intracellular receptors will be dealt with in this same chapter together since they are generally measured in solution, rather than in suspension, as is frequently the case with membrane receptors.

The principles of measuring binding will first be described.

1.1. Measurement of Binding

The principles of binding apply equally well to membrane receptors, though the techniques are different. Some diagrams are taken from membrane receptor binding studies.

    A. The protein must be in solution. In the case of intracellular receptors, this requires preliminary homogenization.
    B. Labeled ligand is added, and binding is allowed to reach equilibrium.
    C. Binding is measured in the presence of increasing concentrations of free ligand.

Such experiments allow for the measurement of saturation binding capacity (n) and of equilibrium dissociation constant (K_d).

Algebraic analyses (1–4) are based upon the law of mass action and in all cases involve infinitely long extrapolations (2). Furthermore, they assume that the protein is pure and that when it binds a ligand it does not undergo conformational changes that modify its binding affinity.

If these assumptions are correct, a plot of bound against free ligand approximates to a portion of a rectangular hyperbola (Fig. 1), which rises asymptotically to the saturation binding capacity (n), and reaches half saturation at a free ligand concentration of K_d. A double reciprocal plot (1/bound ligand vs 1/free ligand) approximates to a straight line (Fig. 2), the intercepts of which yield the reciprocals of n and K_d. A plot of bound ligand/free ligand (a Scatchard plot) is also rectilinear (Fig. 3). Intercepts yield n/K_d and n, respectively.

These graphs may be distorted if (a) the protein is impure or, (b) it undergoes conformational changes when it binds to a ligand.

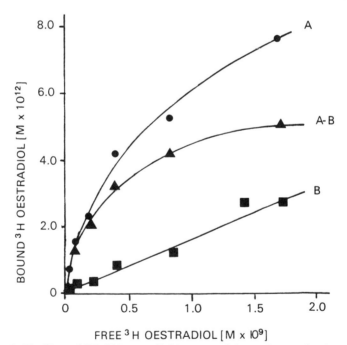

Fig. 1. Binding of ^{3}H-17β-estradiol in cytosol from the cerebral cortex at 30°C. (A) Total binding; (B) Low-affinity, high-capacity binding; (A−B) High-affinity, limited-capacity binding.

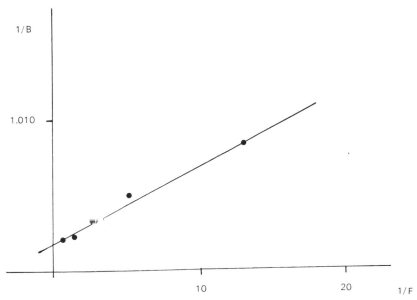

Fig. 2. A double reciprocal plot of the data for "specific" binding shown in Fig. 1.

1.2. Effects of Impurities

If the protein is impure, other proteins may sequester the ligand. We assume that they do so with a lower affinity. This will lead to graphs such as those shown in Fig. 1, curve A and Fig. 3, curve A, which are hard to extrapolate to saturation binding capacity (n) and thence to K_d. Assuming that the aim is to characterize the highest affinity component present, that the other binding sites have much lower affinities, and that the ligand is labeled in some way, the problem can be solved by carrying out control experiments in the presence of an excess of unlabeled ligand (or, better, its closest known congener) and estimating high-affinity binding by difference (Figs. 1 and 3). The appropriate concentration of unlabeled competitor is determined by preliminary experiments (Fig. 4).

1.3. Conformational Changes

When a protein binds a ligand, it may undergo conformational changes: If these reduce its affinity for subsequent molecules of the ligand, binding curves will be similar in form to those obtained when the protein coexists with other ligand-binding molecules. If

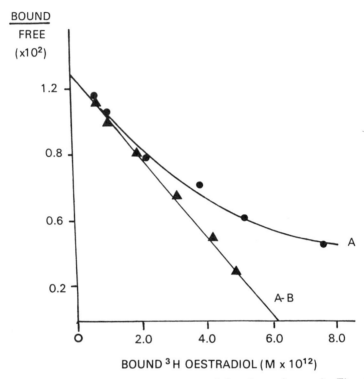

Fig. 3. A Scatchard representation of the data shown in Fig. 1.

the conformational changes enhance the binding of subsequent ligand (5-7), binding curves will be distorted in a different way. Graphs of $[L]_b$ versus $[L]_f$ will be sigmoid, double reciprocal plots will be concave toward the $[L]_b$ axis (Fig. 5), and Scatchard plots will be concave toward the bound axis.

1.4. The Hill Plot

This method of graphical analysis distinguishes between cooperative and other classes of binding reaction. It is explained in detail in refs. *8* and *9*.

Use of the Hill Plot depends upon prior knowledge of the saturation binding capacity (n). The logarithm of the fractional saturation is plotted against the logarithm of the free ligand concentration. At extremes of free ligand concentration (approaching zero and approaching infinity), the slope of the graph approximates to

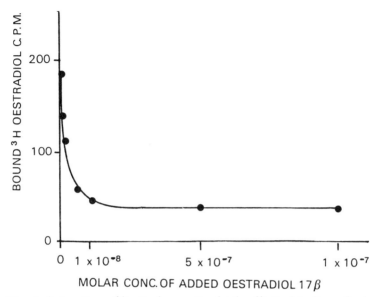

Fig. 4. Saturation of limited capacity, high-affinity binding of estradiol 17β in cytosol from amygdaloid region, in the presence of excess 17β-estradiol. Each incubate contained cytosol and 2,4,6,7-[³H]-estradiol-17β ($1 \times 10^{-9}M$) plus unlabeled 17β-estradiol at the concentrations given on the abcissa.

unity. The slope is unity throughout in the case of a simple biomolecular reaction (Fig. 6). It exceeds one in the middle (around free ligand concentrations that approximate to K_d) if the reaction of binding of ligand enhances subsequent affinity. It is less than one if lower affinity components are present or if binding reduces subsequent affinity.

1.5. Rate Constants

The equilibrium dissociation constant (K_d) is a ratio of the rate constant of association (k_{+1}) and the rate constant of dissociation (k_{-1}). In principle it is possible to measure these rate constants as well (10) using "concentration jump" experiments whereby the concentration of free ligand is suddenly altered and the time taken to approach a new equilibrium is estimated. In practice this seldom works and rate constants are not often quoted in the literature.

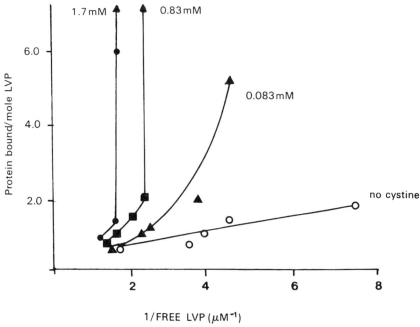

Fig. 5. The effects of increasing concentrations of an allosteric inhibitor upon the binding of an intracellular carrier protein to its ligand (double-reciprocal plots).

1.6. Specificity

The relative binding affinity of a variety of different ligands to a protein is also important. We assume that we are measuring their binding affinities in relationship to the binding affinity of a known, labeled ligand that has an affinity defined as K_d. The affinity of an inhibitor is defined as K_i. The K_i value can be measured in two ways. In the first we assume that, from physiological or other data, we have a rough estimate of K_i. We add this concentration of the putative competitor to our reagents, calculate its effects upon K_d from measurement of the alteration of the slopes of Scatchard plots, and thence calculate K_i (Fig. 7) (11).

In the second method we make no assumptions about the order of K_i. We take a minimal concentration of protein and a minimal concentration of labeled ligand. To these we add increasing concentrations (logarithmic scale) of putative inhibitor. Binding of label

Binding, Carrier, and Receptor Proteins

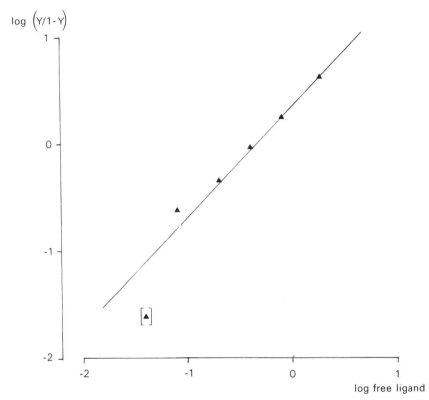

Fig. 6. A Hill plot of the "specific" binding data described in Fig. 1. Note that the slope approximates to unity. The point in parentheses is the one with the highest degree of error.

decreases as the concentration of inhibitor increases (Fig. 8), and the concentration of inhibitor at which the labeled binding is reduced by 50% (IC_{50}) approaches K_i as the concentrations of protein and labeled ligand approach zero. Detailed accounts of the algebra involved are given in ref. 12.

Knowledge of concentrations of inhibitor for which the curves approximate to the obverse of a plateau allows us to determine the appropriate concentration of inhibitor to control for "nonspecific," i.e., "low affinity binding," components in the system (*see* section 1.2).

It will be seen that the measurement of binding parameters involves several interdependent variables and that results should be interpreted with this in mind.

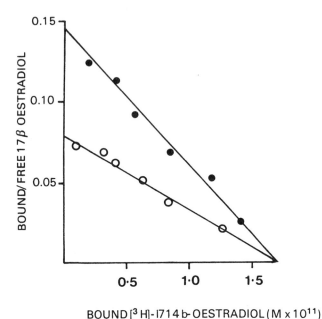

Fig. 7. Inhibition of 17β-estradiol binding to hypothalamic cytosol by unlabeled inhibitor (15).

1.7. Practical Approaches

Binding must be measured without perturbing the equilibrium (except where rate constants are being measured). The methods have been summarized elsewhere (13). They include equilibrium dialysis, ultrafiltration, and, where k_{-1} is very small, sequestration of free ligand by activated charcoal. Other methods, including electrophoresis, have also been described (14). This must perturb the equilibrium. Methods involved in the special case of the measurement of the binding of steroids to their receptors (where dissociation is very slow indeed) are described later. Binding to membrane receptors is often measured with the receptor in suspension rather than solution. This will be considered in detail in chapter 9.

2. Examples of Carrier Proteins

This discussion is restricted to three examples: estrogen-binding proteins in the blood of perinatal rats, steroid-binding proteins in

Binding, Carrier, and Receptor Proteins 197

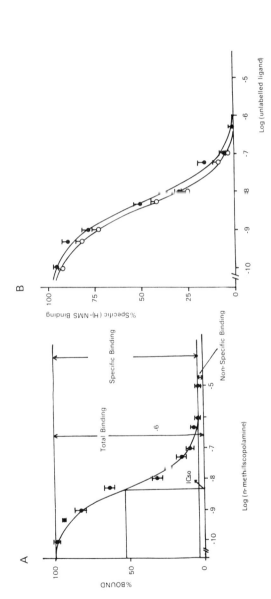

Fig. 8. Competition of unlabeled methylscopolamine for 0.5 nM ^{3}H-n-methylscopolamine bound to 0.2 mg/mL adult male rat "cortical crude tissue preparation." Results are mean ± SEM of four independent experiments carried out in triplicate. Displacement of specific ^{3}H-n-methylscopolamine to crude tissue preparation of adult male rat cortex by atropine (solid circles) and methylscopolamine (hollow circles). Each point is the mean ± SEM of 6–7 independent experiments carried out in triplicate. It should be noted that binding is to a membrane, rather than a soluble protein fraction. This affects the methods of separation, but not the principles involved.

the blood of adult mammals, and a specific lipoprotein, low-density lipoprotein (LDL). These are chosen as the main themes of this chapter, as well as the effects of internalization of hydrophobic ligands.

I shall comment upon the evidence that these proteins may get into cells (internalization) and conceptual problems this may pose.

2.1. Rodent α-Fetoprotein

Rodent α-fetoprotein is secreted by the fetal liver. Its synthesis is shut off immediately after birth (15). It sequesters the ovarian steroid, 17β-estradiol, with a relatively high affinity (K_d in the order of $10^{-8}M$), but has relatively low affinities for other estrogens and nonestrogenic steroids (15).

Thus it alters the bioavailability of 17β-estradiol during perinatal and prepubertal development. This is of great developmental importance since the phenotypic gender of rodent and, maybe, human brains (16,17) is not directly under genetic control, but is conferred by steroids acting at a critical period of development (18). (It should be noted that human and fetoprotein does not bind estradiol.)

The resting state of the brain is female. Masculinization is brought about by perinatal androgens. These act only after being converted to estrogens within the brain (15). It appears from physiological evidence that those estrogens that are not bound by the fetoprotein are more potent masculinizers (mole for mole) than those that are (15). Thus the fetoprotein protects the developing brain against maternal estrogens (Fig. 9). How a protein with a K_d of $10^{-8}M$ prevents activation of a receptor that binds with a K_d of 10^{-10} remains to be seen (15).

Other authors have described the effects of the fetoprotein in prolonging the existence of maternal estradiol within the bloodstream (19) and have suggested that this prolongation might have effects upon the development of the phenotypically female brain (20).

2.1.1. The Properties of α-Fetoprotein

α-Fetoprotein is a glycoprotein and exists in four different forms (21) (Table 1; Fig. 10). It is stated that it can move into cells (internalization). If this were true, it could help to solve one problem: How do steroids (hydrophobic molecules) leave the inner surface of the cell membrane and associate with receptors that are probably associated with the nuclear envelope (22,23). We assume that its function is to facilitate steroid entry. This may be true, but if so

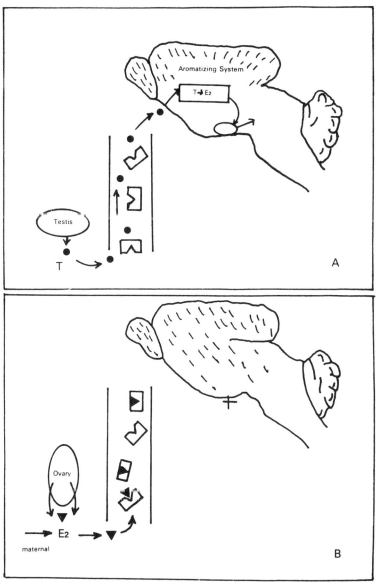

Fig. 9. (A) In males, testosterone (T), secreted from the testis, reaches the brain and is converted to estradiol (E_2) in the hypothalamus by the action of aromatizing enzymes E_2 so produced, then masculinizes the brain. (B) In females, estrogen-binding proteins (α-fetoprotein), represented in the figure by ⬛, protect the brain against the maternal and its own estradiol. These proteins bind estrogens, but not androgens, and prevent them from reaching the brain. The unaffected brain then develops a female pattern of differentiation.

Table 1
Key Properties of Rodent α-Fetoprotein
(MW, 70,000; Sediments at ~4.5S)[a]

Estrogenic potency	Binding affinity, K_d (M)	Compound
+ +	10^{-8}	17β-estradiol
+	10^{-8}	Estrone
+	10^{-6}	Estriol
+ +	10^{-6}	Diethyl stilbestrol
+ +	10^{-6}	17-methoxy, 17α-ethynyl-17β-estriol
−	10^{-6}	Progesterone
−	10^{-6}	Cortisol
−	10^{-6}	Testosterone

Subclasses	K_d1 (M)	% Total sites	K_d2	% Total sites
SLO, LO	3×10^{-8}	45	0.5×10^{-10}	0.45
FAST, LO	2.5×10^{-8}	14	—	—
SLO, HI	0.8×10^{-7}	39	—	—
FAST, HI	10^{-7}	0.7	14×10^{-9}	0.05

[a]Overall binding affinity for 17 β-estradiol (K_d, 10^{-9}M). Specificity for steroids.

[b]Abbreviations: SLO, slow electrophoretic mobility; FAST, fast electrophoretic mobility (similar to SLO); LO, relatively low carbohydrate content; HI, relatively high carbohydrate content; One cannot help but feel that the high-affinity binding components (k2) are so rare as to be ignored.

it poses another problem. If the protein facilitates the entry of estrogens into the cell, why does its presence prevent masculinization (15): The overwhelming physiological evidence shows that estrogens that are relatively weakly bound to it (in comparison to their affinity for estrogen receptors) are much more potent masculinizers than are those that are strongly bound by the fetoprotein.

2.2. Steroid Binding Proteins in Adult Blood

2.2.1. Serum Albumin

Albumin is here defined as a steroid binding protein, since steroids are the main link I use to bring coherence to this chapter,

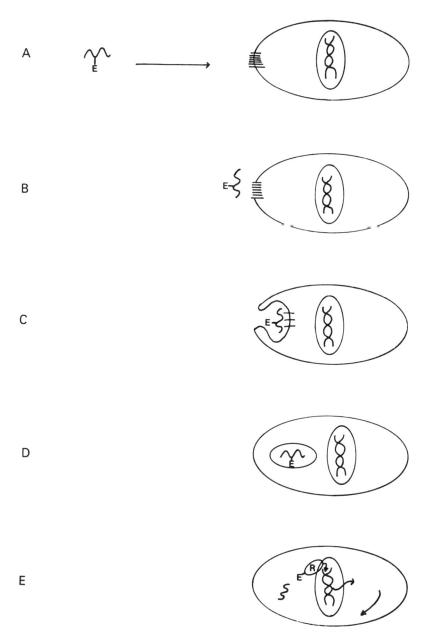

Fig. 10. Are there receptors for α-fetoproteins? An hypothesis that conflicts with some physiological data (see Fig. 8).

but it must be remembered that the binding of steroids may be one of the least important properties of albumin. It binds a large number of other ligands, both naturally occurring and synthetic: These ligands differ enormously in structure and function. They include phenylbutazone, warfarin, clofibrate, congo red, tryptophan, benzodiazepines, fluorescine, sulfonamides, phenytoin methyl red, thyroid hormones, salicylates, and bilirubin. (24,25). It is also important in maintaining the osmotic properties of the blood.

As far as its ability to bind steroids is concerned, its equilibrium dissociation constants (K_d) for glucocorticoids, androgens, estrogens, and progestogens are all in the order of 10^{-3} to $10^{-4}M$ (14). Thus it is relatively nonspecific and of low affinity. Its physiological importance in modifying the bioavailability of drugs is dependent upon its high concentration in the blood.

The fact that its concentration is reduced in old age and in hepatic and renal disease is important in assessing the appropriate doses of drugs that may be bound to it.

2.2.2. Steroid-Binding Proteins of Greater Affinity and Specificity

Steroid binding proteins of greater affinity and specificity are ranked in a roughly increasing order of specificity and affinity. Data are obtained from ref. 14.

α-1-acid-Glycoprotein has a marginally higher affinity for steroids than does albumin and has a slightly higher capacity to distinguish between them.

Corticosteroid binding globulin is also a glycoprotein and has a relatively high affinity (K_d between 10^{-9} and $10^{-8}M$) for glucocorticoids, progestogens, and testosterone, but has relatively low affinity for estradiol (K_d around $10^{-5}M$).

Sex hormone-binding globulin binds most naturally occurring androgens and estrogens with K_d values in the order of 10^{-9} to 10^{-5} M and glucocorticoids, progestogens, some relatively minor estrogens, and the mineralocorticoids, with progressively lower affinity.

More specific is an estrogen-binding protein that is found in human plasma and binds 17β-estradiol and the synthetic, nonsteroidal estrogen, diethyl stilbestrol, with K_d values in the order of $10^{-10}M$, but has little affinity for progesterone and cortisol.

A calciferol binding protein has also been described. It is a glycoprotein (MW ~52,000).

There is evidence that some of these extracellular steroid carrier proteins may be internalized (taken into cells) in company with their ligands (14) (see Fig. 10). This leads to the concept of a con-

Binding, Carrier, and Receptor Proteins

tinuum whereby proteins might carry ligands across the cell membrane, allowing them to come into contact with intracellular receptors or exert other effects within the cell. It also suggests that, if this be so, certain cells have membrane receptors for carrier proteins. Except in the case of LDL (see below), I have found no evidence in the literature for this, but suggest that if more work was carried out we might develop a useful way of bringing potent drugs to the cells where we wish them to act.

2.2.3. Low Density Lipoprotein (LDL)

LDL consists of a protein moiety (the hydrophobic apoprotein B-100) that is mainly associated with cholesteryl esters and long-chain phospholipids, the latter being so orientated that their hydrophilic heads are on the outside, making the total LDL molecule water soluble.

The apoprotein B-100 moiety is recognized by cell surface receptors and the complex is internalized by invagination via coated pits. Once inside the cell the complex is broken down and the cholesterol utilized. Though high levels of plasma cholesterol are epidemiologically associated with vascular disease, it must be remembered that cholesterol is an essential intermediate in many vital synthetic pathways; for example, the formation of steroids, bile acids, and cell surface membranes.

Cells need enough cholesterol, but only enough: though they have the capacity to synthesize some themselves, extrahepatic cells normally take in most of their cholesterol from LDL. If they exceed their requirements by taking in excess LDL, several things happen.

   a. They reduce their own capacity to synthesize cholesterol.
   b. They synthesize fewer LDL receptors.
   c. They store the excess cholesterol as cholesteryl esters.

Thus diets high in cholesterol down-regulate LDL receptors. This causes LDL to accumulate in the plasma. This increases correlates with an increase in the incidence of atheroschlerosis and thence myocardial infarctions and peripheral and cerebral vascular disease.

The current view about the development of atherosclerosis is summarized below.

The vascular endothelium may become damaged by hydrodynamic stress. This stress is maximal where arteries (or arterioles) branch or bifurcate. This damage may become worse in the presence of excess LDL. The excess LDL crosses the damaged endothelium. Macrophages, which are scavenging cells derived from blood-borne monocytes, ingest it by an unknown mechanism; the fatty parts remain in the cells, which become expanded and are then known as "foam cells." Smooth cells may also be converted into "foam cells."

The presence of "foam cells" may lead to the development of atherosclerosis. Thus, high levels of circulating cholesterol and LDL are harmful.

For further information on the nature of LDL, *see* refs. *26–28*.

3. Intracellular Receptors

3.1. Steroid Receptors

I have dealt with the concept of intracellular receptors in Chapter 6 and so will be relatively brief here.

The original story was that steroid hormones entered target cells and were bound by high-affinity cytosolic receptors. The complex then underwent a conformational change, entered the nucleus, associated with the chromatin, and thence induced modifications in genetic expression.

This simple concept has recently been questioned but not fundamentally modified; estrogen receptors have been purified and antibodies have been raised and it has been suggested that the so-called cytosol receptors are found on the outer surface of the nuclear membrane in vivo rather than free within the cytoplasm (22,23). This makes no real difference to the concept: Steroid binding and subsequent activation of the genome is still a two-stage process and casts no doubt upon the validity of the original work describing "cytosol" receptors. The words "cytosol" and "cytoplasm" are not synonymous. The former describes the soluble fraction of the cell after homogenization in a buffered medium: The latter describes the nonparticulate portion of the cell in vivo. If the process of homogenization tears a protein from the nuclear membrane and transfer it to the soluble fraction, it is still properly described as a "cytosol" protein.

In this chapter I shall concentrate upon estrogen receptors. Their location has been described elsewhere: estrogen target tissues, including the brain (*see* refs. *13,15,18*). Here I shall concentrate upon their measurement, their potential interactions with those second messengers that they have in common with neurotransmitters, and some of the proteins that they induce. The reader interested in other steroid receptors is referred to ref. *13*.

Binding, Carrier, and Receptor Proteins

3.1.1. Measurement of Estrogen Receptors

Measurement of estrogen receptors depends upon the fact that they have a very high affinity and that their rate of dissociation is very slow. To all intents and purposes, binding at 0°C can be considered irreversible.

3.1.1.1. Methods

(i) Ultracentrifugation.
(a) The "cytosol" receptor.

Most of the earlier work on estrogen receptors was carried out using this technique. The preformed complex between receptor and radiolabeled steroid [together with all the other proteins that are in the high-speed supernatant (13) fraction of the cellular homogenate] are placed on top of a sucrose density gradient. This is then centrifuged at either 100,000g (or, using a titanium rotor, at 200,000g) for up to 24 h (13). The contents of the centrifuge tube are sampled sequentially and radioactivity is estimated. This permits measurement of the amount of steroid bound and thence the sedimentation coefficient ($S = 1$ Svedberg unit) of the proteins that sequester it. Tubes containing additional unlabeled estrogen (preferably diethyl stilboestrol, DES) serve as controls.

It would be laborious to measure K_d and n by this method, but it has the advantage of giving rough estimates of molecular size. The "cytosol" receptor sediments at $\sim 8S$ in low salt concentrations and at $\sim 4S$ in the presence of 0.4M KCl (Fig. 11).

This can cause practical problems in the newborn rat since α-fetoprotein is present in vast excess and sediments at $\sim 4.5S$ (Fig. 11).

(b) The Nuclear Receptor

This can be extracted from the nucleus by incubation with 0.4M KCl (13). It sediments at $\sim 5S$.

The facts that the "cytosol" receptor can be transformed to 5S form after incubation under appropriate conditions, that the numbers of "cytosol receptors" available in the absence of estradiol approximate the numbers of "nuclear receptors" present after incubation with estradiol, and that the "nuclear receptors" can only be formed after incubation of nuclei in the presence of both steroid and cytosol (13) lead to the two-step concept that states that the "nuclear receptor" is formed by translocation of the "cytosol receptor." This concept is unaffected by modern doubts (22,23) about the in vivo location of the "cytosol receptor," but raises a question about the translocation of the steroid through the cytoplasm to the nuclear membrane: It has been suggested earlier in this

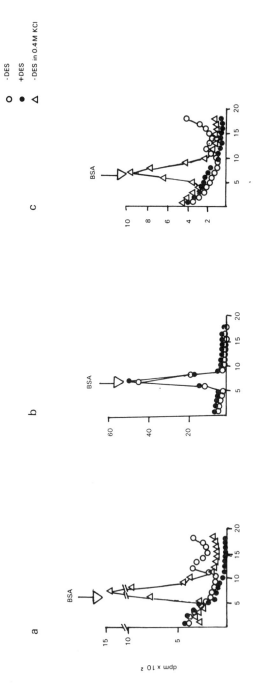

Fig. 11. The binding in the crude cytosol (a) is masked by a large 4S peak of α-fetoprotein. This peak remains in the ammonium sulfate supernatant (b), but is not present in the redissolved ammonium sulfate precipitate, where the 4S receptor peak becomes apparent (c) in the presence of 0.4M KCl. BSA = bovine serum albumin a marker protein that sediments at 4S.

Binding, Carrier, and Receptor Proteins

chapter that the internalization of carrier proteins might be involved. Whether this is so remains to be seen.

With reference to the apparent location of the primary estrogen receptor on the nuclear envelope, and making the assumption that this observation is extrapolable to other steroid receptors, it is worth noting that the enzyme, 5α-reductase, which catalyses the conversion of testosterone into its active metabolite, 5α-dihydrotestosterone, is located upon the nuclear envelope (13).

(ii) Precipitation with Polycations

Protamine sulfate precipitates the "cytosol" estrogen–receptor complex without causing its dissociation (29). The precipitate can be readily removed by low-speed centrifugation and thence estimates of n and K_d obtained.

(iii) Dextran Coated Charcoal

The "cytosol" estrogen–receptor complex can be measured by removal of free steroid by contact with dextran-coated charcoal, which is then removed using a bench centrifuge (13).

(iv) Sephadex LH20

The above methods give high blank values in the presence of lower affinity binding components. An improved method is available (30).

Small columns of Sephadex LH20, ~5–7 cm long × ~0.5 cm diameter [please note that there is a typographical error in the original paper (30) where the diameter is quoted as 0.5 mm], are prepared in an aqueous buffered medium and maintained at between 0° and 4°C. The complex, together with the free hormone, and hormone bound to lower affinity components (such as α-fetoprotein) is applied and allowed to remain in contact with the gel for periods of up to 90 min, and then eluted directly into liquid scintillation vials. Free is bound by the gel: the α-fetoprotein–estrogen complex present in homogenates obtained from newborn rats dissociates and the estradiol so liberated also remains in the columns. The size of the columns is not critical and preliminary experiments determine the volumes to be applied and used for elution. One need not take care to prevent the columns from running dry: If they end up in fine-bore plastic tubing, capillary action prevents this. Incubates containing an excess of unlabeled diethyl stilbestrol are used as controls. Blank values are minimal and are dependent only upon the purity of the radiolabeled estradiol applied. The method is also applicable to other steroid hormone–receptor interactions (31,32).

Care must be taken in measuring K_d. Free concentrations of hormone are measured by difference and if low-affinity components are present in the initial incubate, estimates will be spuriously high (33). This came to light when estrogen receptors in the newborn brain were discovered (33). Affinities for 17β-estradiol and estrone were apparently lower than in the adult. This difference disappeared after partial purification of the receptor (34,35).

3.1.2. Measurement of Unlabeled "Nuclear Receptors"

So far the tacit assumption has been made that the complexes to be measured are radiolabeled. There may be occasion to find out whether unlabeled estradiol binds to "nuclear receptors." This can be done by exchange assay. Cells or tissues are incubated in the presence of unlabeled estradiol. The estrogen–receptor complex moves to the nucleus. Tissues are homogenized and nuclei isolated by differential centrifugation. These nuclei are then incubated in the presence of radiolabeled estradiol. The labeled and unlabeled estradiol equilibrate. The labeled "nuclear receptor" is then extracted and measured. It was by use of this technique that the importance (as far as the sexual differentiation of the brain is concerned) of the conversion of testosterone to estradiol in the hypothalamus was demonstrated (36). Testosterone was administered to newborn rats. It was found (by exchange assay) to have been converted to estradiol and that this estradiol was associated with "nuclear receptors" in their hypothalami. The importance of this in the sexual differentiation of the brain has already been dealt with.

In summary, because of low rates of dissociation, methods are available for measuring steroid–receptor complexes that are not necessarily applicable to other protein–ligand binding reactions.

3.2. Aftermath of Steroid-Receptor Binding

3.2.1. Steroid-Induced Proteins

We have shown that the action of steroids is initially a two-step action. Other steps follow. The steroid–receptor complex associates with the chromatin (DNA + its associated proteins) and modulates the production of messenger RNA (mRNA). This, in turn, leads to the production of many new proteins by different steroids in different organs (37–40). I shall deal here only with the hormone's early effects upon protein biosynthesis and upon those proteins that might be expected to interact with pathways modulated by neurotransmitters in the brain.

Binding, Carrier, and Receptor Proteins

Hormonal induction of proteins can be measured by the inclusion of radiolabeled amino acids into the incubate (15) and measuring their incorporation into proteins either by double labeling (15) or fluorography (38).

Estradiol is a slow-acting hormone. Its effects only become overt many hours (the female sexual cycle) to years (the sexual differentiation of the human brain) (15,41) after its application. However, it induces some proteins rapidly.

Within about an hour after its administration to estrogen target tissues, a protein is produced. It has recently been identified as creatine kinase BB (42), in both the uterus and the brain.

This is interesting since creatine kinase is involved in the metabolism of ATP. ATP is the precursor of cAMP, a substance that acts as a second messenger for many neropeptides and neurotransmitters and also has long-term effects upon cell development, promoting differentiation at the expense of growth (43,44). Since there is an interaction between some neurotransmitters and some steroids in conferring sexual differentiation upon the brain, it could be at the cyclic nucleotide level where this occurs.

We have shown (43,44) (Figs. 12 and 13) that physiological concentrations of estradiol reduce the activity of adenylate cyclase within estrogen target areas of the brain and increase that of guanylate cyclase. There is evidence that cAMP and cGMP play opposing roles both as short-term intracellular messengers for neurotransmitters and as agents mediating cellular development: the former promotes differentiation at the expense of growth, the latter does the reverse (43,44). These effects of estradiol can be inhibited by concurrent administration of cycloheximide, suggesting, though not proving (44) that the action involves protein biosynthesis. Its synthesis and/or actions are dependent upon the presence of Ca^{2+}, an action of the ion that is independent of the effects of Ca^{2+} upon the enzymes. Attempts to purify the estrogen-induced

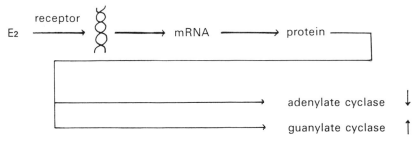

Fig. 12. The effects of estradiol on adenylate and guanylate cyclase activities.

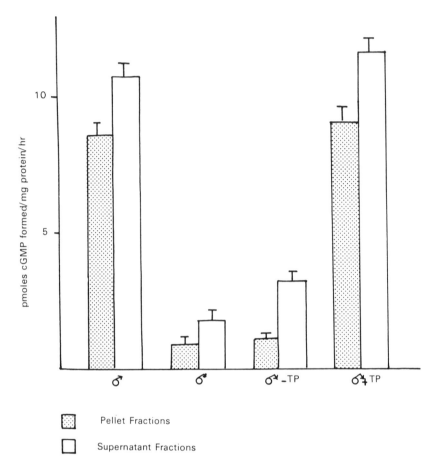

Fig. 13. Guanylate cyclase activity in the anterior hypothalamus/preoptic area of normal adult male, castrated adult male, castrated adult male injected with ethyl oleate, and castrated adult male treated with testosterone propionate (TP). The enzyme activity was significantly higher in normal adult males and lower in untreated castrates. Administration of testosterone propionate (TP) restored the enzyme activity to approximately normal levels. Values represent the mean SE of at least four separate experiments ($p < 0.001$) in triplicate.

protein that modulates cyclase activities are in hand using its effects upon the cyclase enzymes as markers for biological activity. Preliminary results indicate that it is a small peptide, but large enough to be excluded from Sephadex G-10. Any relationships between the estrogen-induced protein described above and creatine kinase BB remain to be seen (Figs. 12 and 13).

It is worth mentioning that 17β-estradiol appears to have no effects on the breakdown of either cGMP or cAMP (45).

It appears that estradiol effects cyclic nucleotides in the uterus in a similar way to its actions upon the brain. 17β-estradiol has been reported to enhance the concentrations of cGMP and to reduce those of cAMP in the uterus. The effects of estrogens on cGMP were prevented by concurrent administration of progesterone (46).

This is interesting since estradiol initiates the proliferative phase of the uterine sexual cycle, and cGMP promotes growth at the expense of differentiation and cAMP does the reverse. The fact that similar results are obtained in brain and uterus (both of which are target organs for estradiol and upon both of which the hormone may have long-term effects—permanent in the case of the developing brain) suggests that the phenomenon might be generalizable to other estrogen target tissues.

If this were so in the case of some estrogen-dependent tumors it might provide a new and useful prognostic tool. It should be mentioned that estrogen receptors in certain breast tumors relate to cyclic nucleotides in a manner analogous to that in the uterus (47) and that an estrogen-dependent 28,000 molecular weight protein has been described in some breast cancers (48).

It is worth noting that the effects of estradiol upon adenylate cyclase may be biphasic, since very high ($2 \times 10^{-6}M$: greatly in excess of the physiological range) concentrations of the hormone are reported to enhance adenylate cyclase activity in the brain (49).

We have seen that the binding of estradiol to its receptors may modulate cyclic nucleotide levels. Is there a reciprocal arrangement? Do cyclic nucleotides have any effect upon the binding of 17β-estradiol to its receptors?

cGMP increases the binding of 17β-estradiol to its receptors: cAMP has the reverse effect (50), at least in the uterus. It has been suggested that these effects are modulated via cyclic nucleotide-dependent protein kinases and subsequent effects upon the phosphorylation of the receptor. At first sight this might seem to lead us into an unstable loop—"positive feedback." Estrogen increases cGMP levels: cGMP increases estrogen binding, and vice versa for cAMP.

The loop, however, is not necessarily unstable. Estrogen-induced cyclic nucleotides do not necessarily come into contact with estrogen receptors in vivo (consider subcellular compartmentalization), and the time courses of the estrogen → cyclic nucleotide and cyclic nucleotide → estrogen receptor effects may be different.

This section on estrogen-induced proteins would be incomplete without mentioning that estrogens can induce the synthesis of the progesterone receptor (a useful marker in determining the estrogen responsiveness of breast cancer) and without comments upon the possible interactions (via common intracellular intermediates) of estrogens and other compounds at the cell surface receptor level. I therefore make some remarks about catecholoestrogens.

3.3. Catecholoestrogens

Estrogens act via intracellular receptors. They also act as substrates for enzymes that hydroxylate them in positions 2 and/or 4 (ring A). These hydroxylated derivatives are referred to as catecholoestrogens. They have a short life in vivo and possess both a catechol structure (they can be recognized as catecholamines) and the C-18 structure of estrogens (they can be recognized as estrogens). In addition to their estrogenic actions, they are substrates for catechol-O-methyl transferase (51), inhibit tyrosine hydroxylase activity (52), and can bind to dopamine and noradrenaline receptors (53–55).

Thus steroids not only share some common intracellular intermediates with cell surface receptor-active drugs, but can, on occasion, be metabolized so interfere with some of the actions of the latter class of drugs at receptor, biosynthesis, or degradation levels.

4. Other Intracellular Receptors

The classical steroid hormones are estrogens, androgens, progestogens, glucocorticoids, and mineralocorticoids. They all act in analogous ways and will not be commented upon further: The reader is referred to refs. *13* and *56* for information.

There are, however, two other classes of intracellular receptors: those for Vitamin D and those for thyroid hormones.

4.1. Vitamin D (Cholecalciferol and Ergocalciferol) Receptors

Vitamin D prevents rickets. The current model of their mode of action appears to be analogous to that of the classical steroid hormones. We have seen earlier that they may be bound in the bloodstream. They enter cells, are bound to intracellular receptors, are translocated to the nucleus, and induce new messenger RNA and thence protein biosynthesis (56). The only difference appears to be in the nature of the induced protein(s). Vitamin D-induced proteins enhance the entry of Ca^{2+} into cells. Classical steroid-induced proteins do other things.

4.2. Thyroid Hormones and Their Receptors

Thyroid hormones, like steroid hormones, have both long and short actions. An excess, or a deficiency, in infancy can lead to permanent brain damage. A balance is also required for normal metabolic activity in the adult.

Current views about the mode of action of thyroid hormones are to some extent analogous to those about steroids and vitamin D, but have certain further complications. Thyroid hormones can act via intracellular receptors to induce mRNA and thence protein biosynthesis. They can also bind to mitochondrial receptors to produce rapid changes in cellular metabolic activity. There is also evidence that there exist cell surface receptors for thyroid hormones (56). Thus thyroid hormones, like estrogens, can act on the cell at many different levels and at different tempi.

5. Summary

The evidence is so far conflicting. On the one hand it appears that serum-binding proteins can reduce and prolong the bioavailability of their ligands. On the other hand it appears that they may be able to carry their ligands into cells and introduce them to receptors. Certain hormones (or their metabolites) can act either internally and/or at the cell membrane or can react with other chemical messengers either at the metabolic or the intracellular level. When these interactions are understood, we may be better equipped to prevent or cure a number of serious diseases.

References

1. I. M. Klotz, in *The Proteins* (H. Neurath and K. Bailey, eds.) Academic, New York (1953), pp. 727–806.
2. I. M. Klotz, *Science* **217**, 1247–1249 (1982).
3. G. Scatchard, *Ann. NY Acad. Sci.* **51**, 660–672 (1949).

4. J. T. Edsall and J. Wyman, *Biophysical Chemistry* vol. 1, Academic, New York (1958).
5. J. Monod, J. Wyman, and J-P Changeux, *J. Mol. Biol.* **12**, 88–118 (1965).
6. G. D. Burford, M. Ginsburg, and P. J. Thomas, *J. Physiol.* **205**, 635–646 (1969).
7. L. W. Nichol, W. J. H. Nichol, and D. J. Winzor, *Biochemistry* **6**, 2249–2456 (1967).
8. J. Wyman, in *Cold Spring Harbor Symposia for Quantitative Biology* **28**, 483–489 (1963).
9. J. R. Bennett, in *Neurotransmitter Receptor Binding* (H. J. Yanamura, S. Enna, and M. J. Kuhar, eds.) Raven, New York (1978), pp. 57–90.
10. D. Colquhoun, in *Towards Understanding Receptors* (J. W. Lamble, ed.) Elsevier/North Holland, Amsterdam (1981), pp. 16–27.
11. M. Ginsburg, N. J. McLusky, I. D. Morris, and P. J. Thomas, *Br. J. Pharmacol.* **59**, 397–402 (1977).
12. Y. C. Cheng and W. H. Prusoff, *Biochem. Pharmacol.* **22**, 3099–3108 (1973).
13. P. J. Thomas, *J. Endocrinol.* **57**, 333–359 (1973).
14. P. Englebienne, *Molec. Aspects Med.* **7**, 313–396 (1984).
15. P. J. Thomas, *Molec. Aspects Med.* **5**, 1–61 (1982).
16. R. T. Rubin, J. M. Reinisch, and R. F. Haskett, *Science* **211**, 1318–1324 (1981).
17. A. A. Erhard and H. F. L. Meyer-Balburg, *Science* **211**, 1312–1317 (1981).
18. P. J. Thomas and A. Knight, in *Current Studies of Hypothalamic Function* (K. Lederis and W. L. Veale, eds.) vol. 1 Karger A. G., Basel (1978), pp. 192–203.
19. P. C. B. McKinnon, E. Puig-Duran, and R. Laynes, *J. Reprod. Fertil.* **52**, 401–412 (1978).
20. K. D. Dohler, *Trends Neurosci.* **5**, 138–140 (1978).
21. G. Vallette, C. Benassayag, L. Belanger, E. A. Nunez, and M. F. Jayle, *Steroids* **29**, 277–289 (1976).
22. W. J. King and G. L. Green, *Nature* **307**, 745–747 (1984).
23. W. V. Welshons, M. E. Lieberman, and J. Gorski, *Nature* **307**, 747–749 (1984).
24. A. L. Lehninger, *Biochemistry* Worth Publishers, New York (1978).
25. A. Goth, *Medical Pharmacology* Mosby, St. Louis, Toronto, London (1981).
26. M. S. Brown and J. L. Goldstein, *Sci. Am.* **251**, 52–60 (1984).
27. Steinberg d. *Arteriosclerosis* **3**, 283–301 (1983).
28. Mahley R. W. and T. L. Inneratity, *Biochem. Biophys. Acta* **737**, 197–224 (1983).
29. A. W. Steggles and R. J. B. King, *Biochem. J.* **118**, 695–701 (1970).
30. M. Ginsburg, B. D. Greenstein, N. J. McLusky, I. D. Morris, and P. J. Thomas, *Steroids* **23**, 773–792 (1974).

31. M. Ginsburg, B. D. Greenstein, N. J. McLusky, I. D. Morris, and P. J. Thomas, *J. Steroid Biochem.* **6**, 989–991 (1975).
32. J. Barley, M. Ginsburg, B. D. Greenstein, N. J. McLusky, and P. J. Thomas, *Brain Res.* **100**, 383–393 (1975).
33. J. Barley, M. Ginsburg, B. D. Greenstein, N. J. McLusky, and P. J. Thomas, *Nature* **252**, 259–260 (1974).
34. D. F. Salaman, P. J. Thomas, and B. R. Westley, *Ann. Biol. Anim. Biochem. Biophys.* **16**, 479–490 (1976).
35. B. R. Westley, P. J. Thomas, D. F. Salaman, A. Knight, and J. Barley, *Brain Res.* **113**, 441–447 (1976).
36. B. R. Westley and D. F. Salaman, *Nature* **262**, 407–409 (1976).
37. A. M. Kaye, in *Regulation of Gene Expression by Hormones* (K. W. McKerns, ed.) Plenum, New York (1983).
38. L. J. Wangh and J. Knowland, *Proc. Natl. Acad. Sci. usA* **72**, 3173–3178 (1975).
39. E. H. Davidson, *Sci. Am.* **212**, 36–40 (1965).
40. B. W. O'Malley, *Biochemistry* **6**, 2546–2551 (1965).
41. B. S. McEwen, *Trends Pharmacol. Sci.* **6**, 22–26 (1985).
42. A. M. Kaye, *J. Steroid Biochem.* **19**, 33–40 (1983).
43. M. Ani, P. J. Butterworth, and P. J. Thomas, *Brain Res.* **183**, 341–353 (1980).
44. O. A. Amechi, P. J. Butterworth, and P. J. Thomas, *Brain Res.* **342**, 158–161 (1985).
45. G. Karakiulakis, P. J. Thomas, P. G. Papaioannidon, and A. G. Paradelis, *Meth. findings Clin. Exp. Pharmacol.* **6**, 437–443 (1984).
46. F. A. Kuehl, Jr., E. A. Ham, M. E. Zanetti, C. H. Sandford, S. E. Nichol, and N. D. Goldberg, *Proc. Natl. Acad. Sci. USA* **71**, 1866–1870 (1974).
47. H. Fleming, R. Blumenthal, and E. Gurdipe, *J. Steroid Biochem.* **20**, 5–9 (1984).
48. W. I. McGuire, D. J. Adams, and D. P. Edwards, *J. Steroid Biochem.* **20**, 73–75 (1984).
49. S. M. Paul and P. Skolnick, *Nature* **266**, 559–561 (1977).
50. H. Fleming, R. Blumenthal, and E. Gurpide, *Endocrinology* **111**, 1671–1673 (1982).
51. P. Ball, R. Knuppen, M. Haupt, and H. Brauer, *J. Clin. Endocrinol. Metab.* **34**, 736–746 (1981).
52. T. Lloyd and J. Weisz, *J. Biol. Chem.* **253**, 4841–4843 (1978).
53. J. M. Schaffer and A. J. W. Hseuh, *J. Biochem.* **254**, 5606–5608 (1979).
54. C. M. Paden, B. S. McEwen, J. Fishman, L. Snyder, and V. De Groot, *J. Neurochem.* **39**, 512–516 (1982).
55. M. Inaba and K. Kamatu, *J. Steroid Biochem.* **11**, 1491–1497 (1979).
56. M. E. Hadley, *Endocrinology* Prentice-Hall, New Jersey (1984).

Chapter 8
Membrane Receptors

P. J. Thomas

Most receptors for biologically active molecules are found upon the cell surface. They recognize their appropriate ligands and bind them. This binding leads to a change in receptor conformation and thence to intracellular signals and responses.

In some cases (hormone receptors), the sequence is of chemical activation through to a chemically mediated biological response, which may involve alterations in the electrical potential of the target cell.

In the case of neurotransmitters the situation is slightly more complicated: A wave of depolarization passes down a nerve process (axon), causes the release of the appropriate chemical messenger that binds to a receptor, and may produce either an immediate (electrical) or a longer-term (chemical through to morphological) response.

Cell surface receptors are many and varied. I shall describe only a few. In the case of those receptors that I describe, I shall comment upon their intracellular linkages and give my views about the relationship between their intracellular messengers and those that are induced by chemical messengers that act upon the genome.

1. Measurement of a Hitherto Unknown Cell Surface Receptor

Measurement of this type requires several steps.
In the beginning we have to show that the putative chemical messenger affects the cell or tissue in question and causes a biologi-

cal response. If this response is receptor mediated it should (a) be saturable, and (b) be prevented by excess concentrations of compounds that are chemically related to the putative messenger, but that, acting alone, produce no effects (antagonist): This inhibition should be dose related. These preliminary steps are essential, but are not the theme of this chapter: The reader is referred to ref. *1* for further information.

Those chemicals that produce a response are referred to as agonists; those that are related, but inhibit response, are termed "antagonists."

The second step is to characterize the putative receptor. This involves many subroutines.

By definition a cell surface receptor exists in intimate contact with membrane phospholipids. It is possible that this phospholipid environment might affect its conformation, so we start by measuring its binding affinities and specificity while it is still membrane bound.

This requires preparation of membrane fragments. A typical procedure is described below. The tissue is homogenized and centrifuged at $1000g$ for 10 min. The precipitate (P) is discarded and the supernatant (S_2) recentrifuged at $17,500g$ for 20 min. The supernatant (S_2) is discarded and the pellet (P_2) is suspended in buffer for assay. In the case of the muscarinic receptor, for example, it has been shown that the bulk of the binding is in the P_2 fraction (*2,3*), and albeit that the gravitational forces described in the literature vary slightly from paper to paper, this appears to be a general truth for receptors for neurotransmitters (*4–6*).

The membrane fragments are almost ready for assay: It must be remembered, however, that they probably still contain enzymes that may metabolize the labeled ligand, that they are associated with nerve terminals that contain neurotransmitters that may compete with the labeled ligand for the binding sites, and that these same nerve terminals may be able to actively incorporate the labeled ligand.

All this could distort measurements of receptor binding. To prevent this distortion, other preliminary steps may have to be taken.

Removal of putative endogenous competitors can be done by preliminary washing (followed by centrifugation) of the membrane fragments (*7*). Only two such washes are normally needed to ensure removal.

The principle is the same in all cases for removal of ligand metabolizing enzymatic activity. In detail it can be done in several ways, depending upon the radioligand of choice.

For example, preincubation with pargyline destroys monoamine oxidase activity and so inhibits the breakdown of catecholamines (7). In all cases it is prudent to check the purity of the radioligand both before and after the assay by methods such as HPLC.

Many, but not all, neurotransmitters can be sequestered by the nerve terminals whence they came: Since these nerve terminals are present in P_2 fractions, this can distort receptor binding measurements in vitro.

This is both easy to deal with and easy to forget. Presynaptic uptake, as opposed to receptor binding, is an active process and requires the activity of sodium dependent ATPase: Leave Na^+ out of the incubation medium and it will not take place. It can also be prevented by the addition of drugs such as ouabain.

It must be emphasized that the above is but a bald summary of the metabolic and competitive problems involved in the measurement of membrane receptor binding and that I may have forgotten certain problems. The reader is referred to the references cited in this chapter for full information.

1.1. Choice of Ligand

The next step is to incubate the P_2 fraction with a radiolabeled ligand until equilibrium is approached. The first and most obvious choice is the naturally occurring agonist: This, however, has potential disadvantages, and neurotransmitters per se should only be used if (a) there is no alternative, or (b) one is looking at the effects of a particular treatment upon a neurotransmitter pathway, but has no prior reason to know which subclass of receptor might be involved (8).

1.2. Receptor Subclass

Receptors for neurotransmitters (at least) appear not to be uniform. Consider an archetypical neurotransmitter: acetylcholine. It is the compound that, when released from appropriate nerve terminals, causes skeletal muscle to contract. It also acts as a neurotransmitter in the brain. It has at least two effects; these are dependent upon dose and location. Some of its actions can be mimicked by nicotine and some by muscarinic compounds.

Observations such as this lead to the concept that neurotransmitters might bind to more than one class of receptor and, because of these receptor subclasses, might give rise to different effects in different areas of the body.

Subclasses of receptors for most other neurotransmitters have also been described.

Neurotransmitters are believed to be relatively nonspecific agents: exactitude occurs at the recognition (receptor) site.

Certain exogenous compounds (be they agonists that mimic some of the effects of the naturally occurring neurotransmitter, or antagonists, that prevent neurotransmitter action) are more selective and also tend to have higher affinities for the receptors. In general, antagonists are the drugs of choice since those that are known tend to have higher affinities for the receptors than do agonists.

1.3. Separation of Bound from Free Receptors

Separation of bound from free receptors is much easier than in the case of soluble proteins. The receptor in the P_2 fraction is membrane-bound and so is effectively very large. It can be separated from the incubation medium by either filtration (followed by superficial washing in buffer) or by centrifugation followed, once again, by rinsing (without disruption) of the pellet. In the latter case it is important to ensure that the gravitational forces involved are at least as great as those that were used to prepare the original P_2 fragments.

It must be reemphasized (referring to chapter 6) that the specificity of binding is all important if one is attempting to define receptors, and that it is improper to classify "receptor" or a receptor subclass merely upon binding data: Many different authors have used many different incubation media to define receptor subclasses in the brain, and it is difficult to compare their results: The ionic composition of the different buffers used might have effects upon measurable binding.

In summary, up to this point, receptors can be defined by their relative binding affinities.

2. Purification and Isolation of Receptors

In principle, purification and isolation of receptors is a problem. They are membrane-bound proteins and so have hydrophobic interfaces: Binding activities are normally measured in an aqueous environment and solubilization with detergents might alter receptor property. Nevertheless, progress has been made. Several membrane receptors have been purified, and in at least one case have been artificially introduced into cells that do not normally contain them. Some of their properties will be described below.

2.1. Dopamine Receptors

Dopamine is a typical example of a neurotransmitter. There is, however, a complication when one tries to measure its receptors, and I have chosen it as my first example for this very reason. Dopamine is not only a neurotransmitter, but can also be converted into another neurotransmitter (noradrenaline or norepinephrine), a compound that is chemically very similar to it. There is an overlap in receptor specificity: noradrenaline can bind to dopamine receptors in the brain and vice versa. Presumably this is not a problem in the normal animal in vivo (anatomical specificity and chemical specificity can summate: If nerve terminal secretes dopamine, but not noradrenaline, the adjacent receptors will only be exposed to the former compound. If, however, one is measuring receptor–ligand interactions in the P_2 fraction, the anatomical specificity is lost).

Dopamine receptors have been described as existing in four subclasses (9), classified as D_{1-4}. The major distinction between D_1 and D_{2-4} lies in their relationship with intracellular messengers. D_1 receptors stimulate adenylate cyclase: The rest do not, but are said to be distinguished by their relative binding affinities for dopamine and neuroleptics in vitro. Whether the distinction between subclasses D_3 and D_4 will prove to be of physiological significance or to be an experimental artifact remains to be seen.

I shall comment upon the D_2 receptor. Attempts have been made to purify it. These have been based upon the use of radiolabeled spirperone (spiroperidol) as a specific D_2 receptor marker (10). Detergents were used to solubilize the putative receptor and it has been stated that the receptor, in solution, has a similar pharmacological profile (as far as binding is concerned) to the membrane bound receptor. It appears, however, that any link with guanyl regulatory proteins has been lost during solubilization. This is hardly surprising. The solubilized D_2 receptor sediments between 11 and 16S (10).

2.2. Effects of Use Upon Dopamine Receptors

We have seen earlier (chapters 6 and 7; ref. 11) that use can modify the selective stabilization of synapses. It is therefore not surprising to learn that prolonged treatment with neuroleptics can modify the characteristics of dopamine receptors in the brain (12), and that dopamine receptor binding in the brains of normal and schizophrenic patients differs (13).

It is also worth repeating that cyclic nucleotides may be involved in the selective stabilization of synapses (14–16), that their activities are modulated by steroids, as well as by drugs acting upon cell surface receptors, and that it would be expected that modulation of their breakdown would affect synaptic development. It has recently been shown (17,18) that perinatal administration of a compound (isobutyl methyl xanthine), which inhibits cAMP phosphodiesterase activity (19), prevents some of the estrogen-induced effects (20) upon long-term dopamine binding in the brain.

In summary, dopamine receptors are heterogenous and some are linked to adenylate cyclase and some are not. Their development may be affected by the exposure of the brain to several different stimuli and progress has been made toward the purification of one of the subclasses.

2.3. Nicotinic Acetylcholine Receptors

The more readily available a protein is, the better is one's chance of successfully purifying it. This may explain why a great deal of work has been done on the nicotinic receptor (one of the subclasses of the acetylcholine receptor). The nicotinic receptor has another advantage: It can be readily and firmly labeled by a specific ligand (the snake venom, bungarotoxin).

The nicotinic receptor is found in high concentrations in the electric organ of an eel, *Torpedo californica*. It has been solublized using detergents (the principle is similar to that quoted above for the D_2 receptor), and not only has it been purified and partially sequenced (21,22), but antibodies have been raised against it and it has been shown that these can induce experimental mysathenenia gravis in rabbits (23). This can be reversed by administration of partially dentured nicotinic receptor, suggesting that the disease is brought about by immunological damage (autoimmune disease) to the receptor and can be relieved by the administration of a substance that can compete with the receptor for the antibody.

(Myasthenia gravis is a muscle wasting disease that can occur in humans. Acetylcholine is the neurotransmitter that stimulates the neuromuscular junction; if its receptors are missing, the muscle cannot contract and will fade away).

The nicotinic receptor is composed of subunits and undergoes allosteric transformations when it sequestors its ligand. It appears to contain an ion channel, which can be opened or closed, depending upon what is bound to the receptor and the opening and closing of which is essential to the cellular response to acetylcholine (21,22). The receptor does not appear to be linked to cyclic nucleotide-forming enzymes.

Purification of the nicotinic receptor has lead to experiments whereby the mRNA that codes it is isolated and injected into oocytes of the amphibian, *Xenopus*, which do not normally have nicotinic receptors upon their outer membranes. After injection, such receptors were developed and, moreover, these receptors contained drug-dependent ion channels (24), as well as having binding characteristics consistent with native nicotinic receptors.

In summary, membrane receptors may be purified, albeit that they are found in vivo in a mixed hydrophobic/hydrophilic environment and such purification may lead to quantum advances in knowledge of cell biology.

2.4. Insulin Receptors

Insulin is a protein. Its main function is to regulate glucose metabolism within the body. It does this in synergism with other hormones, such as the protein, glucagon, and the steroidal, glucocorticoids. The aim is homeostasis.

Insulin may yet be found to have other functions. It and its putative receptors are found in the brain as well as in the periphery (25–27). Insulin cannot pass through the blood–brain barrier, so presumably that insulin found within the brain is endogenous: brain insulin levels do not fluctuate with circulating concentrations of insulin—in the brain it may act as a neuromodulator.

Insulin receptors have been isolated and quantified in almost as much detail as have nicotinic receptors (28,29), but interpretation is more complex. There is evidence that receptors change their conformation when they bind to insulin and that the hormone–receptor complex can be internalized. This latter observation may explain why their immediate second messenger (or messengers) has yet to be identified unequivocally: candidates are cAMP, cGMP, calcium ions, and, also, perhaps uniquely, a direct effect of the complex upon cAMP phosphodiesterase (28). There is evidence that the binding of insulin to the A subunit of the receptor activates protein kinase and thence phosphorylation of the B subunit, which leads in turn to the phosphorylation of other membrane proteins and thence to the release of mediator peptides and activation of intracellular protein kinases and phosphatases. Thus it is not surprising that activation of the receptor should have rapid effects upon a large number of intracellular intermediates (28,30).

2.4.1. Structure of Insulin Receptors

The insulin receptor consists of four subunits, (2 × A and 2 × B) that are interlinked by disulfide bonds. Each of the subunits is a glycoprotein (A with a molecular weight of 130,000, and B with a molecular weight of 95,000). They serve different functions, the A subunits containing the binding sites and the B subunits being affected by the binding of the hormone to the A site in such a way as to modify the protein kinase activity associated with it and thence the sequences of phosphorylation and dephosphorylation mentioned above. Thus the A site is the recognition site and the B site is involved in the early stages of the second messenger signaling (28,22).

2.5. Other Receptors

We have discussed three receptors in some detail and have touched upon their linkage to intracellular messengers. It would be virtually impossible to detail all the receptors that have been described in the literature within a reasonable compass, so I shall restrict myself to receptors to two classes of compounds that illustrate how knowledge of the effects of an exogenous compound upon the body can lead to further understanding of the body's basic function. I shall briefly discuss opioid and benzodiazepine receptors.

2.5.1. Opioid Receptors

Opiates are exonenous substances that have long been used and abused by mankind. They are drugs of addiction and dependence and cause, among other things, analgesia and euphoria. The pharmacological evidence suggested that they acted, in the first instance, by binding to receptors. This posed a problem. Why did the body contain receptors for alien compounds?

Membrane Receptors

The problem was solved by Kosterlitz and Hughes, who in 1973 suggested that there might be endogenous morphine-like compounds that have physiological roles in the brain, and that exogenous opiates might have sufficient chemical similarity to them to recognize their receptors (for references to this beautiful and productive work and its followup, *see* refs. *30* and *31*).

In essence, brains were homogenized and extracts (in various states of purification) tested for their activity upon opiate receptors. This led to the isolation and purification of the enkephalins, endorphins, and dynorphins, all of which are endogenous opioid peptides, and to a partial description of their receptors.

Opioid receptors exist in at least three (mu, delta, and kappa) subclasses (*32*) that may be distinguished by their relative binding affinities to different ligands and may yet be shown to be further subdivided.

It is of interest to note that the development of morphine tolerance/dependence appears to be linked to changes in the adenylate cyclase–cAMP–phosphodiesterase axis (*33–35*) and that the phenomenon involves a reduction of sensitivity of cells to the drug, suggesting a modification of synapse/receptor number or efficacy. I link this to my views (*15–20*) on the effects of estrogens, cyclic nucleotides, and neurotransmitters upon the selective stabilization of synapses and suggest that cyclic nucleotides are common intermediates both in the development of drug tolerance and in normal brain development. I suggest that the developing (and adult) brain adapts to specific chemical stimuli by selectively reducing its capacity to respond to them and that drugs of addiction merely mimic the normal developmental response of the brain to endogenous messengers.

2.5.2. Benzodiazapine Receptors

Benzodiazapines are synthetic compounds that are only too commonly prescribed as tranquilizers without sufficient consideration being given to their possible long-term effects.

We have seen above that exogenous ligands might exert their effects by mimicking the binding activities of endogenously produced biologically active compounds.

This general observation appears to be extrapolable to benzodiazapines, but the interaction appears to be more complex. They appear to interact with receptors for an intrinsic neurotransmitter, gamma-aminobutyric acid (GABA), but do not bind to the same recognition site as does GABA. Instead, their binding appears to

3. Intracellular Messengers

We dealt with the concept of intracellular messengers in chapter 6. In the present chapter, I shall be selective.

Neurotransmitters, neuromodulators, and hormones act at different places at different rates. When they are fast acting, they appear to work by affecting ion channels, thus modulating the rapid ingress or egress of inorganic ions to or from the cell. (Ion channels are believed to be proteins that extend from the outer to the inner surface of the cell membrane, are specific to individual ions, and are linked to certain subclasses of individual receptors.) Ions such as Ca^{2+} may therefore be considered to be fast second messengers (39).

3.1. Cyclic Nucleotides

Cyclic nucleotides are slower second messengers (39) that also have morphogenic effects (15,16). In this chapter I shall concentrate upon their measurement and upon the measurement of the enzymes that catalyze their synthesis and degradation.

The major cyclic nucleotides are cyclic AMP and cyclic GMP. They are formed from ATP and GTP, respectively, by the enzymes adenylate cyclase and guanylate cyclase. Since I have mentioned their linkages to receptors earlier (chapter 6), I shall restrict myself here to their measurement.

3.1.1. Measurement of the Cyclase Enzymes

Enzymes convert substrate into product. The only proper way to measure their activity is to measure the amount of product that is newly formed from a known concentration of substrate.

Adenylate or guanylate cyclase activities should be measured by incubating labeled ATP or GTP with the putative enzyme and separating labeled product from the excess of labeled substrate and then measuring it (15,16).

The composition of the buffer in which enzymatic activity is measured is important. It must be at physiological pH and must contain (a) ions that are essential for enzymatic activity, (b) substances that prevent the breakdown of substrate by ATPase and by GTPase, (c) phosphodiesterase inhibitors to prevent breakdown of the cyclic nucleotides, and (d) a small excess of unlabeled product in order to further reduce any enzymatic breakdown of the labeled product and act as a carrier for the labeled product during purification.

Typical components are detailed below (for concentrations, *see* refs. *15* and *16*). Tris-HCl, pH 7.5, maintains the hydrogen ion composition of the buffer. Phosphoenolpyruvate, pyruvate kinase, and myokinase prevent substrate breakdown. Magnesium chloride provides Mn^{2+}, which is essential for cyclase activity. A phosphodiesterase inhibitor (aminophylline, theophylline, isobutyl-methylxanthine, or 3-isobutyl-1-methylxanthine) should also be included: the different compounds mentioned here have somewhat different effects upon cyclic AMP and cyclic GMP phosphodiesterases, and these vary from tissue to tissue; the compound used should be selected with care. An excess of unlabeled product (cyclic AMP or cyclic GMP, as appropriate) further reduces any breakdown and also acts as a carrier for the labeled product during isolation. In addition, ammonium sulfate is added to the adenylate cyclase assay and bovine serum albumin is added to both assays.

The next step is to stop the reaction. The enzyme is killed by the addition of sodium pyrophosphate followed by boiling, and the products separated by thin layer chromatography. The newly formed (radiolabeled) cyclic nucleotides are estimated by liquid scintilation counting (*15,16*). Under appropriate experimental conditions, both V_{max} and K_m can be estimated by classical methods (*40*), the algebra of which is analogous to that used to determine ligand binding to a receptor with the qualification that measurements are not made under equilibrium conditions.

Methods such as those described above show that adenylate cyclase is predominantly found in the membrane fractions of cells and that its activity in vitro may be affected by neurotransmitters (*15*). The activity of guanylate cyclase, on the other hand, is evenly distributed between membrane and cytosolic ($100,000g$ for 1 h) components and cannot be affected by neurotransmitters in broken cell preparations, possibly because of its less direct linkage to recognition sites for extracellular messengers.

There is another, less direct, method of measuring adenylate and guanylate cyclases. This involves measuring the products by competitive protein binding assays in the presence of appropriate stimuli and subtracting the amount of cyclase found in the absence of such stimuli and assuming that the difference obtained represents newly formed cyclic nucleotides (41). The method is, by definition, indirect, and it must be shown that the enzyme is active by the concomitant administration of activators or inhibitors. The method therefore suffers from a fundamental weakness, but, that having been said, is much more sensitive and alleged to be quicker to carry out than is the direct measurement of the enzymes.

3.2. Phosphodiesterases

Phosphodiesterases must also be assayed. The problem is that they are many and varied: assay is difficult if one attempts to measure a specific phosphodiesterase to the exclusion of others. The problem becomes simple when approaches from a physiological viewpoint, but more complex if one takes a biochemical point of view.

3.2.1. A Physiological Approach

To my knowledge, the breakdown products (5'-AMP and 5'-GMP) of the cyclic nucleotides have no biological activity. The aim is merely to see how rapidly the substrate is destroyed; not to measure the rate of product formation. This can be done by sequestration of cyclic AMP by alumina gel (42) and measurement of enzymatic activity by difference.

3.2.2. A Biochemical Approach

A biochemical approach is more complicated since there are several different classes of phosphodiesterases, a high-affinity cAMP phosphodiesterase (K_m 5 μM) and cyclic AMP specific; a high-affinity (K_m 10 μM) cyclic GMP phosphodiesterase that has a low affinity (K_m 200 μM) for cyclic AMP, but which has a high V_{max} for that compound. A cyclic nucleotide phosphodiesterase that catalyzes the breakdown of both cyclic AMP and cyclic GMP has also been described (43).

Thus it is difficult to measure all the phosphodiesterases at any one time. As an example, I summarize a recipe below; the measurement of a brain cycle GMP phosphodiesterase with a K_m of ~20 μM (44,45). The substrate is incubated with enzyme to produce 5'-GMP. The reaction is stopped by boiling and a second reaction is started (it is not so easy to measure 5'-GMP; it is relatively easy to measure its breakdown product, guanosine) by incubation with a 5' nucleotidase derived from snake venom. The guanosine so formed is purified and measured, thus giving an estimate of phosphodiesterase activity.

3.3. Cyclic Nucleotide-Dependent Protein Kinases

As mentioned earlier, cyclic nucleotide-dependent protein kinases consist of two subunits, regulatory and catalytic. The principles of measuring the activities of the catalytic subunit are in essence the same as those for the measurement of any other enzyme and will therefore not be described further. Measurement of the regulatory subunit involves a new principle: photoaffinity labeling, which may prove very useful in the future and is summarized below.

3.3.1. Photoaffinity Labeling of the Regulatory Subunit of Protein Kinase

A problem involved in measuring any binding is that it obeys the laws of mass action and separation of bound from free ligand can perturb the equilibrium and so distort the results; how much better it would be if binding was covalent. In certain circumstances it can be made to be so (46).

Cyclic AMP can be converted into 8-azido cyclic AMP by treatment with bromine and then with nitrogen. 8-Azido cyclic AMP has physiological properties similar to those of the native compound, but with an important experimental difference. If, once it has bound to the regulatory subunit of protein kinase, in the dark, the complex is exposed to UV light, the complex becomes covalent and can be easily measured. It was by this means that it was shown that the regulatory subunit of protein kinase could move on occasion to the cell nucleus and behave therein in a manner analogous to that of steroid hormone receptor complexes, modifying genetic expression.

Whether it is proper to refer to the regulatory subunits of cyclic nucleotide-dependent protein kinases as enzymes or as intracellular ligand binding proteins is hardly a matter of importance.

The concept of photoaffinity labeling will, in the future, be applied to many more binding reactions and will eventually lead to the purification of important proteins that are relatively unconcentrated in the body and thus to a greater understanding of their physiological role. (I have in mind membrane proteins that are involved in the transfer of ions between the intracellular and the extracellular environment.)

In summary, binding proteins are measured by the separation of bound from free ligand without perturbing the equilibrium. This is relatively easy if the dissociation rate of the complex is slow: if it is fast, the problem can sometimes be solved by photoaffinity labeling. Enzymes are measured by their ability to convert substrate. Proteins whose function is intermediate (for example calmodulin and the guanyl binding proteins that form an activatory or inhibitory link between the neurotransmitter or hormone reception site) have not been mentioned in detail in this chapter. Their measurement is by a combination of binding ability and their effects upon enzymes (for references on calmodulin see ref. 43, and for the guanyl binding proteins and their relationships with forskolin cholera toxin and islet activating protein see refs. 44–46; for a summary of the various interactions between the different pathways, see chapter 6).

References

1. A Goth, *Medical Pharmacology* Mosby, St. Louis (1981).
2. E. C. Hulme, N. J. M. Birdsall, and A. S. V. Burgen, *Mol. Pharmacol.* **14**, 737–750 (1978).
3. N. J. M. Birdsall, A. S. V. Burgen, and E. C. Hulme, *Mol. Pharmacol.* **14**, 732–736 (1978).
4. S. J. Hill and J. M. Young, *Br. J. Pharmacol.* **68**, 687–696 (1980).
5. Y. Clement-Cormier and R. George, *Life Sci.* **23**, 539–545 (1978).
6. E. J. Hartley and P. Seeman, *Life Sci.* **23**, 513–518 (1978).
7. d. R. Burt, S. J. Enna, I. Creese, and S. H. Snyder, *Mol. Pharmacol.* **12**, 631–638 (1975).
8. M. H. Jalilian-Tehrani, G. Karakiulakis, R. Powell, C. B. LeBlond, and P. J. Thomas, *Br. J. Pharmacol.* **75**, 37–48 (1982).
9. P. Seeman, *Pharmacol. Rev.* **32**, 230–313 (1980).
10. P. G. Strange, in *Cell Surface Receptors* (P. G. Strange, ed.) John Wiley, New York (1983) pp. 82–100.
11. P. J. Thomas and M. Ani, *Mol. Aspects Med.* **5**, 1–61 (1982).
12. P. Jenner and C. D. Marsden, in *Cell Surface Receptors* (P. G. Strange, ed.) John Wiley, New York (1983), pp. 142–162.
13. F. Owen, A. Cross, and T. J. Crow, in *Cell Surface Receptors* (P. G. Strange, ed.) John Wiley, New york (1983), pp. 163–173.
14. P. J. Thomas and M. Ani, *Mol. Aspects Med.* **5**, 1
61. (1982).
15. M. Ani, P. J. Butterworth, and P. J. Thomas, *Brain Res.* **183**, 341–353 (1980).
16. O. A. Amechi, P. J. Butterworth, and P. J. Thomas, *Brain Res.* **342**, 158–161 (1985).
17. M. Jalilian-Tehrani, PhD thesis, University of London, london, UK (1983).
18. M. Jalilian-Tehrani, A. Bittles, and P. J. Thomas, *Meth. Find. Exp. Clin. Pharmacol.* **8**, 197–201 (1986).

Membrane Receptors

19. F. W. Smellie, C. W. Davis, J. W. Daly, and J. N. Wells, *Life Sci.* **24**, 2475–2483 (1979).
20. M. Jalilian-Tehrani, G. Karakiulakis, R. Powell, C. B. LeBlond, and P. J. Thomas, *Br. J. Pharmacol.* **75**, 37–48 (1982).
21. J-L. Popot and J-P. Changeux, *Physiol. Rev.* **64**, 1162–1239 (1984).
22. J-P. Changeux, A. Devillers-Thiery, and p. Chemonilli, *Science* **225**, 1335–13345 (1984).
23. S. Fuchs and D. Bartfeld, in *Cell Surface Receptors* (P. Strange, ed.) John Wiley, New York (1983), pp. 126–141.
24. K. Sumikawa, R. Miledi, M. Houghton, and E. A. Barnard, in *Cell Surface Receptors* (P. G. Strange, ed.) John Wiley, New York (1983) pp. 249–269.
25. J. Havrankova, D. Schemchel, J. Roth, and M. Brownstein, *Proc. Natl. Acad. Sci. USA* **75**, 5737–5741 (1978).
26. M. L. Barbaccia, D. M. Chuang, and E. Costa, in *Regulating Peptides, From Molecular Biology to Function* (E. Costa and M. Trabucchi, ed.) Raven, New York (1982), pp. 511–518.
27. J. Havrankova, M. Brownstein, and J. Roth, *Diabetologica* **20**, 268–273 (1981).
28. S. Gammeltoff, *Physiol. Rev.* **74**, 1321–1378 (1984).
29. M. P. Czech, *Am. J. Med.* **70**, 142–150 (1981).
30. M. E. Hadley, *Endocrinology* Prentice Hall, New Jersey (1984).
31. J. Hughes, *Opioids—Past, Prsent and Future* (J. Hughes, H. O. J. Collier, M. J. Rance, and M. B. Tyers, eds.) Taylor and Francis, London (1984).
32. E. J. Simon and J. M. Hillier, in *Opioids—Past, Present and Future* (J. Hughes, H. O. J. Collier, M. J. Rance, and M. B. Tyers, eds.) Taylor and Francis, London (1984), pp. 33–52.
33. K. Kuriyama, K. Nakagawa, K. Naito, and M. Muramatsu, *Japan J. Pharmacol.* **28**, 73–84 (1985).
34. H. O. J. Collier, N. J. Cuthbert, and D. L. Francis, *Fed. Proc.* **40**, 1513–1518 (1981).
35. H. O. J. Collier, in *Opioids—Past, Present and Future* (J. Hughes, H. O. J. Collier, M. J. Rance, and M. B. Tyers, eds.) Taylor and Francis, London (1984), pp. 109–125.
36. M. K. Ticku, *Neuropharmacology* **22**, 1459–1470 (1983).
37. A. J. Turner and S. R. Whittle, *Biochem. J.* **209**, 29–41 (1983).
38. I. L. Martin, *Trends Pharmacol. Sci.* **60**, 343–347 (1984).
39. J. Altman, *Nature* **315**, 537–539 (1985).
40. A. L. Lehninger, *Biochemistry* Worth, New York (1970).
41. D. J. Judson, S. Pay, and K. D. Bhoola, *J. Endocrinol.* **87**, 153–159 (1980).
42. A. Gadd, PhD thesis, University of Bristol, Bristol, UK (1985).
43. A. Nathanson and G. Kebabian, *Handbook of Experimental Pharmacology* Springer/Verlag, New York (1982).
44. G. Karakiulakis, P. J. Thomas, P. G. Papaiannidou, and A. G. Paradelis, *Meth. Find. Exp. Clin. Pharmacol.* **6**, 437–443 (1984).

45. W. J. Thompson and M. M. Appleman, *Biochemistry* **10**, 311–316 (1971).
46. J. Kallos, *Nature* **265**, 705–709 (1977).

Chapter 9

Protein and Peptide Hormones

P. J. Thomas

1. Introduction

The cyclical control of protein and peptide hormone synthesis and release and the existence of prohormones has been considered in chapter 6, as has the existence of neuromodulators that sometimes are chemically identical to classical hormones, though subserving presumably different functions. These problems will not be dealt with in the present chapter, in which the old-fashioned definition of hormones (chemical messengers acting at a distance) will be accepted by default: It is true more often than not.

There are so many protein and peptide hormones that it would be out of the question to describe or even comment upon them all. I shall restrict myself to a few examples: some of those hormones secreted by clearly defined (in an anatomical sense) endocrine organs. This chapter should only be read after due consideration of chapters 6 and 8. When possible I shall, in my bibliography, refer to standard textbooks in which the material is more than 5 yr old. Other references are given in chapter 6.

2. Release and Release-Inhibiting Factors Secreted by the Hypothalamus

Release and release-inhibiting factors secreted by the hypothalamus play a part in the control of the function of the anterior pituitary. They are sometimes referred to as release or release-inhibiting hormones. The term hormone is used when a substance has been identified, beyond reasonable doubt, as having specific physiological effects upon a given pituitary hormone. They affect the synthesis and release of the protein hormones of the anterior pituitary, either in a positive or negative manner, as is implied by their names, and reach that organ in relatively high concentrations by way of the pituitary portal blood vessels.

Their discovery arose from the observation that, although the adenohypophysis is not part of the brain, its anatomical position (adjacent to the median eminence of the hypothalamus) is essential to its function, and that a vascular system (the pituitary portal system) carries blood from the median eminence to the anterior pituitary.

In order to qualify for consideration as a potential releasing (or release-inhibiting factor), a substance must ideally (a) have relatively specific effects upon an individual pituitary hormone, (b) be present in the neurons of the median eminence, (c) be demonstrable in the portal blood, and (d) fluctuate in concentration in relationship to its physiological usage (1).

Release and release-inhibiting factors include thyrotrophin releasing hormone (TRH), somatostatin, somatotrophin releasing hormone, gonadotrophin releasing hormone (GRH), corticotrophin releasing hormone (CRH), melanostatin release inhibiting factor, melanostatin (MIF), and prolactin release inhibiting factor (PIF). The latter two are currently referred to as factors rather than hormones, since their current chemical nature remains debatable. Their names define their roles.

Release and release-inhibiting hormones (or factors) are generally polypeptides consisting of 50 or fewer amino acid residues. The smallest one to be described is the tripeptide, TRH (pyroglutamyl-histidyl-proline-amide) (2). The structures of some of the others are described below.

> Somatostatin: Ala-Gly-Cys-Lys-Asn-Phe-Phe-Trp-Lys-Thr-Phe-Thr-Ser-Cys
> Somatotrophin releasing hormone: Tyr-Ala-Asp-Ala-Ile-Phe-Thr-Asn-Ser-Tyr-Arg-Lys-Val-Leu-Gly-Gln-Leu-Ser-Ala-Arg-Lys-Leu-Leu-Gln-Asp-Ile-Met-Ser-Arg-Gln-Gln-Gly-Glu-Ser-Asn-Gln-Glu-Arg-Gly-Ala-Arg-Ala-Arg-Leu-NH_2
> Gonadotrophin releasing hormone (GRH): (pyro)Glu-His-Trp-Ser-Tyr-Gly-Leu-Arg-Pro-Gly-NH_2
> Corticotrophin releasing hormone (CRH): Ser-Gln-Glu-Pro-Pro-Ile-Ser-Leu-Asp-Leu-Thr-Phe-His-Leu-Leu-Arg-Glu-Val-Leu-Glu-Met-Thr-Lys-Ala-Asp-Gln-Leu-Ala-Gln-Gln-Ala-His-Ser-Asn-Arg-Lys-Leu-Leu-Asp-Ile-Ala-NH_2

Protein and Peptide Hormones 235

For references to the elucidation of their structure, *see* ref. 2.

There may be an exception to the rule that hypothalamic release and/or release-inhibiting hormones are peptides. I draw attention to this because it illustrates the interactions between peptides and neurotransmitters within the brain and also the difficulty of distinguishing between the naturally occurring effects of substances that are present in high concentrations in a given area and the effects of the artificial introduction of biologically active compounds into that same area. It might also illustrate the potential strengths and weaknesses of immunohistochemistry.

For a long time, PIF was thought to be dopamine (1,2). Dopamine is present in the hypothalamus (it is a neurotransmitter) and, when applied to the adenohypophysis, inhibits the release of prolactin. This does not necessarily mean that it is PIF.

It has recently been suggested (3–5) that PIF might be a 55-amino-acid residue peptide that is identical to one derived from a prohormone secreted by the human placenta, which can be located by immunohistofluorescence in appropriate neurons within the hypothalamus. This is exciting; it suggests that tissues as diverse as placenta and hypothalamus can secrete the same hormones. It should, however, be treated with caution until its anatomical interrelationships with hypothalamic dopamine-secreting neurons are better understood. The reader is also asked to bear in mind the concept (chapter 6) that peripheral hormones can also function as cotransmitters and that CRF, LHRH, somatostatin, and TRH have been found in nerve terminals in the central nervous system that are not known to be associated with endocrine or neuroendocrine function. It should also be emphasized that immunohistochemical and biological identities are not necessarily synonymous.

3. Protein Hormones of the Anterior Pituitary (Adenohypophysis)

It seems to be a natural progression to move from release and release-inhibiting hormones of hypothalamic origin to a discussion of the trophic hormones of the pituitary: of these, however, I shall have little to say. Their structure and function is textbook knowledge (2). It suffices to say that they are relatively large proteins: some (thyrotrophin, TSH; follicle stimulating hormone, FSH; and luteinizing hormone, LH) are two-chain glycoproteins maintained by disulfide bonds. It appears that only one of these chains confers receptor specificity, but that the presence of the other chain is essential for biological activity. Others, somatotrophin corticotrophin and prolactin, for example, are single-chain polypeptides.

Their approximate molecular weights, to the nearest 1000, and their principle actions, are summarized below.

Hormone	~MW, × 10^{-3}	Actions
TRH	28	Stimulates synthesis and secretion of thyroid hormones
ACTH	5	Enhances the secretion of glucocorticoids by the adrenal cortex
LH	28	LH and FSH are known collectively as the gonadotrophins, with synergistic actions upon the ovaries and testes. In the former case, FSH stimulates follicular growth; LH precipitates ovulation
FSH	34	
PRL	21	Stimulates milk secretion
GH (growth hormone: somatotrophin)	21	Stimulates somatic growth via somatomedins. Immediate effects are insulin-like: long-term effects are anti-insulin like.

3.1. Multiple Functions of Individual Anterior Pituitary Hormones

It has been shown (chapter 6) that individual proteins may have more than one function.

3.1.1. ACTH

Corticotrophin is a most interesting protein in this respect since it was one of the first hormones to be shown to have different functions and that the coding for these different function lies in different parts of the peptide chain.

The complete hormone consists of 39 amino acid residues (aar); namely, Ser-Tyr-Ser-Met-Glu-His-Phe-Arg-Trp-Gly-Lys-pro-Val-Gly-Lys-Lys-Arg-Arg-Pro-Val-Lys-Val-Tyr-Pro-Asn-Gly-Ala-Glu-Asp-Glu-ser-Ala-Glu-Ala-Phe-Pro-Leu-Glu-Phe. This sequence is for human ACTH. In other species (cattle, sheep, and pigs), substitutions are found in positions 31 and 33.

ACTH shares similarities with a MSH, a melanocyte stimulating hormone that is secreted by the pars intermedia of the pituitary and causes darkening of the skin. It is therefore not surprising that one of the side effects of ACTH involves an increase in skin pigmentation: the sequence of both hormones is included in that of β-lipotrophin, and in that of its precursor, proopiomelanocortin, which contains not only the sequences of MSH and ACTH, but also those of the endogenous opioids.

A much more interesting side effect of ACTH involves the brain. It is said to enhance memory. I will not explore the complex psychopharmacological experiments involved in analyzing the potential effects of drugs upon behavior, nor do I propose to comment upon their validity. I restrict myself to saying that animals can be trained to perform statistically quantifiable tasks (running through mazes, pressing levers to obtain reward, taking action to avoid predicted and unpleasant stimuli), and that the effects of drug administration upon performance can be monitored.

In general, rats learn to perform such tasks more effectively when ACTH is present than when it is absent. (6–8) This is unaffected by adrenalectomy and so would appear to be an effect of ACTH as a molecule rather than of ACTH as a trophic hormone. Furthermore, it has been indicated that the residue ACTH (4–10) is the active principle, insofar as memory is concerned, and that ACTH (4–10) acts by competing with β-LPH fragments, has a direct effect upon neurons (whatever that may mean), or works by influencing neurotransmitter levels remains to be seen.

Other functions of ACTH fragments are summarized below. The numbers refer to the amino acid residues.

1–24	Full agonist
5–24	Full agonist, steroidogenesis, partial agonist, lipolysis
6–24	Partial agonist, steroidogenesis, partial agonist, adenylate cyclase stimulation
7–24	Partial agonist, adenylate cyclase
11–24	Antagonist, Mg^{2+} accumulation, antagonist-steroidogenesis (9)

In summary, hormones may have more than one effect, some of which are mediated by different biologically available amino acid sequences of the entire hormone.

It also should be pointed out that hormones can have overlapping effects and peripheral organs (for example, the placenta) can secrete hormones that are normally believed to be of pituitary origin (3–5). It should also be reemphasized that ACTH, GH, and β-lipotrophin are among the hormones that function independently as cotransmitters.

The hormones of the pars intermedia will not be dealt with in this chapter, except for the comments above and to point out that α-MSH appears to have cotransmitter as well as melanotrophic activities (10).

4. The Posterior Pituitary (Neurohypophysis)

The posterior pituitary, unlike the anterior pituitary, is a part of the brain. It consists of neuronal endings that are in close juxtaposition to the anterior pituitary, but that, as far as I know, have no physiological contact with the anterior pituitary.

Though the posterior pituitary is part of the brain (the gland itself is located in the pituitary fossa in juxtaposition to the anterior pituitary), the relationship ends there. Whereas the secretory cells in the anterior pituitary are wholly contained within that organ, those of the posterior pituitary are not. The posterior pituitary consists of nonsecretory pituocytes (which are analogous to neuroglia) and the swollen nerve endings (axons) of neurons whose cell bodies are found within the supraoptic and paraventricular nuclei of the hypothalamus (the word nucleus has two meanings in neuroanatomy: first, that portion of a cell that contains the genetic material, and, second, a distinct cluster of neurons within the brain: the supraoptic and paraventricular nuclei are distinguished by the latter sense).

The posterior pituitary secretes two distinct octapeptide hormones (or nonapeptides; depending upon whether you regard cysteine or cystine as a single amino acid residue) held into a cyclic structure by disulfide bonds. The hormones are oxytocin and vasopressin (alternatively called antidiuretic hormone; ADH). Oxytocin causes milk ejection in lactating mothers and also stimulates the contractions of the estrogen- and progesterone-primed uterus, aiding parturition. It has no known effects upon milk secretion. That is the function of prolactin. Vasopressin is an antiduretic hormone in physiological concentrations. At higher concentrations it raises blood pressure, as its name implies. The two hormones are distinguished by differences in two amino acid residues, of which the most interesting is in position eight, where vasopressin may have either an arginyl or lysyl residue. Most mammals secrete 8-arginine vasopressin: Pigs secrete 8-lysine vasopressin; the South American wild pigs (peccaries) secrete both 8-arginine and 8-lysine vasopressin, both of which subserve much the same function. For an analysis of the evolution of posterior pituitary hormones, *see* refs. *11,12*.

Protein and Peptide Hormones

I also draw attention to the late Hans Hellers contribution to this field 13. In essence he showed that the classical Darwinian evolutionary tree could be paralleled by a molecular one.

In essence, the pituitaries of "lower" vertebrates, such as fish, contain neither oxytocin or vasopressin: instead, they contain arginine vasotocin, which, in evolutionary and embryological terms, is a precursor to the adult mammalian hormones.

Oxytocin and vasopressin differ from one another in positions 3 and 8.

Oxytocin: Cys-Tyr-Ile-Gln-Asn-Cys-Pro-Leu-Gly-(NH$_2$)

AVP: Cys-Tyr-Phe-Gln-Asn-Cys-Pro-Arg-Gly-(NH$_2$)

LVP: Cys-Tyr-Phe-Gln-Asn-Cys-Pro-Lys-Gly-(NH$_2$)

Arginine vasotocin (AVT): Cys-Tyr-Ile-Gln-Asn-Cys-Pro-Arg--Gly-(NH$_2$)

It will be seen that AVT is an intermediate between oxytocin and the vasopressins. Whether the evolutionary changes between AVT and the mammalian neurohypophyseal hormones are the results of successive single-point mutations or not remains (to my knowledge) to be worked out in detail.

Other "lower" species secrete other analogs of the posterior pituitary hormones. These include ichthyotocin, mesotocin, glumitocin, valitocin, and aspargtocin.

4.1. Neurophysins

In mammals, oxytocin and vasopressin are not released alone. They are stored in vesicles together with proteins, neurophysins, to which they are bound with a relatively high affinity (K_d values in the μM range) and are released together by exocytosis. There are at least two distinct neurophysins, one of which is preferentially associated with oxytocin (and is secreted by oxytocin-synthesizing neurons and preferentially sequesters the latter hormone. They have monomer molecular weights in the range of a thousand.

It would appear that the neurophysins and the active posterior pituitary hormones are synthesized as single long chains and are separated by proteolytic cleavage as they pass from the supraoptic and paraventricular nuclei to the pars distalis. Their mutual conformation is such that they remain in intimate juxtaposition until they are released and separated by dilution in the bloodstream. The speed of their movement from brain to pituitary and the fact

4.2. Some Effects of Neurohypohyseal Hormones Upon the Brain

We have seen earlier in this chapter that ACTH and certain of its fragments have not only adrenocorticotrophic effects, but also modulate brain function. Neurohypopyseal hormones also have dual functions. Vasopressin enhances memory. So does des-glycinamide vasopressin (vasopressin from which the C-terminal glycinamide residue has been removed). des-Glycinamide vasopressin has no classical hormonal functions: it still works upon the brain (7,15). This should be considered in relation to the neuromodulators referred to in chapter 6, as well as in relationship to the multiple functions of ACTH.

In summary, peptide hormones may have more than one function: The amino acid sequences that are critical to the different functions may not be the same.

5. Hormones of the Periphery

I will deal here with a few examples of hormones of the periphery and try to relate them to both their classical actions and to their possible effects upon the brain. I shall also comment upon their relationship to prohormones.

5.1. Insulin

The archetypical peripheral peptide hormone is insulin. Its classical actions involve glucose metabolism: with too little, diabetes results, with too much, hypoglycemic coma and death result. It is mainly secreted by the β-cells of the islets of Langerhans of the pancreas: As we shall see later, it is also secreted by the brain.

Insulin's synthesis involves the secretion of a prohormone (proinsulin). Proinsulin consists of a single chain of 81 amino acid residues. Proteolytic cleavage yields a two-chain residue.

The structure of proinsulin is described below. Only the key amino acids are given their classical abbreviations. The rest are numbered (N-terminal = 1). The points of cleavage are indicated by arrows.

Protein and Peptide Hormones

Proinsulin

1-6-Cys-8-18-Cys-20-30-[Arg-Arg]-33-58[Lys-Arg]-61-65-Cys-
Cys-68-70-Cys-72-79-Cys-Asn

Insulin consists of the fragments, A chain; 61–68 and B chain; 1–30. The intermediate (after cleavage) C chain makes up the remainder.

The insulin is another example of a hormone-prohormone relationship, with the additional complication that proteolytic cleavage converts a single-chain precursor into a two-chain (held together by disulfide bonds) hormone and a single-chain (C-fragment) nonhormonal residue.

The functions of insulin upon glucose and cellular metabolism are too well known to warrant further description. I shall therefore restrict myself to pointing out that insulin, like other protein hormones, varies slightly with species in its amino acid sequence, that it may subserve a completely different function in the brain, and that it has been stated that the rat can secrete two different molecular forms of insulin.

5.2. Insulin in the Brain

Insulin cannot pass the so called blood–brain barrier. In other words, it cannot enter the brain from the blood stream. Nevertheless, insulin is found in the brain (16). The concentrations of insulin that are found in the periphery; they are not effected by hypo- or hyperglycemia. Thus it would seem possible that insulin is synthesized within the brain and that it might fulfill functions other than those that it has in the periphery (17). This concept should be considered together with the multiple roles of ACTH fragments and with the effects of vasopressin and des-glycinamide vasopressin upon the brain. It should also be considered with the concept of neuropeptides (neuromodulators) that was outlined in chapter 6.

In summary, I have described or discussed a few protein hormones: I have omitted many others. I have shown that current dogma postulates that they are synthesized as long-chain precursors, that they are activated by proteolytic cleavage, and that they, or some of their precursor remnants, can fulfill more than one function.

Other protein or peptide hormones not mentioned here include: chalones, the hormones of the intermediate lobe of the pituitary, parathormone, calcitonin, gastrin, secretin, cholecystokinin, gastric inhibitory peptide, vasoactive inhibitory peptide, substance P, gastric releasing peptide, motilin, glucagon (in essence an insulin antagonist), endorphins and enkephalins, and bombesin. The reader is referred to ref. 2 for further information.

6. Conclusions

Endocrinologists once thought that hormones could be defined as chemical messengers secreted by specific organs that act at a distance. This is an oversimplification, the more so since it has recently been shown that individual cells (and indeed individual vesicles) can secrete more than one hormone. CRF and vasopressin, for example, have been found together in nerve terminals in the median eminence of the hypothalamus (18).

Thus, many proteins and peptides that were originally thought to be classical hormones (chemical messengers acting at a distance) may also have neuromodulatory role. As time goes on, it may be shown that many other proteins also have more than one function, depending upon their location, production, and site of action in the body.

References

1. B. T. Donovan, *Mammalian Neuroendocrinology* McGraw Hill, New York (1970).
2. M. E. Hadley, *Endocrinology* Prentice-Hall, New Jersey (1984).
3. K. Nikolics, A. J. Mason, E. Szoni, J. Ramachadran, and P. M. Seeburg, *Nature* **316**, 511–517 (1985).
4. H. Phillips, K. Nicolics, D. Braunton, and P. H. Seeburg, *Nature* **316**, 542–545 (1985).
5. G. Fink, *Nature* **316**, 487–488 (1985).
6. D. de Weid, in *Centrally Acting Peptides*, (J. Hughes, ed.) MacMillan, London (1978), pp. 241–251.
7. T. B. van Wimersman and D. de Weid, *Pharmacol. Biochem. Behav.* **5** (suppl. 1), 29–34 (1976).
8. H. Ritger and H. van Riezen, *Physiol Behav.* **14**, 563–566 (1975).
9. R. Schwyzer, in *Towards Understanding Receptors* (J. W. Lamble, ed.) Elsevier/North Holland (1981), pp. 139–146.

Protein and Peptide Hormones

10. L. Iversen, *Trends Neurosci.* **7**, 293–294 (1983).
11. R. Acher, *Proc. Roy. Soc. Lond.* (B), **210**, 21–43 (1980).
12. W. H. Sawyer, *Am. Zool.* **17**, 727–737 (1977).
13. Ciba Foundation Study Group, **39**, Ciba, London (1971).
14. B. T. Pickering, *Essays Biochem.* **14**, 45–81 (1978).
15. J. T. Martin and T. B. van Greidanus, *Psychoneuroendocrinology* **3**, 261–269 (1979).
16. K. Katika, S. Giddings, and M. A. Permutt, *Proc. Natl. Acad. Sci. USA* **79**, 2803–2807 (1982).
17. J. Havrankova, M. M. Brownstein, and J. Roth, *Diabetologica* **20**, 268–273 (1981).
18. M. H. Whitnall, E. Mezey, and H. Gainer, *Nature* **317**, 248–249 (1985).

Chapter 10

Analysis of Amino Acids and Small Peptides

Colin Simpson

1. Introduction

Proteins are composed of amino acids, and the nature and composition of the various constituent amino acids confer upon the protein its characteristic properties. Of even greater importance is the order in which the constituent amino acids are arranged in the protein molecule and the tertiary structure the protein molecule may adopt, although we may at this stage be unable to relate structure to protein function. Thus there are two different analyses to be performed: (1) the proportions of the various naturally occurring amino acids that are present in a given protein, and (2) the determination of the order in which the constituent amino acids are arranged in the protein molecule. Alternatively, the straight amino acid composition of proteinaceous material can provide information as to the source of a particular food for human consumption or animal feed, or indicate the nature of a protein from the constituent amino acids present, e.g., whether the protein is fibrous or globular in character. Thus amino acid analysis is an important tool in the biochemists armory, indeed one that is undertaken routinely on a day-to-day basis.

2. Amino Acid Analysis

Various means have been applied to the separation and quantification of amino acids in protein hydrolysates; these may be divided into chromatographic and electrophoretic methods and suitable combinations of the two. It would be useful to consider the merits of the various analytical procedures available, because even though some of the modern methods may enable us to separate and detect the naturally occurring amino acids at the picomolar level or less in about 10–15 min, they may not necessarily be capable of separating di- and tripeptides on the same time scale and at the same levels of detection.

2.1. Paper Chromatography

Amino acid analysis stems from the classical experiments of Consdon, Gordon and Martin (1) and subsequently Dent (2). This work catalyzed the production of a large number of papers by a variety of workers for the separation of amino acids by paper chromatography, e.g., Kowkabany and Cassidey (3) investigated the characteristics of some 70 different papers with different types of mobile phase, whereas McFarren (4) investigated the effect of buffer type over a wide pH range on the migration of the amino acids.

Paper chromatography is a three-phase system consisting of a mobile phase, a stationary phase adsorbed on the surface of the cellulose fibers of the paper, and a vapor phase, which may or may not be in equilibrium with the mobile phase, which is difficult to ascertain. The rate of migration of the mobile phase is a function of the physical properties of the solvent, e.g., viscosity, surface tension, and density, all of which are dependent on temperature. Further, the rate of migration is also a function of the nature of the paper, its porosity, which controls the capillary action that draws the mobile phase up (or down) the paper, and the lie of the paper itself, i.e., whether the paper fibers lie with the direction of mobile phase flow or across, inhibiting free flow. The rate of migration of the various species is a function of the partition coefficient of the solute between the mobile phase and what may be considered to be a strong solution of a polysaccharide (1). Normal amino acids migrate at a rate that is close to the theoretical rate for a partition

Amino Acids and Small Peptides

process, but for basic amino acids, the ion-exchange process affects migration.

It is usual to express migration rates in terms of the retardation factor R_f, where:

$$R_f = \frac{\text{Distance traveled by the solute zone}}{\text{Distance traveled by the solvent front}}$$

The major disadvantage with the open-bed forms of chromatography lies in nonuniform operating conditions; nevertheless it is a powerful method capable of separating far more components than the naturally occurring amino acids.

In their classical procedure, Consdon et al. (1) used Whatman No. 1 paper and developed in the downward direction first with phenol ammonia and, after drying, in the second direction with collidine. This is a most powerful solvent system, but fell into disuse because the variation in the collidine isomer distribution resulted in changes in R_f. Figure 1 shows the map of some 60 spots obtained using this system, and the power of the separation method is clear. The main disadvantages of two-dimensional paper methods are the excessive time required to develop the chromatogram, only one analysis can be carried out at a time, and it is difficult to obtain quantitative results.

A number of one-dimensional methods have been developed giving near-perfect separation of the common amino acids (5–8), but all suffer from excessively long development times compared with modern HPLC techniques. However, if relatively few analyses are required, it can often provide a simple method of separation.

2.2. High-Voltage Paper Electrophoresis

Amino acids can easily be separated by high-voltage paper electrophoresis (up to 5 kV) using paper impregnated with suitable buffers. Good results may be obtained using an 8% formate buffer, pH 1.6, with which complex mixtures may be separated in a reasonable time, about 20 min. Visualization is normally achieved by spraying with ninhydrin. Not all the amino acids can be separated in a single run, however, and for a more complete separation, two-dimensional separation is preferable. Gross (9) has shown excellent results by running the second dimension in a buffer of pH 9.2 when 28 amino acids were separated in less than 1 h (Fig. 2).

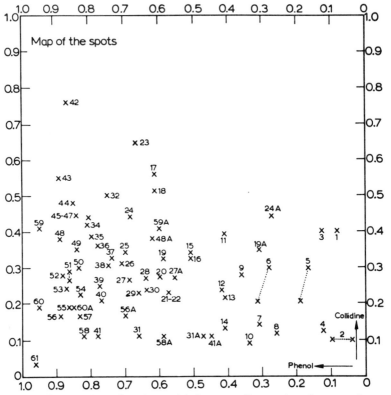

Fig. 1. Separation of amino acids by two-dimensional paper chromatography (from ref. 2). Average positions of spots taken up on phenol-collidone two-dimensional chromatograms. Phenol first solvent. Identifications are given below.

Alanine (20)
Alanine (29)
Allothreonine (15)
Amino-n-butyric acid (26)
Aminoisobutyric acid (37)
Aminobutyric acid (40)
Aminohexanoic acid (55)
Amino-ε-hydroxycaproic acid (38)
Aminooctanoic acid (43)
Aminopentanoic acid (43)
Aminophenylacetic acid (33)
Arginine (56)
Asparagine (13)
Aspartic acid (5)
Carnosine (54)
Citrulline (30)
Cystathionine (7)

Cysteic acid (1)
Djenkolic acid (14)
Ethanolamine (48)
Ethanolamine-phosphoric acid (10)
Glucosamine (24)
Glutamic acid (6)
Glutamine (21)
Glutathione (4)
Glycine (12)
Histamine (27)
Histidine (27)
Homocysteic acid (3)
Hydroxylysine (31)
Hydroxyproline (28)
Lanthionine (8)
Leucine (45)
Isoleucine (46)
Lysine (58)
Methionine (34)

Methionine sulfone (25)
Methionine sulfoxide (39)
α-Methyl-α-amino-n-butyric acid (49)
Norleucine (47)
Norvaline (35)
Orthinine (41)
Phenylaline (44)
Proline (52)
Serine (9)
Serine-phosphoric acid (2)
Taurine (11)
$\beta,\beta,\beta^1,\beta^1$-Tetramethylcystine (19)
Treonine (16)
Thyroxine (42)
Tryptophan (32)
Tyrosine (18)
Valine (36)

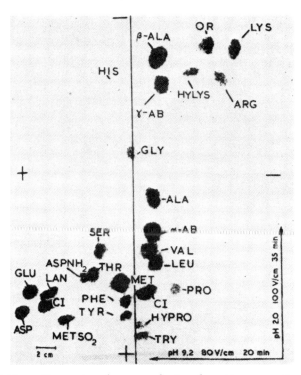

Fig. 2. Two-dimensional paper electrophoretic separation of amino acids (from ref. 9).

2.3. Combined Paper Chromatography and Electrophoresis

An alternative approach to 2D electrophoresis is the combination of electrophoresis followed by chromatography. It is desirable to perform the separation in this order because it automatically desalts the sample and eliminates the presence of residual chromatographic solvents that could impair separation. Figure 3 shows a map of a two-dimensional separation obtained from such a separation. Operating conditions and component identification of the ninhydrin positive solutes are also shown (10).

2.4. Thin-Layer Chromatography

Thin-layer chromatography (TLC) has many advantages over paper as a separation medium for amino acids. One is that the nature of the medium leads to appreciably smaller spots because

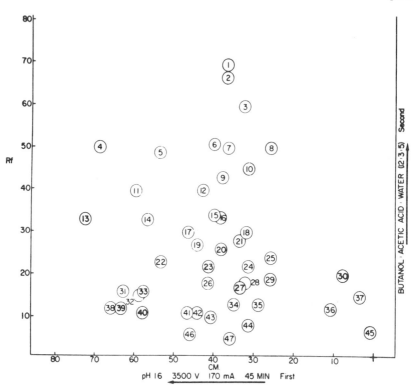

Fig. 3. Electrophoretic and chromatographic paper separation of amino acids and peptides (from ref. *10*). Map of the positions taken by ninhydrin-positive substances in a two-way separation, using electrophoresis for the first separation and butanol-acetic acid-water for the second (chromatographic) separation. The amino acids in a standard solution were applied to the origin (+). Electrophoresis was carried out at pH 1.6, using an 8% formic acid buffer. A potential of 3500 V at a current density of 170 mA during 45 min. All the amino acids migrated toward the cathode under these conditions. Chromatographic development was carried out in the ascending mode. Spot identifications: (1) leucine (2) isoleucine (3) phenylalanine (4) delta-aminovaleric acid (5) beta-aminoisobutyric acid (6) valine (7) methionine (8) tryptophan (9) kynurenine (10) tyrosine (11) gamma-aminobutyric acid (12) alpha-aminobutyric acid (13) ethanolamine (14) beta-alanine (15) proline (16) 3-hydroxykynurenine (17) alanine (18) homocitrulline (19) sarcosine (20) threonine (21) glutamic acid (22) glycine (23) serine (24) hydroxyproline (25) aspartic acid (26) homocystine (27) glutamine (28) citrulline (29) gamma-hydroxyglutamic acid (30) taurine (31) carnisine (32) arginine (33) methylhistidines (34) asparagine (35) penicillamine-penicillamine (36) phosphoethanolamine (37) cysteine sulfinic acid (38) ornithine (39) lysine (40) histidine (41) argininosuccinic acid C, cyclic anhydride, five-membered ring (42) argininosuccinic acid A (43) cystathionine (44) penicillamine-cysteine (45) cysteic acid (46) argininosucinic acid B, cyclic anhydride, six-membered ring (47) cystine.

Amino Acids and Small Peptides

diffusion is more restricted. This in turn means that the separation is sharper and hence a smaller surface area is required to effect the separation, and, as a consequence, analysis time is shortened and the level of detection is about an order lower. Separation may be effected using either buffered or unbuffered silica gel plates with a variety of different mobile phases. Brenner and Niederwieser (11) have studied the separation of 25 amino acids using six solvent mixtures and obtained good separation of a large proportion. These two workers (12) have also investigated two-dimensional separations and obtained excellent resolution (Fig. 4).

High-performance TLC (HPTLC) has been developed over the last 10 yr with considerable savings in analysis time. It is now possi-

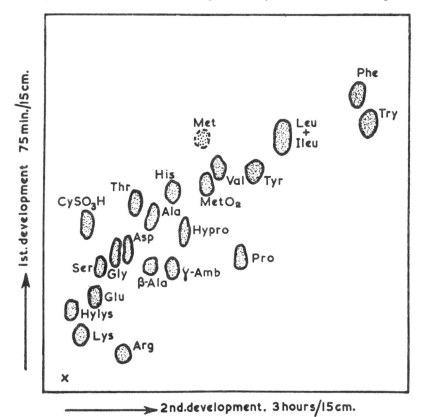

Fig. 4. Thin-layer separation of amino acids (from ref. 12). Two-dimensional TLC separation of some of the commoner amino acids on silica gel G. 1st dimension: chloroform/methanol/17% ammonia, (2:2:1); 2nd dimension: phenol/water (3:1); Load: 0.5 micrograms of each acid. Development times for 15 cm: 1st: 75 min; 2nd: 3 h.

ble to develop a 10 × 10 cm plate and separate 60 components with quantitative densitometry in less than 1 h. The principal reason for this enhanced performance lies in the manufacturing ability to produce close particle size range and uniform pore size silica gels. Further, the layer thickness has been reduced from approximately 200–500 to 100 µm. This has the disadvantage that the loading capacity of the plate is markedly reduced; typical sample sizes in the order a few nanoliters must be used for optimum performance. Sample applicators have recently been developed, however, to allow for the application of up to 1000 µL, while maintaining the spot size between 0.1 and 0.5 mm (13). Separations may be carried out using similar mobile phases as in the conventional methods of TLC. Precoated HPTLC plates may be obtained from a number of suppliers. It is of interest to note that amino acids may also be separated on polyamide HPTLC films by 2D separation in about 15 min (14).

As with paper methods, combined electrophoresis/chromatography may be used with thin layer films. Figure 5 shows the separation of some peptides using this technique.

Recently, Grinberg and Weinstein (15) have demonstrated the separation of the enantiomers of some dansylated amino acids by TLC using reversed-phase (RP-18) silicas. The mobile phase used was 8 mM N,N-di-n-propyl-L-alanine (DPA) and 4 mM Copper(II) acetate dissolved in 0.3M sodium acetate in water-acetonitrile (70:30) (apparent pH 7). Figure 6 shows the separations obtained on five dansylated amino acids. The resolution coefficient obtained ranged between 1.3 and 2.0. An interesting feature of this separation is the use of a temperature gradient of 6.25°C/cm up the length of the plate.

2.5. Gas Chromatography

The problem of performing quantitative analyses on amino acids by gas chromatography (GC) depends principally on the ability to quantitatively produce volatile derivatives of all the acids. In terms of sensitivity, the GC method is capable of detection on a routine basis down to the 5–10 ng level with a coefficient of variation in results of 1%. Further, with electron-capturing derivatives, the minimum detection level is of the order of 10 pg, and coupling to a mass spectrometer can give lower levels of detection and relatively simple component identification so that the disadvantage of having to prepare derivatives before analysis is amply recom-

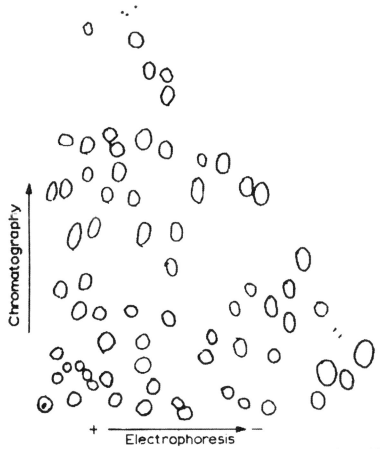

Fig. 5. Electrophoretic and chromatographic separation of peptides on thin-layer plates (from ref. 59). Separation of protein fragments taken from a tryptic digest of myosin. Enzyme to substrate ratio, 1:100. Buffer: 0.1M ammonium carbonate, pH 8.5. Incubation time, 24 h. 1st dimension: Chromatography using chloroform/methanol/ammonium hydroxide (40,40,20) eluent during 60 min; 2nd dimension: Electrophoresis using pyridine/acetic acid/water (1:10:489) carrier buffer. 980 V at 30 mA for 1 h. The separation was carried out on a 200 × 200 mm silica gel G plate. Note the use of chromatography for the 1st dimension.

pensed. It must be recognized that in nearly all methods of amino acid analysis, direct detection is impossible particularly at low concentration.

The first application of GC to the separation of amino acids was reported in 1956 by Hunter et al. (16), who oxidized isoleucine

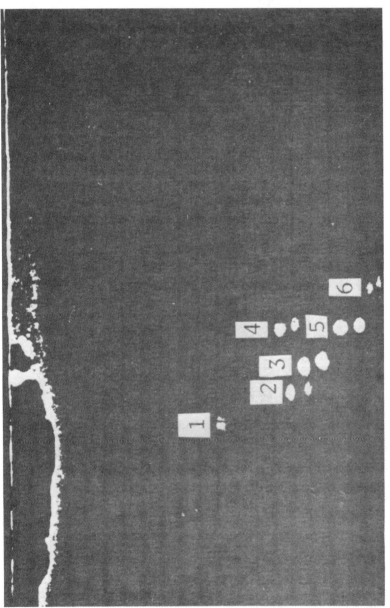

Fig. 6. Resolution of some amino acid enantiomers by chiral thin-layer chromatography (from ref. 15). Reversed-phase thin layer separation of some dansylated amino acids enantiomers using copper complexation competition as the separation method with a temperature gradient. Spot identification: (a) dansyl-OH (2) asparagine (3) serine (4) methionine (5) alanine (6) phenylalanine.

and leucine with ninhydrin to form the aldehydes which were separated on a silicone oil column. Subsequently a search occurred to find suitable derivatives and over 100 have been described in the literature.

Of the more useful derivatives, the propyl, butyl, and amyl esters of trifluoroacetyl amino acids have been extensively investigated, but problems arise resulting from the poor solubility of cystine and the basic amino acids in butanol. However, yields may be improved by transesterification; alternatively, direct esterification may be markedly improved by ultrasonication, which breaks up precipitates and generally improves the rate of solution and reaction. A very recent method is that of MacKenzie and Tenaschuk (*17*), who esterify either directly or indirectly with isobutanol, followed by reaction with heptafluorobutyric anhydride to form the heptafluorobutyryl isobutyric ester derivatives. These may be separated within 30 min on a column packed with 3% SE30 on deactivated support by temperature programming from 90 to 240 °C. Detection may be by flame ionization (FID) or electron capture (ECD).

Clearly the stability of the derivatives formed is important for GC studies, e.g., the *N*-TFA-*n*-amyl ester derivatives of cysteine, hydroxyproline, serine, threonine, and tyrosine are not stable on standing at room temperature, and undergo thermal decomposition in the injection port of the chromatograph, or break down with irreversible adsorption on the chromatographic support.

A recent paper by Husek (*18*) has suggested that a useful alternative to the above methods, in that it eliminates the esterification process, is to react the amino acid with 1,3-dichlorotetrafluoroacetone (DCTFA) to form cyclic derivatives. This has the advantage of reacting both to amino- and carboxyl-functions in one step:

$$RCH[NH_2]COOH + CF_2ClCOCF_2Cl = RC \begin{array}{c} \diagup NH \diagdown \\ CH \\ \diagdown CO \diagup \end{array} C(CF_2Cl)_2$$

The derivatives are formed in one step, and the reaction is rapid at room temperatures (about half that of the esterification method). The derivatives show a high response to the ECD allowing for detection in the picomolar range.

The trimethylsilyl-range of derivatizing agents has been used with some success. Again, this is a one-step, generally room temperature, reaction, although the rate is faster at elevated temperatures (e.g., 150°C). Good reproducibility is obtained with the N,O-bis(trimethylsilyl)acetamide reagent (BSTFA). All the amino acids except glycine undergo monosubstitution on the alpha amino group, whereas mono- or di-TMS-substitution can occur on the omega amino, depending on reaction conditions. For example, 17 amino acids can be reproducibly silylated at 150°C in 15 min, but for glycine, arginine and glutamic acid 2.5 h are required.

It is worth noting here that in the modification to Edman's (19) classic procedure for protein sequencing, which has been used for GC analysis (20), the amino acids are reacted with phenylisothiocyanate at pH 8–9 at about 40°C to form the phenylthiocarbamyl peptide (amino acid). This is then cleaved under suitable conditions to give the cyclic phenylthiohydantoin:

$$C_6H_5N=C=S + RCH[NH_2]CONHCHR_1CO- = C_6H_5NHCS-NH-CHRCONHCHR_1CO- \xrightarrow{H^+}$$

$$\begin{array}{c} CO \\ / \ \backslash \\ S \quad CHR \\ | \quad | \\ C_6H_5NHC-NH^+ \end{array} \longrightarrow \begin{array}{c} RCH-CO \\ | \quad | \\ NH \quad N-C_6H_5 \\ \backslash \ / \\ CS \end{array}$$

The above scheme shows the Edman degradation of a peptide. The final product is the phenylhydantoin (PTH) of the terminal amino acid. Amino acids derivatization follows a similar scheme, except that the final hydrolysis step is omitted.

The amino acid PTHs form three groups of derivatives, according to their volatilities and compatibilities with different stationary phases. Group 1 includes alanine, glycine, valine, leucine, isoleucine, methionine, proline, and phenylalanine; these migrate well on all stationary phases, but the best separation is achieved on nonpolar phases. Group 2 includes asparagine, glutamine, tyrosine, histidine, and tryptophan; these derivatives are more polar and are best separated on polar phases (e.g., DEGS). Group 3 is formed from aspartic acid, glutamic acid, lysine, serine, and threonine, which are difficult to separate. Trimethylsilylation of these deriva-

Amino Acids and Small Peptides

tives improves their migration properties as it does to all of the PTH derivatives. Indeed, the TMS derivatives are commonly used rather than the parent PTH derivatives, and these are generally separated on either an OV 210 or OV 225 stationary phase.

The choice of stationary phase has also been intensively investigated; their polarity, ranging from nonpolar to highly polar, has been used with varying degrees of success, e.g., a mixed phase of diethylene glycol succinate/methyl silicone has successfully separated 20 components.

2.6. Amino Acid Analyzers

Spackman et al. (22) developed the first automatic amino acid analyzer in 1958 for the quantitative analysis of amino acids from protein hydrolysates using ion exchange (see chapter 11) as the separation step. The analysis time was 24 h. Since then many other systems have been devised, cutting the analysis time to about 1 h. These improvements in performance occurred through a better understanding of the kinetics of chromatography, which indicated the use of small regular-size ion exchangers and higher mobile phase flow rates (23). Further improvements arose through the use of single-column systems (24) using gradient elution instead of stepwise gradients, although the time of analysis was not improved. Sensitivity is not a great detection problem in the large number of cases, but the use of ninhydrin as the detection mode has become restrictive when dealing with the much smaller samples now becoming usual from biological studies. Developments in derivatization using fluorescamne (25) and o-phthaldehyde (26), which have taken levels of detection to the nano- and picomolar levels, have been of assistance here. It is interesting to note that sodium citrate buffers proposed by Spackman et al. have remained in use virtually unchanged and, apart from changing the cation to lithium (27), are the preferred buffers for the analysis of neutral and acidic amino acids in physiological fluids. In 1975, Murren et al. (28) produced a versatile buffer system that develops a pH gradient from only two buffers, one acid (0.2M sodium citrate, pH 2.2) and the other basic (0.2M sodium borate, pH 11.5). Complete analyses of all the naturally occurring amino acids from protein hydrolysates can be completed in less than 1 h. Equipment and the methodology is commercially available (29). Another (30) commercially available system has cut analysis times to about half an hour for protein hydrolysates and to less than 15 h for physiological fluids. This system uses a 48 cm long × 0.18 cm id ion exchange column packed with 8 μm spherical beads. The amino acids are eluted by a four-component gradient incorporating buffers with pH values of 3.21, 3.25, 4.25, and 9.45. The temperature is changed during the run from 56 to 65 to 75°C. One disadvantage of this method is that reequilibration to the starting temperature is required between runs, as well as changing back to the starting buffer.

In a relatively recent review, Errser (31) has surveyed the available commercial amino acid analyzers and compared them in terms of their hardware and methodology. He points out that, when using IEC, it is unlikely that an unequivocal separation and identification can be achieved between the various amino acids.

2.7. Liquid Chromatography

The understanding of the extra-column processes that control band broadening in chromatographic systems and that are included in the design of commercial amino acid analyzers has been most important, but of even greater importance has been the construction of stationary phases for column packings, for example, the production of stable reverse phases. However, the highly polar nature of the amino acids does not lend itself to simple separations. Furthermore, the absence of strong UV absorbing chromophores in the majority of amino acids leads to difficulties in detection. This can be alleviated by the production of suitable UV or fluorescent derivatives, which has the added advantage of reducing the highly polar nature of the acids and making them amenable to separation by reverse-phase chromatography (RPC). (Note: In most amino acid analyzers, separation is by IEC, followed by postcolumn derivatization.)

Clearly the stability and ease of preparation of the various types of derivative are important, these factors will now be considered together with the appropriate separation steps.

2.7.1. Amino Acid Derivatization Techniques

Six methods of derivatization have been employed with various degrees of success and levels of detection, and the time scale for the complete analysis has fallen from approximately 1 h to 10 min. The first method is based on the Edman method of protein sequenc-

Amino Acids and Small Peptides

ing with the formation of the PTH derivatives, as indicated above, and is worth considering again in some detail—particularly the sequencing method.

2.7.1.1. Phenylisothiocyanate

The steps in the Edman process (section 2.5) may be called coupling, cleavage, and hydrolysis.

Coupling occurs in basic solution (pH 8–9) at 40 °C. The rate of reaction is a function of the pK_a value of the amino acids, but is essentially complete after 30 min. Reaction is carried out in aqueous pyridine (50%) or dioxane. At completion, the products are extracted with benzene and cyclohexane to remove excess reagent. The aqueous phase is freeze dried, followed by further extraction with ethyl acetate.

Cleavage is performed with anhydrous acid to avoid breakdown of the peptide chain. The shortened peptide is precipitated with ethylene chloride, with the derivatized amino acid remaining in solution. The precipitated protein is returned for further sequencing.

Conversion takes place in aqueous acid solution (e.g., HCl) at 80 °C for 10 min. The product is extracted with ethyl acetate. *Note*: the PTHs of arginine, histidine, and cysteine remain in the aqueous phase.

Preparation of the PTH derivatives of the free amino acids follows a similar path, apart from cleavage. After the benzene extraction, molar HCl is added, which precipitates the phenylthiocarbamyl amino acids that after filtration, are taken up into solution by warming with molar HCl. The PTH derivatives crystallize on cooling to give products with high melting points and high stability. Exceptions to this are threonine, serine, cysteine, and cystine. The derivatives, with three exceptions (arginine, histidine, and cystine), are soluble in organic solvents, show strong UV absorbance with a maximum at 245 nm, and are best stored in the dark at −15 °C.

Separation of the PTH derivatives was originally carried out by TLC on silica or polyamide sheets (section 2.4) or by GC (section 2.5) and LC on silica columns with moderately polar mobile phases, e.g., dichloromethane, dimethylsulfoxide, *tert*-butanol, and chloroform, but the reactive surface of silica gives rise to problems with the elution of the basic derivatives. The preferred method for these derivatives is RPC using octadecylsilyl silicas, (ODS) with ammonium acetate/acetonitrile, and/or methanol as the eluent. Alternatively, acetic acid/tetrahydrofuran/acetonitrile, pH 5.8, and sodium acetate/acetonitrile, pH 4.32, mixtures have been used under isocratic conditions. Gradient systems prepared from the same solvents have also been used with success.

Cyanopropyl silicas have been used as the stationary phase with success as well, and Dimari et al. (32) have reported excellent separations. Cunico et al. (33) have prepared a mixed stationary phase from cyano- and alkyl-silicas in ratios adjusted to give the maximum separation between components using a ternary gradient system with sodium acetate/methanol/acetonitrile as the mobile phase. It has also been shown recently (34) that a wide-pore diphenyl bonded phase will separate all the PTH amino acids in about 35 min using a binary gradient prepared from a TFA acetate/acetate (66 mM, 4 mM, pH 5.8) buffer and 75% acetonitrile/25% TFA acetate (35 mM, pH 3.6) buffer, e.g., Fig. 7.

Detection limits at the picomolar level are easily attainable using PTH derivatives.

A commercial system (the Pico.Tag system) recently announced by Waters Associates (35) is based on this method. Here the analysis time has been reduced to 12 min with 1 pmol level of detection. It is claimed that a complete analysis can be obtained on as little as 100 ng of sample with highly reproducible retention times. As indicated above, the reaction is quantitative for both primary and secondary amino acids with a linear response over the range 10 to 1000 pmol. The system is capable of automatically handling 60 samples/d and consists of a custom-made reverse-phase column, 3.9 × 150 mm thermostated to 38°C. Elution is obtained with a gradient between two solvents: (1) 0.14M sodium acetate, pH 6.40, containing 0.05% triethanolamine, and (2) 60% acetonitrile/40% water. Starting solvent 10% (2) rising to 51%: (2) using a convex gradient (no. 5 on the M680 gradient former) at a flow rate of 1 mL/min. Figures 8 and 9 show the separation obtained at two levels of concentration.

2.7.1.2. Dabsyl Chloride

Dabsyl chloride, dimethylaminoazobenzene-4'-sulfonyl chloride, reacts with amino and imino acids under alkaline conditions to form derivatives that have an absorption maximum at 420 nm. The reaction is rapid and the products are stable for extended periods. The amino acids are dissolved in sodium bicarbonate buffer (pH 9.0) and a solution of dabsyl chloride in acetone added. The reaction mixture is kept at 70°C for about 15 min with occasional shaking, and is diluted with 70% ethanol or 50% acetonitrile before

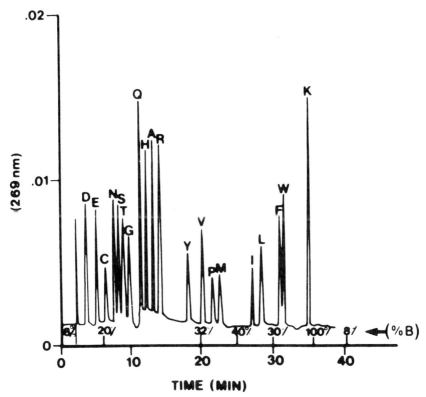

Fig. 7. Separation of PTH amino acids by liquid chromatography (from ref. 60). This separation was obtained on a Bakerbond wide-pore diphenyl bonded phase column, 5 μm particle size, 4.6 × 250 mm. The separation was developed at a flow rate of 1.5 mL/min using a gradient of A. TFA acetate/acetate (66 mM/4 mM) pH 5.5, and B. 75% acetonitrile/2% TFA acetate (35 mM) pH 3.6. See figure for gradient. Detection by UV at 269 nm. Peak identification: (D) aspartic acid (E) glutamic acid (C) cysteine (N) asparagine (S) serine (T) threonine (G) glycine (Q) glutamine (H) histidine (A) alanine (R) arginine (Y) tyrosine (V) valine (P) proline (M) methionine (I) isoleucine (L) leucine (F∼) phenylalanine (W) tryptophan (K) lysine.

separation. It is necessary to control the pH at 9.0; otherwise multiple derivatives of histadine, lysine, tyrosine, and cystine are formed. Under these conditions, excess reagent is hydrolyzed to methyl orange.

Separation is best achieved using gradient elution with phosphate buffer, dimethylformamide and acetonitrile as the mobile

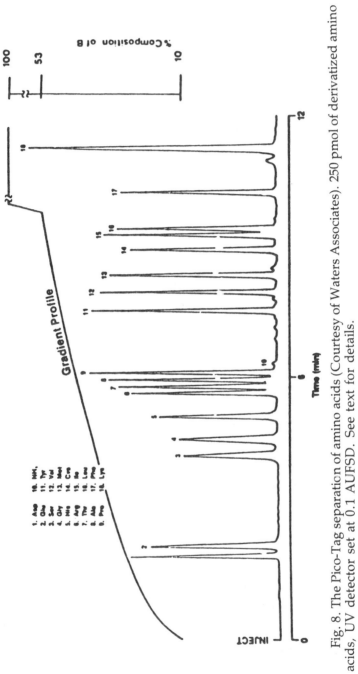

Fig. 8. The Pico-Tag separation of amino acids (Courtesy of Waters Associates). 250 pmol of derivatized amino acids, UV detector set at 0.1 AUFSD. See text for details.

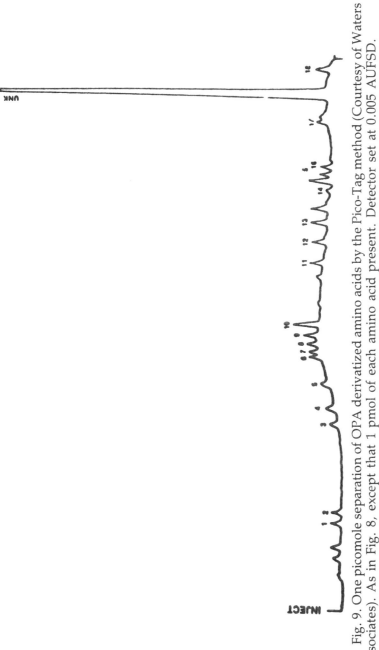

Fig. 9. One picomole separation of OPA derivatized amino acids by the Pico-Tag method (Courtesy of Waters Associates). As in Fig. 8, except that 1 pmol of each amino acid present. Detector set at 0.005 AUFSD.

phase on a RP ODS column thermostatted at 50°C. Chang et al. (36) have obtained a separation of 21 derivatives using this method with about 10 pmol limit of detection.

2.7.1.3. Dansyl Chloride

Dansyl chloride, 1-dimethylaminonaphthalene-5-sulfonyl chloride, reacts with free primary and secondary amino groups to yield fluorescent sulfonamides that are quite stable when stored in the dark. The derivatives are resistant to both acid and alkaline hydrolysis, which makes them particularly suitable for amino acid analysis. Reaction occurs readily with the unprotonated form of the amine and is thus favored by high pH. The reagent, however, is also hydrolyzed to the corresponding sulfonic acid at high pH so that careful control is necessary for derivatization. The optimum pH has been found to be 9.5. Thus the derivization step is usually carried out in bicarbonate buffer. The reagent is only sparingly soluble in water, so it is normally used in acetone solution. Care must be exercised not to use an excess of dansyl chloride in the derivatization step. This causes fragmentation of the derivative, producing a dansyl amide that is controlled by the structure of the amino acid with respect to steric hindrance of the derivative and the excess reagent added. As might be expected, hindered amino acids decompose most easily. Tauphi et al. (37) have examined the various procedures given and concluded that the ratio of amino acid to dansyl chloride is critical. They recommend that the reagent be used at a concentration of 1.5 mg/mL in a 2:1 volume ratio of the amino acids dissolved in lithium carbonate buffer, pH 9.5. Reaction is complete in approximately 30 min at room temperature.

The spectral properties of the dansyl amino acid derivatives depend on their structure, the solvent, and its pH. The maximum emission of primary amines is shorter than the secondary amines, and the quantum yields are similar, but not identical. The wavelength of maximum emission increases and the quantum yield decreases with increasing dielectric constant of the solvent.

Separation of the derivatives has undergone the same developments as the dabsyl derivatives, and is now preferably carried out on RP columns. The retention and selectivity of the separation is a function of the mobile phase pH. At low pH the derivatives elute in order of their amino acid side-chain hydrophobicity in three groups: (1) acid derivatives, (2) neutral aliphatic derivatives, and (3) *bis*-(dansyl) derivatives. Wiedweier et al. (38), using an Ultrasphere ODS column and a gradient between sodium acetate (0.1M,

Amino Acids and Small Peptides

pH 4.18)/THF and acetonitrile/THF have separated 22 derivatives in 55 min, and recently, Levina and Nazimov (39), using a trifluoroacetate buffer, have completed the separation in 26 min.

The levels of detection of the dansyl derivatives are in the 10-pmol range using conventional equipment, but this level may be lowered by using the microbore technique of Oray et al. (40), in which the limit of detection was reduced to the femtomolar range. Another interesting way of improving the sensitivity of the detection system was developed by Kobayashi and Imai (41). They developed a chemiluminescence detector in which the dansyl derivatives are excited by the addition of bis-(2,4,6-trichlorophenyl) oxalate and hydrogen peroxide to give detection limits of about 10 fmol. By combining the technique with microbore methodology, detection limits in the order of 10^{-16} mol were obtained.

2.7.1.4. Fluorescamine

Samejima (42) reported in 1971 that the sensitivity of the ninhydrin reaction to amino acids could be enhanced by the inclusion of certain aldehydes when a strong fluorescent response was obtained. Weigele (43) noted, however, that the oxidizing properties of ninhydrin could impair this reaciton, and synthesized a similar reagent to the ninhydrin/phenylacetaldehyde pair. This was called fluorescamine, 4-phenylspiro[furan-2(^{3}H)-1'phthalan]-3,3'dione, which in acetone solution readily reacts with amino acids in aqueous solution at pH 8.5 to yield in a few seconds a strongly fluorescent emission at 475 nm when excited at 390 nm. Fluorescamine itself is nonfluorescent and hydrolyzes to nonfluorescent products. The derivatives formed are stable for several hours.

One of the disadvantages of using this reagent is that it only forms fluorescent derivatives with primary amines, but secondary amines can compete for the reagent if they are present in sufficient quantity. Tertiary amines do not react. The optimum pH for reaction lies between 8.5 and 9.0, and is somewhat lower for aliphatic amines and peptides. This optimum pH range reflects the balance that must be considered between excessive reagent hydrolysis at high pH and protonation of the amino acids at lower pH, which reduces the ability for reaction. Reaction yield is quite low and about 100-fold excess reagent is required to achieve 80% yield. However reagent concentrations in excess of 1 μmol/mL results in UV absorbance that can reduce the fluorescence output and give rise to a nonlinear response. Fluorescence output is substantially constant over the pH range of from 4.5 to 11.0, but is significantly affected by the nature and concentration of the organic cosolvent used. A 1:1 aqueous/organic ratio appears to be optimum.

An acetone solution of fluorescamine has been used chiefly as a post-column reactant in IEC, and is mixed with the column eluent and a controlling buffer of about pH 9 to optimize the reaction conditions. Detection levels at the picomolar levels may be obtained for primary amino acids. A disadvantage, however, is the lack of fluorescent response to secondary amino acids, although they do respond in the UV with a maximum absorption at 320 nm at pH 9.0. The total number of amino acids may be estimated by UV absorption at 380 nm.

2.7.1.5. Orthophthaldehyde

Orthophthaldehyde (OPA) reacts with primary amines in the presence of a reducing agent (ethanethiol or mercaptoethanol) to form highly fluorescent adducts. Secondary amines do not react and must be converted to primary amines prior to derivatization. Detection is at 455 nm and excitation is at 340 nm.

The reaction results in the formation of a 1-alkylthio-2-alkylisoindole (44). The reaction products are unstable and slowly decompose. The reaction proceeds rapidly in alkaline media; the optimum pH lies between 8 and 11, preferably in a borate buffer. Since the reaction products have only a limited stability, it is usual to inject solutions for separation at precise time intervals after mixing. According to Lindroth and Mopper (45), the fluorescence response is linear and independent of reagent composition, providing a 200-fold excess is present. Cooper et al. (46) have studied the stability of the derivatives in the chromatographic column and shown similar stability for all adducts except the di-acidic amino acids, aspartic and glutamic acid, which show a significant loss in fluorescence with time. Since these derivatives elute first, however, no loss in sensitivity is apparent.

The quantum yields of all the derivatives are substantially constant, except for that of lysine. This is because of the presence of two isoindole fluorophors in the molecule, which causes severe quenching. The monosubstituted lysine derivative has a fluorescent lifetime of 16.2 ns, only slightly less than the average (19.2 ns). Addition of a nonionic detergent to the solution eliminates this quenching, allowing the lysine derivative to be observed with the same order of magnitude as the other derivatives.

The inability to prepare OPA derivatives of the secondary amino acids proline and hydroxyproline has led to considerable research.

Amino Acids and Small Peptides

The following procedure is from Cooper et al. (47). The dried protein hydrolysate is dissolved in borate buffer, pH 8.5. Chloramine T in dimethyl sulfoxide and borate buffer are added, and the mixture is held at 60°C for exactly 1 min. A mixture of sodium borohydride and thiolactate in lithium hydroxide is then added. After 10 min, iodoacetic acid in borate buffer is added and an aliquot of the resulting solution is derivatized with OPA.

OPA may be used in both the pre- and postcolumn modes, and amino acid analyzers frequently use this reagent to ninhydrin because the sensitivity of detection is considerably enhanced. It is necessary to ensure that the reagents and mobile phases are carefully purified before use. Separation is carried out by the reverse phase mode when using precolumn derivatization, but a tight time regimen must be followed. Mobile phases may be suitably buffered gradients prepared from THF/methanol, with the buffers phosphate (pH 7.2), acetate (pH 5.9), or citrate (pH 6.5) present. Phosphate buffers with methanol are generally preferred. The time of analysis ranges between 35 and 95 min, depending on the mobile phase used, with a detection limit in the 50–100 fmol range.

An automated precolumn derivatization process using OPA has been recently developed by the Gilson company; using a gradient system, 21 amino acid derivatives may be separated in 11 min. Full details of the conditions are given in Fig. 10.

2.7.1.6. The Halogenonitrobenzofurazans

Two halogen-substituted 4-nitrobenzo-2-oxa-1,3-diazoles, 7-F and 7-Cl, (NBD-F and NBD-Cl) have been used to derivatize both primary and secondary amino acids to form highly fluorescent products. The halogen on the NBD reacts with the amino group of the amino acid to form the NBD–amino acid. The reaction is carried out in aqueous buffer (pH 9.5)/organic solvent mixture with the reagent in methanol, ethanol, or acetonitrile. The reaction mixture is held at 60°C for 20 min. At the completion of the reaction, the pH is adjusted to 1.0 with HCl to reduce the fluorescence blank caused by the hydrolysis product of the reagent (NBD-7-OH).

The reaction of all amino acids with NBD-hal obey a pseudo-first-order rate constant when the reagent is in excess, and although the reaction rate for secondary amines is 10 times faster than for primary amines, the rate of hydrolysis of the reagent is similar to the derivatization rate of the latter. A close study of the derivative formation rates in methanol shows that they are related to the concentration of the solvolysis product, $NBD-OCH_3$, rather than the

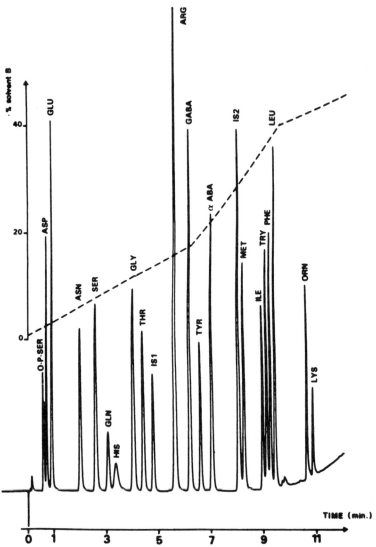

Fig. 10. Automated separation of OPA derivatized amino acids (Courtesy of Gilson Medical Electronics Inc.). Separation conditions. SGE Hypersil C18, 3 µm 100 × 4.0 mm column, flow rate, 1.3 mL/min gradient between (A). 0.02M sodium acetate-methanol-THF (85-11-4 v/v) and (B) acetonitrile. See figure. Fluorescence detection was used, excitation 340 nm, emission 450 nm. Sample: 4 µL of mixture containing two standards IS1: methionine sulfone and IS2: 6-amino-caproic acid. MDL: 0.4 pmol of each compound.

concentration of the reagent, thus suggesting that NBD-OCH$_3$ is a reactive intermediate in the derivatization step. Indeed, this intermediate may well be a more powerful derivatizing agent than the parent compound (48).

The reaction yields appear to be high (95%), and the products are stable when stored in the dark. The fluorescence intensity is stable in the pH range of from 5 to 9, and the quantum efficiencies are similar to the dansyl derivatives in medium polarity solvents, but are drastically reduced in water. Detection levels are about 10 fmol.

The reaction rate of NBD-F is very fast, and it may be used in both the pre- and postcolumn derivatization mode for the detection of all the amino acids, but the reaction rate of NBD-Cl with the primary amino acids is significantly slower than for the secondary. Hence, using short reaction times, proline and hydroxyproline may be selectively determined. Separation may be performed on reverse-phase columns using buffered (pH, 6.0) methanol/THF mobile phases either isocratically or with a gradient.

2.8 Separation of Racemic Amino Acids

Ligand exchange chromatography may be used to separate the enantiomeric amino acids and to separate amino acids as a class from peptides and amino sugars. The technique was introduced in the mid-1960s, but received little attention at that time. Amino acids form complexes with divalent metal ions, e.g., Ni, Cd, Co, Cu, and Zn, according to the scheme:

$$RCH(NH_2)COOH + M^{2+} \leftrightarrow \begin{bmatrix} RC \begin{smallmatrix} COO^- \\ \\ NH_2 \end{smallmatrix} M^{2+} \end{bmatrix}^+$$

and it is this chemical equilibrium that controls the selectivity of the chromatographic system. Copper is the usual cation to use and is adsorbed onto the support. Elution is performed with another ligand, usually ammonia. A discussion of the factors controlling separation by ligand exchange has been reviewed by Doury-Berthod et al. (49). If the eluant ligand is optically active, this method provides a mean of separating the enantiomers of the amino acids.

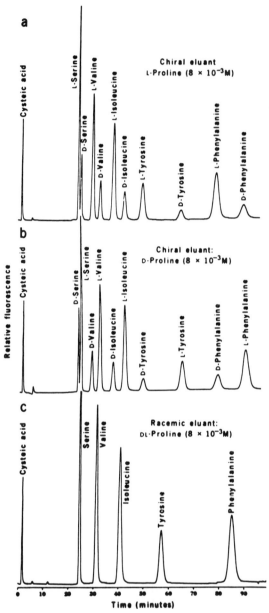

Fig. 11. Chiral separation of some amino acid enantiomers showing the effect of different chiral and achiral mobile phases (from ref. 51). Effect of the chirality of the eluant on the separation of D- and L-amino acid enantiomers by ligand-exchange chromatography. Identical portions of a mixture of five pairs of amino acid enantiomers, each consisting of 0.375 nmol of L form and 0.125 nmol D form, were injected in each run. Sodium

2.8.1. Ligand Exchange Chromatography for Enantiomer Separation

The large majority of enantiomer separations of the amino acids have been obtained by the method indicated above, but it is worth noting that Gil-Av et al. (50) first demonstrated the separation of derivatized amino acid enantiomers using capillary gas chromatography with an optically active stationary phase in 1966.

For enantiomeric separation, there should be a stereoselective interaction between the solute molecules and the stationary phase. The stationary phase may be optically active in its own right, or a chiral reagent may be included in the mobile phase that will interact with the stationary phase (either silica, a reverse-phase, or an ion-exchange resin) to form the chiral stationary phase *in situ*. It is clear that for separation to take place, a differential adsorption between one or other of the enantiomers and the modified support is necessary.

The method is illustrated very elegantly in a paper by Hare and Gil-Av (51) (Fig. 11). Here the mobile phase is a sodium acetate buffer (0.05N, pH 5.5) containing $4 \times 10^{-3}M$ copper sulfate and $8 \times 10^{-3}M$ proline, and the column, 12 × 0.2 cm id, packed with DC4a cation exchange resin. After equilibrium at a flow rate of 10 mL/h at 75°C, the sample is injected. Detection is by fluorescence after postcolumn derivatization with *o*-phthaldehyde. *Note*: Secondary amino acids do not react with the reagent; proline, therefore, does not interfer with the detection method. The details of the separation are given in Fig. 11 and Table 1 gives the adjusted retention times and separation factors for the amino acids investigated.

2.8.2. Other Methods of Enantiomer Separation

A minor modification to the method discussed above involves a column-switching procedure and a different chiral mobile phase,

---

← Fig. 11 (*continued*)
acetate buffer (0.05N, pH 5.5) containing $4 \times 10^{-3}M$ CUSO$_4$ and $8 \times 10^{-3}M$ proline was used as eluant. The chirality of the proline ligand was as indicated. The column was equilibrated with each separate eluant for 15 min before the sample was injected. The column was 12 by 0.2 cm (inside diameter) packed with DC 4a resin. The eluant flow rate was 10 mL/h, the reagent flow rate was 10mL/h, the column pressure was 200 bars, and the column temperature was 75°C. (a) L-Proline effected the separation of all five pairs of enantiomers with the L enantiomers eluting before the corresponding D enantiomers. (b) D-Proline reversed the order of elution. (c) With racemic proline no resolution occurred. The amino acids eluted halfway between the corresponding enantiomeric peaks in (a) and (b).

Table 1
Retention Characteristics of Amino Acid Enantionmers Separated by Ligand Exchange Chromatography Using a Chiral Eluent (51)[a]

Amino acid	Form	t'_R, min	Separation factor, r
(Cysteic acid)		(0)	
Aspartic acid	L	2.2	1.00
	D	2.2	
Glutamic acid	L	3.3	1.00
	D	3.3	
Allothreonine	L	10.0	1.14
	D	11.4	
Glutamine	L	12.9	1.00
	D	12.9	
Serine	L	12.9	1.04
	D	13.4	
Threonine	L	13.0	1.05
	D	13.7	
Asparagine	L	14.0	1.00
	D	14.0	
α-Amino-n-butyric acid	L	16.2	1.02
	D	16.6	
Valine	L	16.3	1.09
	D	17.7	
Alanine	L	17.2	1.00
	D	17.2	
Isoleucine	L	21.5	1.10
	D	23.7	
Norvaline	L	21.5	1.04
	D	22.3	
3,4-Dihydroxyphenylalanine	L	22.3	1.28
	D	28.5	
Alloisoleucine	L	23.5	1.09
	D	25.5	
Methionine	L	23.8	1.03
	D	24.6	
Leucine	L	28.3	1.01
	D	28.6	
Ethionine	L	29.0	1.04
	D	30.3	
Norleucine	L	30.8	1.05
	D	32.3	
Tyrosine	L	30.8	1.28
	D	39.4	
m-Tyrosine	L	34.1	1.21
	D	41.1	

(continued)

Table 1 (continued)
Retention Characteristics of Amino Acid Enantionmers Separated by Ligand Exchange Chromatography Using A Chiral Eluent (51)[a]

Amino acid	Form	t'_R, min	Separation factor, r
o-Tyrosine	L	36.7	1.18
	D	43.2	
Phenylalanine	L	48.7	1.13
	D	55.0	
p-Flurophenylalanine	L	64.0	1.18
	D	75.6	

[a]Mobile phase: 0.1N sodium acetate with $8 \times 10^{-3}M$ $CuSO_4 \cdot 5H_2O$ and 16×10^{-3} L-Proline flow rate, 10 mL per h. The column was a 12×0.2 cm id packed with DC 4a resin. The retention time for cysteic acid is taken as the column void volume.

in which Nimura et al. (52) demonstrate a separation of all the enantiomers of the common protein amino acids. The chiral mobile phase (I) used contained 10 mL of 100 mM N-(p-toluenesulfonyl)-D-phenylglycine-copper(II) [Tos-PhG-Cu(II(] in acetonitrile, 125 mg hydrated copper sulfate, and 75 mg sodium carbonate, made up to 1 L with water. For gradient work, seven volumes of (I) was diluted with three volumes of acetonitrile. Separation was performed on a 200 × 6 mm column packed with Develosil ODS and detection was by derivatization with OPA. Figure 12 shows the separation achieved. Full details of the method are given in ref. 52.

Armstrong and DeMond (53) have demonstrated the separation of beta-naphthylamide derivatives of some amino acids (alanine, methionine) on a 100 mm column of cyclodextrin using methanol/water, 1:1, as the mobile phase. Figure 13 shows the separation obtained on these two amino acids together with the separation of the dansyl derivatives of leucine where the D-leucine is present at very low levels.

Yuasa et al. (54) have also demonstrated the direct separation of the enantiomers of some amino acids on a cellulose column (250 × 0.85 cm) activated with hydrochloric acid. The mobile phase was 0.2M sodium citrate and visualization was by postcolumn reaction with ninhydrin. Figure 14 shows a typical separation in which the amino acids are initially separated as the racemic pair by conventional IEC, derivatized, and transferred to the cellulose column by a column-switching method. It was found that after several runs, the performance of the column deteriorated, but it was possible

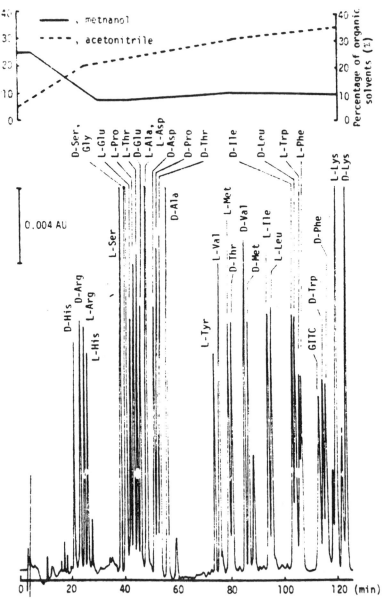

Fig. 12. Reversed-phase chiral separation of all the amino acid enantiomers (from ref. 52). Chromatogram and gradient program of the simultaneous analysis of GITC derivatives of all common protein amino acid enantiomers. Column: Develosil ODS (200 × 6 mm id, particle size 5 μm). Flow-rate: 1.3 mL/min. Each peak corresponds to 100 ng of free amino acid.

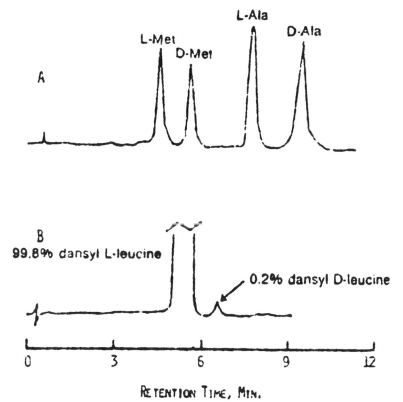

Fig. 13. Resolution of some amino acid enantiomers using cyclodextrin as the stationary phase (from ref. 53). Chromatogram A shows the baseline separation of the enantiomers of β-naphthylamide derivatives of alanine and methionine. Chromatogram B illustrates the potential use of a β-cyclodextrin column to determine optical purity when one of the enantiomers is present at very low levels. Both separations were done on a 10-cm column.

to regenerate the cellulose by simply passing 5 mL of 1.0M hydrochloric acid through the column.

3. Electrophoretic Separation of Amino Acids in Capillaries

3.1. Isotachophoresis

A major advance in the separation of amino acids occurred with the development of isotachophoresis (ITP) (55,56), which provides

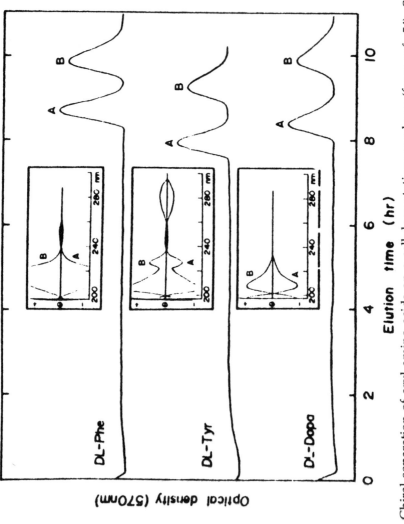

Fig. 14. Chiral separation of aryl amino acids on a cellulose stationary phase (from ref. 54). Separation of DL-aromatic amino acids. DL-Phe, DL-Tyr, and DL-Dopa (500 μg) were eluted with pyridine:ethanol:water (4:1:1, v/v/v). The CD spectra for each DL-amino acid are shown in the insets.

Amino Acids and Small Peptides

a means of quantitatively detecting the amino acids and providing a permanent record of the separation achieved. ITP has as its motive force a constant current supply driven by a varying potential that increases as the resistance of the circuit increases. Analysis is carried out in narrow-bore PTFE capillaries (0.1–0.5 mm) filled initially with a leading electrolyte (LE) that has a higher mobility than the constituents of the sample, which is placed at the junction between the LE and a terminating electrolyte (TE) chosen to have a lower mobility than the sample components. All the components of the system are in the presence of a suitable buffering counter ion. Thus at the start of an analysis the sample is "sandwiched" between the LE and TE, and the mobilities are in the following order:

$$m_{L-} > m_{A-} > m_{B-} > m_{T-}$$

provided the selection of the LE and TE have been correctly made. *Note*: Positive ions may also be analyzed.

When the current is applied to the system, the negatively charged ions will commence migrating to the anode at a velocity, v, dependent on the effective mobility, m, of the LE, according to the equation:

$$v = mE$$

Thus ions with the highest effective mobility will migrate first, followed in turn by ions of lower effective mobilities until the sample constituents achieve a stacked configuration in the order of their effective mobilities. Since there is no carrier electrolyte in the system, in order to preserve conductivity the zones are forced to move in contact with each other. Further, since the current density is constant, it follows that the concentration of the ions in any given zone must be the same as that in the other zones; hence the length of any zone in the capillary is directly proportional to the concentration of the various species present.

After the components of the system have equilibrated in the stacked configuration, they will migrate through the system at constant velocity. Hence it is possible to write the Kohlrausch regulating function in the following form:

$$C_A/C_B = m_A/(m_A + m_R) (m_B + m_R)/m_B$$

where C_A/C_B is the ratio of concentrations of ions A and B, m = ionic mobility ($cm^2/V/s$, and A, B, and R refer to the ions A^-, B^-, and R^+, respectively. This equation gives the condition at the boundary between two ions in the presence of a common counter-ion, when the boundary migrates in an electric field. And since the effective mobilities of the participating ions must be constant under a defined set of conditions, the Kohlrausch equation above can be rewritten in the form:

$$C_A = C_{L^-} \text{ constant}$$

This indicates that in equilibrium conditions, the sample ion concentration is directly proportional to the LE concentration. Thus, as stated above, the concentration of the various ion constituents is a constant, and this is the unique feature of ITP.

Figure 15 shows an example of the separation of some amino acids using ITP and the operating system used. The upper trace is that obtained from the conductivity detector showing steps associated with the presence of different amino acids. The "peaks" in the upper trace are the differentials of the steps and provide a simple means of measuring the zone length for quantitative purposes. The lower trace is the response of the UV detector to the constituents and shows the presence of tyrosine and phenylalanine.

3.2. High-Voltage Zone Electrophoresis

One of the problems of any electrophoretic experiment resides in the Joule heat that is produced on passing an electric current through the system, whatever it may be. If a tube is used to perform separations, there will exist a parabolic temperature profile across the bore of the tube. Since both electrophoretic mobilities and diffusional effects increase with increasing temperature, band broadening will occur because there will be a temperature differential between the axis and the walls of the tube. Clearly this effect can be minimized if the radius of the tube is reduced. Heat dissipation will be increased and an ion will be able to diffuse across the bore of the tube many times during its passage through the tube. Thus such temperature differences as exist will effectively be averaged out.

Jorgenson and Lukacs (57,58) have investigated this proposition using a silica capillary with a bore of 75 μm to dissipate the heat generated during zone electrophoresis using a potential drop of 30 kV. The migrating ions were detected using a fluorescence detector, the cell of which was constructed in the capillary itself. The system is simple to construct and consists of a suitable length of capillary (preferably silica), the ends of which dip into two beakers containing the carrier buffer (e.g., 0.05M phosphate, pH 7.0). Two carbon electrodes provide the electrical connections; one

Amino Acids and Small Peptides

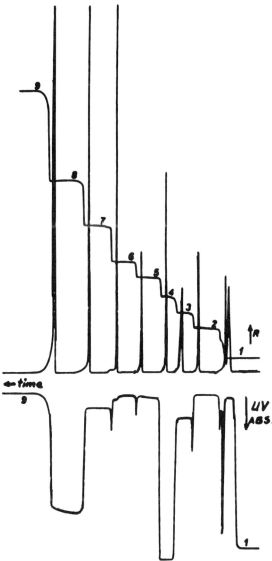

Fig. 15. Isotachophoretic separation of some amino acids (from ref. 61). Operational system at pH 9.2 and 5-bromo-2,4-dihydroxybenzoate (0.004 M) as the leading ion and L-lysine as the counter-ion. Note carbonate, moving in front of L-ASP, clearly visible in the linear trace of the UV-absorption detector. The carbonate has an effective mobility in this system comparable with that of the leading ion. 1 = 5-Bromo-2,4-dihydroxybenzoate; 2 = L-ASP; 3 = L-Cys; 4 = I:-L-Tyr; 5 = L-Asn; 6 = L-Ser; 7 = L-Phe; 8 = DL-Trp; 9 = β-Ala. R = increasing resistance (conductivity detection).

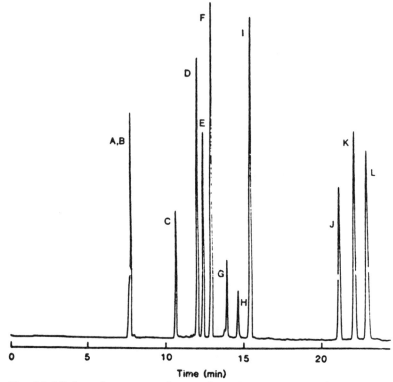

Fig. 16. High-voltage zone electrophoretic separation of some amino acids (from ref. 57).

is connected to the high-voltage power supply (the anode) the other (the cathode) to earth. Samples are applied in the buffer to the anodic end, and the power is applied. Figure 16 shows an electropherogram of some dansyl amino acids, and an excellent separation is obtained in about 25 min. The efficiency expressed in terms of number of theoretical plates was 400,000 for a capillary approximately 60 cm long. An interesting aspect of this separation is that, in the buffer used, the dansyl amino acids would be negatively charged, but they are apparently migrating toward the negative electrode. This anomaly is readily explained by the presence of a strong electroendosmotic flow toward the negative electrode, sufficiently strong to sweep the ions in the opposite direction to their electrophoretic migration, which is necessarily present. This effect

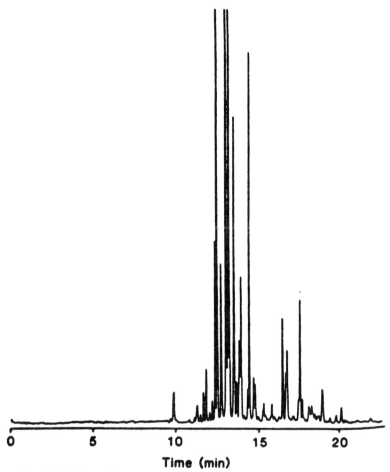

Fig. 17. High-voltage zone electrophoretic separation of the peptides obtained from a tryptic digest of chicken ovalbumin (from ref. 58).

is important in that it allows both positively and negatively charged species to be analyzed in a single run.

Clearly this method of analysis has high potential as a separation method for all charged species. A number of groups are investigating this method and it is likely, when developed, to provide the high efficiencies necessary for complex biochemical mixtures. Figure 17 shows such a complex separation of fluorescamine derivatives of simple peptides from a tryptic digest of egg white lysozyme.

References

1. R. Consdon, A. H. Gordon and A. J. P. Martin, *Biochem. J.* **38**, 224 (1944).
2. C. E. Dent, *Biochem. J.* **43**, 169 (1948).
3. G. L. Kowkabany and H. G. Cassidy, *Anal. Chem.* **22**, 817 (1950).
4. E. F. McFarren, *Anal. Chem.* **23**, 168 (1951).
5. L. Fowden, *Biochem. J.* **50**, 355 (1952).
6. C. S. Hanes, *Can. J. Biochem. Physiol.* **39**, 119 (1961).
7. E. H. M. Wade, A. T. Matheson, and C. S. Hanes, *Can. J. Biochem. Physiol.* **39**, 417 (1961).
8. T. B. Moore and C. G. Baker, *J. Chromatog.* **1**, 513 (1958).
9. D. Gross, *Nature* **184**, 1298 (1955).
10. I. Smith, ed., *Chromatographic and Electrophoretic Techniques* vol. 2, 4th Ed. Heinemann, London (1976).
11. M. Brenner and A. Niederweuser, *Experimentia* **16**, 373 (1960).
12. A. R. Fahmy, A. Niederwieser, G. Pataki, and M. Brenner, *Hely. Chim. Acta* **44**, 2193 (1961).
13. The Nanomat Applicator, Camag, F.D.R.
14. Schleicher and Schuell (Application Notes).
15. N. Grinberg and S. Weinstein, *J. Chromatog.* **303**, 251 (1984).
16. I. R. Hunter, K. P. Dimik, and J. Corse, *Chem. Ind.* 294, (1956).
17. S. L. MacKenzie and D. Tenaschuck, *J. Chromatog.* **97**, 19 (1974).
18. P. Husek, *J. Chromatog.* **234**, 381 (1982).
19. P. Edman, *Acta Chem. Scand.* **4**, 277–283, (1950).
20. Applied Biosystems, Foster City, California.
21. S. Moore, D. H. Spackman, and W. H. Stein, *Anal. Chem.* **30**, 1185 (1958).
22. D. H. Spackman, W. H. Stein, and S. Moore, *Anal. Chem.* **31**, 1190 (1958).
23. P. B. Hamilton, D. Bogue, and R. A. Anderson, *Anal. Chem.* **32**, 1782 (1960).
24. K. A. Piez and L. Morris, *Analyt. Biochem.* **1**, 187 (1960).
25. S. Udenfriend, S. Stein, P. Bohlen, W. Dairman, W. Leimgruber, and M. Wiegele, *Science* **178**, 871 (1972).
26. J. R. Benson and P. E. Hare, *Proc. Natl. Acad. Sci. USA* **72**, 619 (1975).
27. C. Murren, D. Stelling, and G. Felstead, *J. Chromatog.* **115**, 236 (1975).
28. Chromaspek, Rank Hilger, Margate, England.
29. Durrem d500 Amino Acid Analyser. Palo Alto, California.
30. R. Errser, *Lab. Equip. Dig.* **17**, (6) 61 (1979).
31. S. J. DiMari, J. P. Robinson, and J. H. Hash, *J. Chromatog.* **213**, 91 (1981).
32. R. L. Cunico, R. Simpson, L. Correia, and C. T. Wehr, *J. Chromatog.* **336**, 105 (1984).
33. W. G. Kruggel and R. V. Lewis *Anal. Biochem.* (submitted).
34. S. A. Cohen, T. L. Tarvin, and B. A. Biblingmeyer, *Internat. Lab.* Oct., **68** (1984).

35. J-Y. Chang, R. Knecht, and D. G. Braun, *Meth. Enzymol.* **91**, 41 (1981).
36. Y. Tauphi, D. E. Smith, W. Linder, and B. L. Karger, *Anal. Biochem.* **115**, 123 (1981).
37. V. T. Wiedmeier, S. P. Porterfield, and C. E. Hendrich, *J. Chromatog.* **231**, 410 (1982).
38. N. B. Levina and I. V. Nazimov, *J. Chromatog.* **286**, 207 (1984).
39. B. Oray, H. S. Lu, and R. W. Gracy, *J. Chromatog.* **270**, 253 (1983).
40. S. Kobayashi and K. Imai, *Anal. Chem.* **52**, 424 (1980).
41. K. Samejima, W. Dairman, and S. Udenfriend, *Anal. Biochem.* **42**, 222 (1971).
42. M. Weigele, S. L. DeBernardo, J. P. Tengi, and W. Leimgruber, *J. Am. Chem. Soc.* **94** (16), 5927 (1974).
43. M. Roth, *Anal. Chem.* **43**, 880 (1971).
44. P. Lindrith and K. Mopper, *Anal. Chem.* **51**, 1667 (1979).
45. J. D. H. Cooper, G. Ogden, J. McIntosh, and D. C. Turnell, *Anal. Biochem.* **142**, 98 (1984).
46. J. D. H. Cooper, M. T. Lewis, and D. C. Turnell, *J. Chromatog.* **285**, 490 (1984).
47. L. Johnson, S. Lagerkvist, P. Lindroth, M. Ahnoff, and K. Martinsson, *Anal. Chem.* **54**, 939 (1982).
48. M. Doury-Berthod, C. Poitrenaud, and B. Tremillon, *J. Chromatog.* **179**, 37 (1979).
49. E. Gil-Av, B. Feibush, and R. Charles-Sigler, in *Gas Chromatography* A. B. Littlewood, Institute of Petroleum, London (1966). p. 227.
50. P. E. Hare and E. Gil-Av, *Science* **204**, 1226 (1979).
51. N. Nimura, A. Toyama, and T. Kinoshita, *J. Chromatog.* **316**, 547 (1984).
52. D. W. Armstrong and W. DeMond, *J. Chromatogr. Sci.* **22**, 411 (1984).
53. S. Yuasa, M. Itoh, and A. Shimada, *J. Chromatogr. Sci.* **22**, 288 (1984).
54. F. M. Everaerts, J. L. Beckers, and Th. P. E. M. Verheggen, *Isotachophoresis J. Chroma.* vol. 6, Elsevier, Amsterdam (197e).
55. S-G. Hjalmarsson and A. Baldesten, *Capillary Isotachophoresis* CRC (1981).
56. J. W. Jorgenson and K. D. Lukacs, *Anal. Chem.* **53**, 1298 (1981).
57. J. W. Jorgenson and K. D. Lukacs, *J. chromatog.* **218**, 209 (1981).
58. A. G. Georgiadis and J. W. Coffey, *Anal. Boichem.* **56**, 121 (1973).
59. W. G. Kruggel and R. V. Lewis, *Anal. Biochem.*, in press.
60. F. M. Eveaerts, J. L. Beckers, and Th. P. E. M. Verheggen, *J. Chromatog.* **19**, 129 (1976).

Chapter 11
Analytical Liquid Chromatography of Proteins

Colin Simpson

1. Introduction

The application of liquid chromatography (LC) to protein separations has advanced significantly over the last few years, and three of the conventional LC techniques have proven to be valuable in this respect. All LC methodology may be applied to the separation of proteins to a greater or lesser extent; namely, liquid/solid (adsorption) (LSC), liquid/liquid (partition) (LLC), bonded phase (BPC), particularly reversed phase (RPC), hydrophobic interaction (HIC) (effectively a modified form of reversed-phase chromatography), ion exchange (IEC), and size exclusion [SEC (GPC)], together with the "newer" techniques, affinity chromatography (AC), chromatofocusing (CF), and field flow fractionation (FFF). Of the methods mentioned above, four [BPC (RPC), HIC, IEC, and SEC] have proven to be the most useful, possibly because they are simple extensions of known methodologies applied to proteins, but also because they are the "softer" of the techniques, less liable to bring about denaturation of the protein, or easier to control. However, care has to be exercised, particularly with BPC (RPC) and HIC, because the nonpolar nature of the stationary phase can have a deleterious effect on proteins. AC has had many triumphs, and it is capable of highly selective separations, but is a difficult technique to utilize in that the construction of the bound substrate to the matrix is often chemically demanding. Chromatofocusing, a relatively new technique, shows great promise, combining the high resolution of isoelectric focusing (chapter 12) with the convenience of chromatography. Field flow fractionation, a sophisticated technique, has also proved capable of providing good separations; several groups are working on this technique and it is likely that, when fully developed, the method will provide the "softest" method of separation, both on an analytical and preparative scale.

In this chapter the various chromatographic methods of protein separation will be reviewed, and their advantages and disadvantages will be discussed. Virtually all of the methods discussed are capable of being scaled up for preparative purposes, using the techniques indicated in chapter 13.

2. Adsorption Chromatography

Adsorption chromatography (LSC) methods have not been used extensively for the separation of whole proteins, but the method has been used for the separation of protein fragments up to pentapeptides (1). Silica gel is the principle adsorbent used, in conjunction with a suitable multisolvent gradient designed to minimize the high surface energy of silica gel and provide the necessary solvent strength to elute the protein fragments. Figure 1 shows a separation of various peptides of up to five amino acid units in length, and demonstrates the type of separation possible using this technique. It must be realized that the separation is not taking place on the surface of the silica gel, but on a mono-, or bilayer of the mobile phase adsorbed on to the silica surface, which thus deactivates the silica. Separation occurs either by displacement of the adsorbed solvent molecules or by association with the adsorbed solvent molecules, which will have different solvent characteristics from the free solvent (mobile phase) (2). Alternatively, if mobile phase bilayers are produced, the adsorption mechanism could be a combination of the two modes. Indeed it is questionable whether true adsorption actually occurs, particularly with the wide pore size silicas that are preferable to use in protein separations. "Adsorption" methods are rarely, if ever, used, however, for whole protein separations, principally because of the strong adsorption energies involved, even with the solvent-moderated systems available and the strong possibility of surface denaturation of the protein occurring.

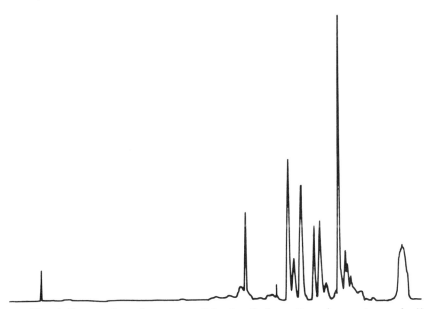

Fig. 1. Separation of pentapeptides by "adsorption chromatography" using a multisolvent gradient (from ref. 1).

3. Partition Chromatography

Partition chromatography (LLC) has not been exploited at all (in the true sense of the technique) for separation of proteins, other than the pseudo-LLC/adsorption system outlined above. Developments in the method are occurring that could, however, lead to the exploitation of LLC in its own right as a separation method that is intrinsically soft in its effect on proteins. The choice of phase system can be diverse and chosen specifically for a given separation. It is important that the nature of the support medium has no interaction with the solutes. Providing the pore size of the (usually silica) support is correctly chosen, true LLC is the operative mechanism, e.g., insulin and its precursers have been separated using *n*-butanol/water as the phase system on 30 nm pore size silica as the support (3). The main restriction of this method lies in the choice of the organic phase, which must not bring about denaturation of the protein. Further developments in this field may be expected.

The technique of "droplet counter-current chromatography," (DCCC) (4), a so-called chromatographic technique, and associated recent developments, which are in fact refined methods of solvent extraction, has been used for peptide separations using the solvent systems indicated in Table 1.

In many ways it is surprising that LLC has not received greater attention in this application. This is probably because it is a relatively difficult technique to control from a technical viewpoint. From the nature of the method, however, it is "gentle," in that it does not allow the protein to come into contact with high-energy surfaces, which can bring about denaturation, and indeed complicate a given separation by causing the tertiary structure of the protein to "unfold," thus presenting different interactive surfaces/groups to the stationary phase. Several groups are investigating the technique in some detail, and novel methods of operation are being investigated that may lead to its application to protein separation in due course.

Table 1
Solvent Systems Used in DCCC of Peptides

Solvent system	Solvent ratios
C_6H_6–$CHCl_3$–MeOH–H_2O	15:15:23:7
C_6H_6–$CHCl_3$–MeOH–0.1N HCl	11:5:10:4
$CHCl_3$–MeOH–0.1N HCl	19:19:12
sec-BuOH–TFAA–H_2O	120:1:160
n-BuOH–AcOH–H_2O	4:1:5

4. Bonded Phase Chromatography

Bonded phase chromatography (BPC), particularly in the reversed-phase mode, has received considerable attention over the past few years, and many elegant separations have been achieved with its use. Bonded phases consist of a silica matrix to which suitable modifying groups are chemically bonded. The nature of the modifying group depends upon the type of separation to be undertaken. For normal phase separations, i.e., where the mobile phase is relatively nonpolar and the stationary phase is polar, the bonded group may contain the functional groups, NH_2, CN, $CH(OH)CH_2OH$, $COOC_2H_5$, and so on, whereas in the reversed-phase mode, the functional group is an alkyl group of varying chain lengths (C_2–C_{24}) of carbon atoms, and the mobile phase is water or aqueous/organic in nature, where the organic part may be alcohols, acetonitrile or tetrahydrofuran, usually at relatively low concentration.

Various chemistries have been used to prepare bonded phases, but in general, two methods are in regular use that produce the "brush"-type and the "bulk"-type phases. For most part the normal phases containing functional groups are reasonably consistent between manufacturers. However, within a group of reversed phases, e.g., the octadecylsilyl silicas (ODS phases), significant differences exist between the products manufactured by different companies, and great care has to be exercised when comparing separations between different sources of ostensibly the same stationary phase type. Further, the difference between the brush and bulk types of phase is evident when comparing their responses to changes in mobile phase composition. The brush types take time to equilibrate, and several column volumes of mobile phase must be passed, but the polymeric phases respond virtually immediately to the change in mobile phase composition. This can have important consequences when performing gradient elution separations (Fig. 2).

Separation in BPC depends on the form of chromatography used, i.e., whether normal or reversed-phase operation is to be used. With normal-phase operation, the mechanism is similar to that of adsorption chromatography; namely, the interaction of the species with the stationary phase depends upon the competition between the solute molecules and the mobile phase for the active sites present. The interaction energies can be quite high; although not as high as those associated with "naked" silica, they are nevertheless sufficient to lead to surface denaturation of some proteins. Further, mobile phases in the "normal" mode of operation are usually organic in character, which can also bring about protein denaturation. Thus this method is not often used. Glycol bonded phases may be used successfully with aqueous solvents in which the molecular interactions are minimal and separations occur principally by the size exclusion mechanism (see later). Obviously, since the surface of the glycol bonded phase is amply covered with hydroxyl functions, it is unlikely that adsorption plays any part in the separation process.

Thus, the reversed-phase mode has been the principle mode investigated. Here the stationary phase is hydrocarbonaceous in character, and although it has been shown that a C_{18} chain can bring about surface denaturation, this is not too common. Because of this possibility, however, the C_8 chain (or shorter) is often used for protein work.

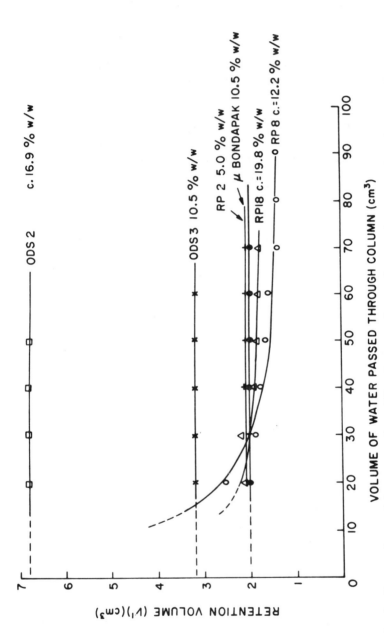

Fig. 2. Comparison of the rates of equilibration to solvent changes between "brush" and polymeric type reversed phases columns initially equilibrated to pure methanol followed by and instantaneous change to pure water. Equilibration followed from the retention volumes of ethanol (from ref. 24).

4.1. Mechanism of Separation

Because of the nature of the stationary phase, retention in RPC must depend on the ability of the solute molecules to interact with the hydrocarbon surface presented, which implies that it must be the hydrophobic part of the protein tertiary structure that is responsible for retention. This view is in accordance with the proposal of Singer (5), who considered that the interaction of a protein with a hydrophobic surface consisted of the membrane protein "floating on a sea of lipids," in which the nonpolar regions of the protein are in contact with the nonpolar lipids present in the membrane, whereas the polar moieties are exposed to the aqueous environment (Fig. 3).

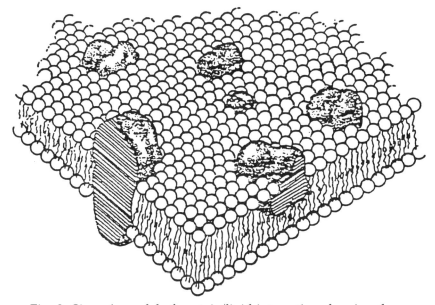

Fig. 3. Singer's model of protein/lipid interaction showing the proteins "floating" in a sea of lipid. Here the solid bodies represent globular proteins; the phospholipid bilayer are represented by the open circles (from ref. 5).

Further studies have suggested that in the presence of lipid, the apoprotein components of a lipoprotein assume regions of high helical content that exhibit faces of opposite polarity. Thus, if these proteins are to interact with the surface of a lipoprotein particle, the nonpolar portion will align itself with the nonpolar lipid components of the particle, whereas the polar groups interact with the polar head groups of cholesterol and phospholipid and the aqueous environment (Fig. 4). Thus the hydrophobic effect plays an important role in these models.

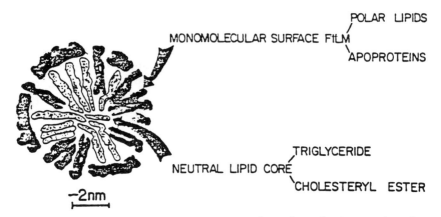

Fig. 4. Model of lipoprotein structure based on the interactions between apolipoproteins and lipid components. The surface monolayer is composed of phospholipids and apolipoproteins. The apoproteins contain regions of a helical nature that are amphipathic in nature. The hydrophobic surface of such amphipathic helices interact with the fatty acid chains of the phospholipid, whereas the hydrophilic portions are exposed to and interact with the aqueous environment (from ref. 25).

In RPC, however, the hydrophobic chains are anchored to the surface of the silica matrix, making it impossible for the hydrophobic alkyl chain to avoid contact with the aqueous environment. If a polar modifier, e.g., an alcohol or acetonitrile, is included in the system, however, this will be adsorbed on to the surface of the RP, thus shielding it from the aqueous environment. (This effect is analogous to the way a polar modifier will be adsorbed from a non-polar solvent in LSC, thus deactivating the silica surface.)

On the basis of these arguments, we can propose a mechanism for the interaction of a protein with a RP in which nonpolar regions of the molecule will displace some of the modifier molecules from the surface of the RP and be themselves adsorbed. It is unlikely that the protein will penetrate far into the body of the RP for two reasons. (1) The pore structure of the RP prevents penetration, (unless the RP has been prepared with wide pore silica; these are now available), and (2) the relatively rapid change from hydrophobic to hydrophilic nature in the protein would prevent "wetting." Thus the preferred mechanism is adsorption.

This mechanism has been substantiated by Horvath (6) during a study of the hydrophobic effect in aqueous-organic systems (solvophobic theory). It was shown that the dominant interactions were between the mobile and stationary phases, and between the mobile phase and sample molecules. The driving force in *both* interactions was the shielding of a nonpolar region of *either* the solute or stationary phase from the aqueous environment (Fig. 5).

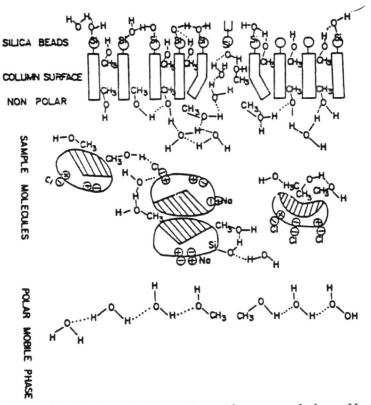

Fig. 5. Model of protein interaction with a reversed phase. Here the hydrophobic regions of the sampe molecules displace solvent molecules from the adsorbed layer of solvent on the reversed phase (from ref. 26).

On the basis of this theory, it is possible to explain the effect of an organic solvent gradient in displacing nonpolar samples from the surface of the stationary phase. The effect of the addition of the organic solvent is to reduce the surface tension of the water molecules, which will decrease the energy required to produce a cavity in the matrix of water molecules. Hence, if this decrease is sufficiently large, the sample molecule will not favor the adsorbed state and will pass back into the mobile phase and, in due course, be eluted. Simplistically, on the basis of this theory, the retention of a protein molecule depends upon a balance of interaction energies, which will either tend to favor the adsorbed state or the solution in the mobile phase. This is basically a function of the relative strength of the molecular interactions of the protein molecule between the stationary and mobile phases. Thus, those proteins that favor the mobile phase will pass through the column system rapidly, and those favoring the stationary phase will pass through more slowly (which is a description of the chromatographic process).

The theory enables us to predict that the order of elution of various proteins will be related to the nonpolar surface areas exposed that are capable of interaction with the number of nonpolar moieties on the surface of the stationary phase that are also available. It is possible to restrict the number of exchange sites available to a protein by including in the mobile phase appropriate moderators, e.g., alcohols, nitriles, and so on. These will be strongly adsorbed on the reverse phase, the strength of adsorption dependent on the hydrocarbon chain length of the modifying species (7). The longer the hydrocarbon chain, the stronger the modifier is adsorbed and the less likely the protein is to interact, hence its retention will be reduced. Thus it has been possible to predict the order of elution of some apolipoproteins, C-1, C-11, and C-111, on an alkyl-phenyl column with an acetonitrile/water gradient. The observed elution order (Fig. 6) is consistent with the known polarity of these compounds, with the apolipoprotein with the greatest nonpolar surface area being retained the longest.

Calculations have been carried out by Nice and O'Hare (8) that provide a good correlation between the number of hydrophobic sites and retention for a series of peptides up to 15 units long. Longer protein chains also follow this correlation, probably because the molecules possess the ordered three-dimensional structure that "buries" some of the hydrophobic sites and render them incapable of interacting.

Great care must be exercised when evaluating a separation from a reversed-phase column, however. For example, Molnar (9) showed that by size exclusion and ion exchange methods, a given protein was substantially pure. The same sample when chromatographed on a reversed-phase column, however, showed a considerable number of components, a fact ascribed to impurities in the

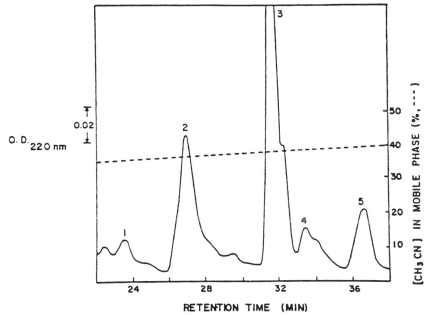

Fig. 6. Separation of C-apolipoproteins. Conditions: Bondapak alkylphenyl column, mobile phase 1% triethylammonium phosphate, pH 3.2 with an acetonitrile gradient (see figure) at 1.5 mL/min flow rate. Peak identity confirmed by amino acid analysis. 1 = apoLP C-I, 2 = apoLP C-III$_2$, 3 = apoLP C-III$_1$, 4 = apoLP C-III$_0$, 5 = apoLP C-III (from ref. 27).

sample. This seems a very unlikely event, even though SEC and IEC may not be the most powerful of separation methods for proteins (see later). In view of what was stated above (i.e., the retention of a protein is a function of the extent of the hydrophobic area available to interact with the reversed phase), it is possible to postulate that a given protein, in the presence of the reversed-phase surface, may unfold from its tertiary structure, thus presenting additional hydrophobic surface for interaction. Further, the protein may unfold in several different positions, thus presenting a series of different hydrophobic surface areas, each possessing its own unique retention on the reversed phase. Hence a protein sample may appear to be impure, whereas in fact it is just many "structural" isomers of the one protein. This view is also held by Hearn (10).

It is also interesting to consider the strength by which proteins interact with the surface, and how small differences in mobile phase composition can bring about either irreversible adsorption on the one hand or rapid elution on increasing the organic solvent content of the mobile phase by as little 1 or 2%. For example, BSA can be eluted in three column volumes from a reverse phase column by 34% volume *n*-propanol in water, but 32% of the alcohol results in irreversible adsorption. Increasing the propanol content to 34% again brings about elution. This tendency for minor changes in mobile phase composition to bring about a drastic change in retention appears to be characteristic of proteins.

5. Hydrophobic Interaction Chromatography

Hydrophobic interaction chromatography (HIC) is performed using a reversed-phase column identical to those used in reversed-phase separations. The eluting solvent, however, consists of aqueous salt solutions, in contrast to aqueous organic mobile phases normally used in RPC. Retention in HIC is inversely proportional to the salt concentration, and the nature of the salt used can change the selectivity (*11*) of the separation. In many ways, HIC is to be preferred to RPC because the mobile phase is most unlikely to bring about denaturation of the protein, which is a real possibility in RPC, in which organic solvents are used. Further, the pH of the mobile phase is generally at physiological levels. The method of development in HIC suggests that the order of elution of some of the components of a sample will be reversed compared to the equivalent separation carried out by RPC. This is illustrated in Fig. 7, which compares the separation of four proteins—myoglobin, ribonuclease, lysozyme, and chymotrypsinogen A—on a TSK gel phenyl-5PW column.

6. Ion Exchange Chromatography (Electrostatic Interaction Chromatography, EIC)

Separation by IEC may be viewed as the simple electrostatic interaction of the various charged sites on a molecule with the fixed charged ligands on the stationary phase. This simple view is satisfactory for simple molecules, but for proteins the situation is more complex insofar as a protein possesses many sites where inter-

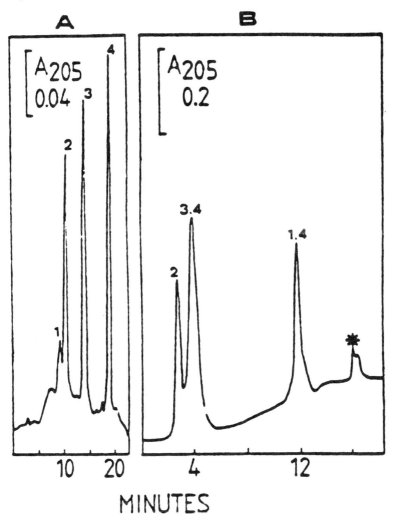

Fig. 7. Comparison of reversed phase and hydrophobic interaction chromatographic separations of some protein standards. Note the appearance of two peaks for chymotrypsinogen by the reversed-phase method. Conditions: HIC on a TSK gel Phenyl-5 PW column. Gradient: from 1.5 to 0M ammonium phosphate in 0.1M sodium phosphate, pH 7.0, 1.0 mL/min. Peak identity: 1, myoglobin; 2, ribonuclease; 3, lysozyme; 4, chymotrypsinogen. (A) 0.5 microgram of each component. RP on the same column, conditions: 15 min gradient from 0 to 60% acetonitrile in 0.1%TFA. 1.0 mL/min, 25 μg of each component, same identification. Asterisk, impurity (from ref. 28).

action can occur, and in addition there are areas of hydrophobicity that can interact with the support of the exchanger by hydrophobic interaction chromatography. Thus these two mechanisms can be superimposed, complicating the simple ion exchange mechanism, but providing additional means of varying selectivity. A further advantage of IEC arises from the relatively "soft" nature of the stationary phase. Thus structural changes in proteins are rare.

6.1. The Nature of the Stationary Phase

A variety of different stationary phases are available for IEC that are characterized by the nature of the functional group bonded to the support matrix. These may be divided into two classes, anionic and cationic phases, with a further subdivision dependent on whether the bonded group is strong or weak in nature; thus we have the data shown in Table 2.

Table 2
Characteristics of Ion Exchangers

Type	Character	Functional group
Cationic	Strong	SO_3
	Weak	Carboxyethyl, pK 3.5
		Sulfopropyl, pK 2.0
Anionic	Strong	$N(CH_3)_4$, pK 12
	Weak	Diethylaminoethyl, pK 9.5

The pH ranges wherein the various types of exchanger exist in an ionized form and hence are capable of undergoing exchange are shown in Fig. 8. It will be seen that the strong exchangers are fully ionized at the extremes of the pH range (i.e., pH 2 for cationic, pH 11 for anionic), whereas the weak exchangers ionize between pH values of 5–9 and 4–7 for the cationic and anionic types, respectively, and the weak exchanger is particularly useful for protein separation, the cationic for basic, the anionic for acidic proteins. With cationic exchangers, the basic proteins are adsorbed below their isoelectric points (cf. chapter 12) when they possess a positive charge (pH 7–8), acidic and neutral proteins (and the majority fall into this class) are usually chromatographed above their isoelectric points, and hence are negatively charged and will be adsorbed on to a positively charged anion exchanger. IEC is a powerful method of protein separation and small changes in operating con-

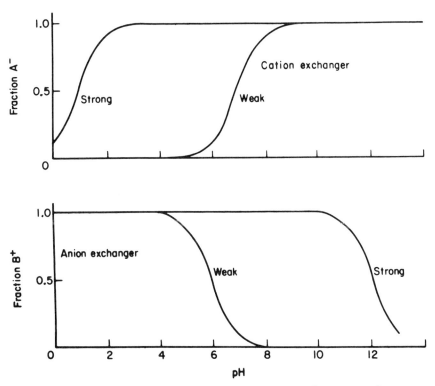

Fig. 8. pH ranges in which ion exchangers are in the appropriate state to undergo ion exchange.

ditions brought about by changing the buffer molarity can bring about large differences in selectivity possibly by invoking HIC. Many of these operating parameters give predictable changes; hence it would be advantageous to consider them in detail.

6.2. Operating Parameters in IEC

The ion exchange mechanism is a function of the competition between the solute ions and counter-ions (present to preserve electroneutrality of the system) for the fixed oppositely charged sites on the surface of the exchanger. This means that the solute ions must displace one or more counter-ions that are associated with the fixed functional groups in order to be adsorbed themselves.

For simple monovalent solute and counter-ions the process may be described by the simple equilibria:

Anion exchange: $M-B^+Y^- + X^- = M-B^+X^- + Y^-$

Cation exchange: $M-A^-R^+ + NH^+ = M-A^-NH^+ + R^+$

where M represents the substrate to which the functional group is bound, X^- and NH^+ represents the solutes for the two processes, Y^- and R^+ are the adsorbed counter-ions, and B^+ and A^- are the bound functional groups. The equilibrium constants for the two processes are:

Cation exchange: K_{NR}; Anion exchange: K_{XY}

and these are dependent upon a variety of parameters for which only sparse data are available.

It is possible, however, to write equations describing the ionic equilibria that pertains; namely, K_a

For acids: $HX = H^+ + X^-$

For bases: $NH^+ = N + H^+$

where K_a is the ionization constant for the two equilibria.

If it is assumed that only the charged form of the solute is adsorbed on to the surface of the exchanger, which will be true providing the only mechanism operative is that of ion exchange, then it is possible to derive relationships that describe the distribution coefficient, D_i, of a solute, i, in terms of the ratio of the concentration of the solute adsorbed on to the stationary phase to the total concentration (ionized + nonionized) of solute in the mobile phase. Thus we get:

Anion exchange: $D_x = [M-B^+X^-]/\{[X^-] + [HX]\}$

$= K_{XY}[M-B^+Y^-]/[Y^-]l/\{1 + [H^+]/K_a\}$

Cation exchange: $D_N = [M-A^-NH^+]/\{[NH^+] + [N]\}$

$= K_{NM}[A^-M^+]/[M^+]l/\{1 + K_a/[H^+]\}$

Analytical LC of Proteins

These equations show that, to a first approximation, the distribution coefficient is proportional to the number of exchange sites that are available on the surface of the exchanger. The equilibrium constant for the exchange process, D_i, is inversely proportional to the counter-ion concentration in the mobile phase, the solute pK_a values, and the pH of the mobile phase. These latter parameters allow us to vary the distribution process in a predictable way.

6.3. Various Other Parameters That Control Separation by IEC

1. The ion exchange capacity of the exchanger (expressed as mEq/g of dry exchanger), which determines the sample size that may be chromatographed without overloading the exchanger, affects only the absolute retention of the solute, and has no effect on the selectivity of the separation.
2. The pore size of the exchanger controls the kinetics of the exchange process. For proteins this should be as large as possible concomitant with structural strength, i.e., the ability to tolerate high mobile phase flows without compression of the bed. This is particularly important for protein separations because of the large size of the protein molecule. Because of this large size and the attendant low diffusivity, however, it is preferable to operate the column at relatively low flow rates. The two effects are therefore effectively self-canceling.
3. pH is a very valuable parameter for adjusting the retention of proteins. Figures 9 and 10 show the effect of pH on the distribution coefficient of solutes of differing pK_a values on strong and weak anion and cation exchangers. Here it will be seen that, on strong exchangers, for weak acids and bases a significant change in D_i occurs at about pH = pK_a. Thus when separating proteins of differing pK_a values, pH is most useful in adjusting selectivity through changing the effective charge on the molecule. With weak exchangers, the situation is somewhat more complex in that pH also affects the ion exchange capacity of the exchanger. Thus for weakly acidic or basic solutes, the D_i passes through a maximum at a certain pH, which is a function of both the exchanger and the solute molecules, and falls to zero at low and high pH.
4. The concentration and charge of the counter-ion can be used to change the selectivity of separations predictably, as indicated by the above equations suitably modified to take into

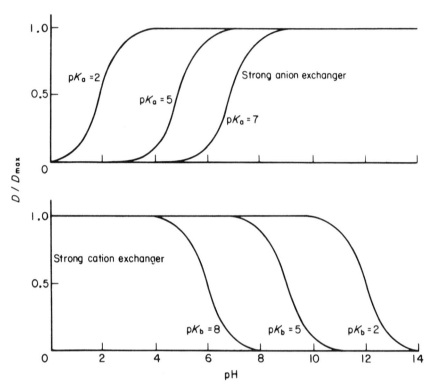

Fig. 9. Influence of pH on the distribution coefficient of acids and bases with different pK_a and pK_b values on strong ion exchangers.

    account the different equilibria involved. In general, $\log D_i$ is linearly related to the ratio of the charge on the solute to that on the counter-ion with increasing counter-ion concentration when both species are fully ionized.

5. The nature of the counter-ion can also affect selectivity, i.e., the affinity of the solute vs counter-ion. In general, the exchanger shows strongest affinity for (1) ions with the largest charge, (2) ions with the smallest hydrated volume, and (3) ions with the largest polarizability. Here through the complex nature of the protein molecule can restrict the predictability of the separation in that the charge on the protein is not localized.

Ion exchange may not be the only mechanism controlling separation. In certain circumstances the hydrophobic moieties on the protein molecule exposed to the aqueous environment can in-

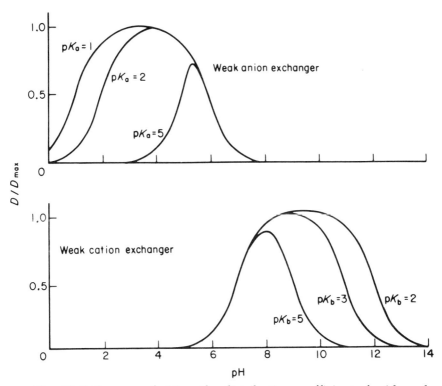

Fig. 10. Influences of pH on the distribution coefficient of acids and bass with different pK_a and pK_b values on weak ion exchangers.

teract with the support matrix of the exchanger and hence modify the separation selectivity. This form of interaction can be enhanced by increasing the salt concentration of the mobile phase or by performing a decreasing salt gradient to exploit the hydrophobic effect.

Temperature may be used to modify the exchange kinetics; elevated temperatures will increase the diffusivity of the solute molecules and lower the mobile phase viscosity, thus lowering the pressure drop across the column. Further it has the effect of reducing the strength of the electrostatic and hydrophobic interactions with a concomitant decrease in retention. Clearly this variable is restricted in range of applicability, dependent on the denaturation temperature of the proteins present. Additionally, in HIC, conformational changes may occur that can be enhanced by increase in temperature. This will have the effect of increasing the hydrophobic contact area with an increase in retention with temperature.

Finally the nature of the support matrix can have a profound effect on the separations possible, and this is a function of the degree of crosslinking of the substrate used. A high level of crosslinking will exclude the protein molecules from the pores, although this will have good flow characteristics with little if any compression of the packing in the column, whereas a low level of cross-linking will allow good penetration into the pore structure of the matrix, but this gives a compressible packing with poor flow characteristics. The following Table 3 lists the exclusion limits for various levels of crosslinking in a polystyrene resin.

A variety of suitable weak ion exchangers for protein separation are commercially available from a number of laboratory suppliers, e.g., Pharmacia, L.K.B., Biorad, Serva, Toya-Soda, and so on.

Table 3
Approximate Exclusion Limits at Different Levels of Crosslinking

Crosslinking, %	Exclusion limit, mw
2	<2000
4	<1000
8	<500

6.4. Other Operating Methods

6.4.1. Mixed Bed Columns

Mixtures of anionic and cationic exchangers may be used with some success for analytical monitoring purposes. These enable both positive and negatively charged analytes to be separated at one time (Fig. 11). These mixed bed columns have relatively low retentions and loading capacities, but are useful for scouting purposes.

There are two modes of operation of mixed bed ion exchangers: (1) where the components of the bed are additive in their retention and so the net retention is the sum of the retentions of the weighted components of the mixed bed, i.e., predictable retentions and (2) where the retention of the individual resins are nonadditive with concomitant unpredictable results.

6.4.2. Tandem Columns

Here columns of individual ion exchangers are arranged in sequence. The order and dimensions of the component columns may

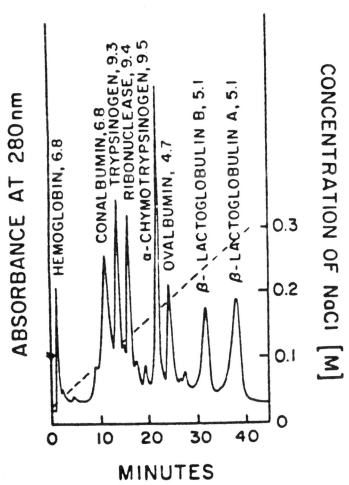

Fig. 11. Separation of some protein standards using a mixed bed of 40/60% wt: strong anion exchanger/strong cation exchangers and a NaCl molarity gradient.

be arranged to provide the optimum separation, and the separation may be carried out under gradient conditions if required. Care has to be exercised in these circumstances in that one column can affect the gradient that the second column experiences with a modification of the separation obtained (Fig. 12).

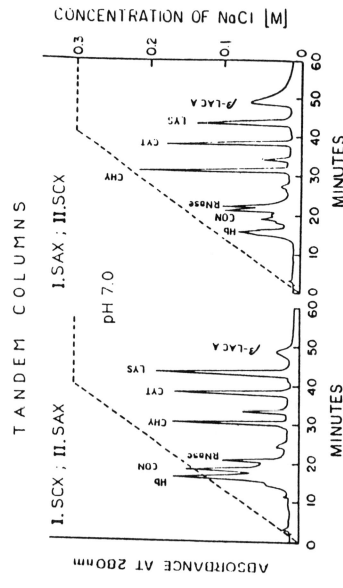

Fig. 12. Comparison between the separation of protein standards using tandem columns at pH 7.0. Order of columns, left: I,SCX, II,SAX; right: I,SAX, II,SCX.

6.5. Metal Chelate Interaction Chromatography

Metal chelate interaction chromatography (MCIC) is a method of separation in which the selectivity of a given separation may be modified by altering the nature of the metal ion on the surface of the support matrix.

This matrix may be any microparticulate rigid porous support with a "soft" surface capable of chemically binding suitable ligands. For example, iminodiacetic acid (IDA), tris-(carboxymethyl)-ethylene diamine (TED), and these bound ligands can themselves be used for EIC and HIC at appropriate pH values. Alternatively the selectivity of the "naked" phase can be moderated by immobilizing metals. Table 4 compares the relative retentions of some amino acids on "naked" IDA vs various immobilized metals.

Table 4
Relative Retentions of Some Amino Acids on Metal-IDA vs IDA

Amino acid	Relative retention at pH 6.0				
	Zn^{2+}	Ni^{2+}	Cu^{2+}	Mg^{2+}	Fe^{2+}
Phenylalanine	0.65	0.73	8.23	0.65	0.82
Tryptophan	1.00	1.00	19.6	0.92	1.21
Histidine	1.85	5.67	No elution	1.05	4.53
Lysine	0.63	0.60	3.76	1.00	1.91
Arginine	0.61	0.64	4.29	1.00	1.75

These results show that there is an order of "hardness" associated with the various metals used where:

soft = titanium (III)

intermediate = zinc(II), magnesium(II), nickel(II)

hard = copper(II), iron(III)

The strength of the metal complex varies inversely with temperature and it is usual to operate the system with increasing salt gradients.

Fig. 13 shows the effect of changing the metal ion on the separation obtainable between some protein standards at pH 5.0. Here it will be seen that the use of a "softer" metal gives better resolution between the components of the mixture and also illustrates the change in the order of elution and changing the nature of the metal. The separations shown demonstrate a markedly better

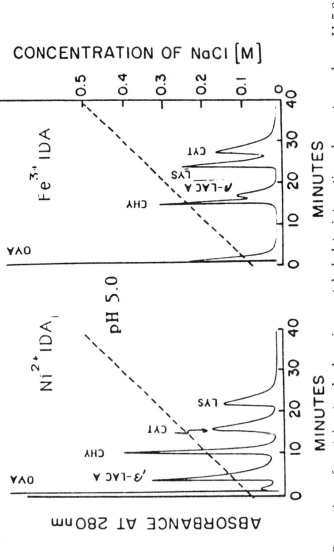

Fig. 13. Separation of protein standards using metal chelate interaction chromatography, pH 5.0. Left, Ni(II)IDA; right, Fe(III)IDA.

separation for these components when compared with the "naked" IDA (not shown).

7. Chromatofocusing

Isoelectric focusing (chapter 12) is a powerful method of protein separation and a novel way of operating ion exchange columns that incorporates the focusing principle, was first described by Sluyterman and coworkers in 1978 (*12-14*).

They proposed that a pH gradient could be set up in an ion-exchange column by taking advantage of the buffering action of the charged groups of an ion exchanger. If an ion exchanger is initially equilibrated at high pH using a suitable buffer, and a second buffer of low pH is run through the column, a pH gradient would be formed within the column. If such a mobile phase was used to elute proteins adsorbed on the exchanger, the proteins would elute in order of their isoelectric points. In addition, focusing would take place if the rate of movement of the mobile phase was faster than the rate of change of the pH gradient. This would result in band sharpening with associated sample concentration and hence very high resolution between components of the mixture.

This proposition was examined both theoretically (*12*) and experimentally (*13*), and was shown to be a powerful method for protein separation.

Initially a suitable anion exchange resin is equilibrated at high pH (e.g., 10.5) and packed into a suitable column. The dialysed sample is dissolved in the lowest pH buffer to be used, and applied to the column. Elution is allowed to proceed with this buffer. The acid–base exchange between the elution buffer and the ion exchanger produces a pH gradient between the limits of the two initial pH values, and this slowly proceeds down the column as the volume of the eluent buffer added increases. The elution buffer is prepared from suitable ampholytes with the desired pH and consists of a variety of differently charged species that, during migration down the column, bind to the exchanger, thus forming the gradient. The pH of the eluent initially leaving the column is that of the starting pH, but as elution proceeds, the pH of the column becomes more and more acidic until eventually the eluting buffer breaks through.

When a protein is applied to the column, its charged state depends upon the pH present in the column at that point. Thus, when the pH is less than its isoelectric point, (pI), the protein will be positively charged and will migrate with the eluting buffer. When it reaches a point at which the pH of the buffer in contact with the protein has increased so that the pH is greater than pI, the following occurs: The protein reverses its charge and is adsorbed to the exchanger, where it remains until the developing pH gradient causes the instantaneous pH at the point of adsorption to drop below the pI of the adsorbed protein, which again undergoes a reversal in charge and is carried through the system until again the pH drops; this process is repeated many times. Thus the protein travels through the column at the same rate as the pH gradient forms the condition pH = pI within the column, i.e., as the appropriate pH front moves down through the column. Hence the protein elutes from the column at its pI, and this is possibly one of the main disadvantages of this method, for at its pI, the protein is at its minimum solubility in the eluent and could precipitate if the applied sample size is too great. Obviously proteins with different pI values will migrate at different rates through the column, and the order of elution will be in order of decreasing pI.

7.1. Factors Controlling Resolution

The resolution obtained in chromatofocusing depends upon several factors, as indicated below.

7.1.1. Slope of the pH Gradient

The best resolution between bands is obtained with a fairly flat gradient, which may be accomplished by simply using dilute buffer concentrations that yield good separations with a narrow band width for the eluting components. Too flat a gradient, however, will result in large volume bands, and protein concentration will be low. This can be of advantage if the solubility of the protein is low. Thus a good compromise is a gradient volume of between 10 and 15 column volumes.

7.1.2. Column Packing

Since chromatofocusing produces very narrow bands, the quality of the packed bed should clearly be good. The particle size distribution of the exchanger should be as narrow as possible and evenly packed into the column.

7.1.3. Nature of the Ion Exchanger

In isoelectric focusing, the applied potential controls the resolution and speed of focusing. In CF, therefore, the difference in charge between the exchanger and the eluting buffer contributes to the sharpness of the zones produced. Thus an exchanger with a high ion exchange capacity is required when operated with dilute buffers (but note section 7.1.1).

Chromatofocusing is supported by Pharmacia, who supply reagents for undertaking the method. Hutchens (15,16), however, has undertaken high-performance studies using modified Pharmacia "polybuffers" to good effect, comparing the separations obtained with those of size exclusion chromatography (see later). Examples of separations obtained by CF are shown in Figs. 14 and 15.

8. Affinity Chromatography

Affinity chromatography (AC) is a highly selective form of separation technique that depends for its efficacy upon the specific interactions of biological molecules with their substrates, without which chaos would reign in biological systems. For example, an enzyme has strong affinity for its substrate and occasionally for its cofactor or effector, an antibody for the antigen that stimulated its response, a hormone for its carrier or receptor, and so on. These interactions have a very high level of specificity, which arises from the precise "fit" between the interacting species. Thus a binding site for an enzyme may incorporate ionic, hydrogen bonded, and hydrophobic sites, all spacially orientated so that the substrate is held in precisely the right position for enzymic reaction to occur. It is this property of specific interactions that is utilized in affinity chromatography.

Basically the technique involves the binding of an appropriate ligand to a water-insoluble matrix to form a tailor-made material capable of adsorbing from a mixture only those components of interest. Other materials present pass through the system, leaving the adsorbed components to be selectively desorbed by some suitable change in conditions, e.g., pH, increased salt concentration, and so on. There are spacial constraints that must also be considered. For example, the egg-white protein avidin has a huge affinity for biotin (approximately 10^{-15} mol/L). In spite of this affinity, avidin is scarcely retained by biotinyl-cellulose, but if a small spacer

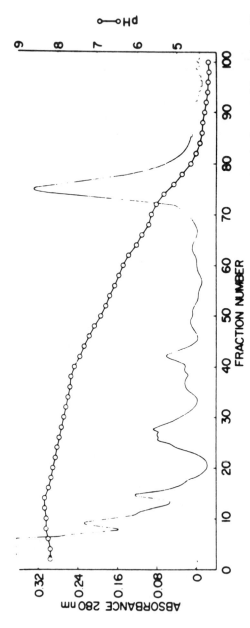

Fig. 14. HPLC chromatofocusing of cytosolic proteins from rat mammary gland after 14 d lactation. Conditions: 2–5 mg of cystolic proteins injected and eluted from the AX-300 column with a 70/30 mixture of Polybuffers 96 and 74, diluted 1:10 with 20% glycerol at pH 4.5 (from ref. 15).

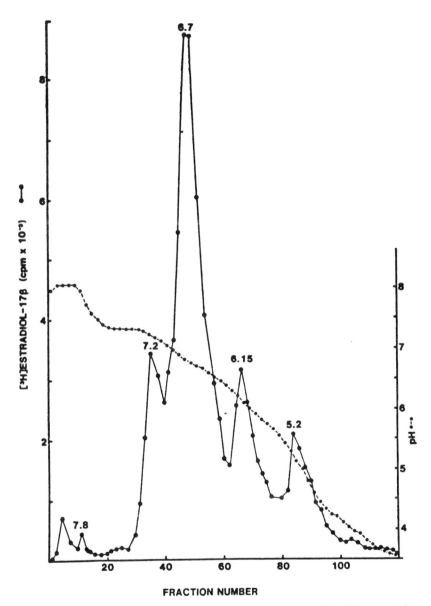

Fig. 15. Separation of ³H-estradiol-labeled estrogen binding proteins from the human uterus by high performance chromatofocusing (from ref. 16).

group is included between the matrix and the biotin, as in biotinyl-L-lysyl-agarose, the binding strength of the avidin is so strong that it can only be desorbed using 6M guanidium hydrochloride, pH 1.5! It is important to note that the specificity of the ligand may be altered by the spacer and the matrix, so that it is unwise to assume that the magnitude of an interaction between, for example, an enzyme and its substrate will be the same when an adsorbent is constructed from that substrate.

8.1. Support Matrices and Coupling

A wide variety of materials has been used as the support matrix for preparing the specific adsorbent (for example, cellulose, cross-linked dextran, acrylic copolymers, agarose, and glasses). The principal requirement is that the matrix contain reactive groups capable of undergoing chemical reaction, and the most favored group is the hydroxyl function.

The most important coupling method involves activation of the matrix, e.g., agarose, with cyanogen bromide, followed by reaction with a primary amine. This leads to the formation of cyclic or acyclic imidocarbonates that are able to react with primary amino groups such as lysyl residues of proteins to immobilize ligands. This process is pH dependent, but a pH of about 11 is a good compromise between optimizing the reaction and keeping the product intact. *Note*: Activated agaroses are now commercially available from suppliers such as Pharmacia, Biorad, and so on.

Although agarose-based supports are most widely used, they have some disadvantages, e.g., biodegradability and limited physical stability. Glass beads offer a viable alternative in that they are stable to extremes of pH and temperature, and are mechanically strong. They may be activated by reactions similar to the production of chemically bonded phases, e.g., by the use of aminopropyltriethoxysilane, when further chemistry to attach the appropriate ligand is relatively simple to perform. Figure 16 show some typical pathways.

8.2. Spacers

When the ligand is small, it is often necessary to introduce a spacer molecule to provide spacial accessibility when constructing the adsorbent. There are two approaches that may be used: (1) The modular, and (2) the systematic.

Analytical LC of Proteins

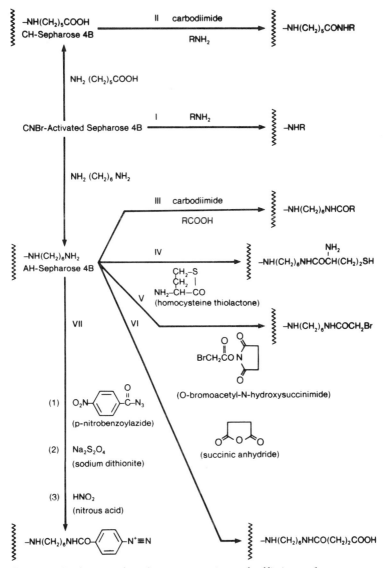

Fig. 16. Pathways for the preparation of affinity substrates.

A typical example of the modular method is the activation of glasses, as indicated above, which may be directly coupled to the ligand by one of the schemes indicated in Fig. 16. The disadvantage to this method lies in the large number of spacer arms available, which it may be impossible to completely substitute with ligands, thus leaving unwanted polar or nonpolar groups on the matrix.

In the systematic method, the ligand is constructed with the spacer arm, which is then reacted with the matrix. The advantage is that the product has no extra spacers present, although the chemistry involved may be very difficult. This is the preferred method when practicable, however. It is worth noting at this stage that the production of a suitable adsorbent for affinity separations should only be considered if the desired product cannot be obtained by any other means, or if considerable quantities of a protein are required. Very often the mere act of immobilizing a ligand can change its affinity for the protein, but when a satisfactory product is obtained, it is invaluable.

8.3. Operation

If the specificity of the assembled adsorbent is high, the separation of the component of interest may be achieved by simply incubating a slurry of the adsorbent with the mixture containing the component of interest, filtering through a sinter, and washing and desorbing the protein by washing with a suitable buffer or changing the pH of the wash.

Should several components have affinity for the adsorbent, then it may be packed into a column and elution conditions chosen to selectively elute successive fractions. This may be performed by a step-wise change in conditions or by performing a gradient (pH or ionic strength). Alternatively, denaturants such as urea or guanidine hydrochloride may be used to desorb strongly bound proteins that are capable of regeneration (Fig. 17).

Further details of the use of affinity chromatography may be found in refs. *17* and *18*.

All of the methods discussed above rely on the molecular interactions of the protein molecule with some suitable adsorbent. Now we consider the separation of proteins by methods that depend upon the molecular size of the molecules.

9. Gel Permeation Chromatography

Gel permeation chromatography (GPC) has many synonyms, notably size exclusion chromatography or gel filtration, but all operate by the same mechanism—the ability of molecules of dif-

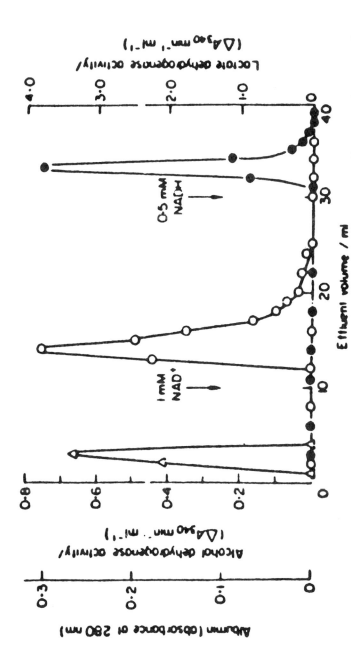

Fig. 17. Affinity separation of a mixture of proteins using two different eluting buffers. Conditions: column, Sepharose, bound NAD$^+$ analog. Order of displacement (1) bovine serum albumin (void volume), (2) horse liver alcohol dehydrogenase, (3) lactate dehydrogenase. Mobile phase changes at arrows, phosphate buffer to phosphate buffer containing 1 mM NAD$^+$ and phosphate buffer containing 0.5 mM NADH (from ref. 29).

ferent molecular dimensions to penetrate the pore structure of the chromatographic bed. Those molecules that are unable to penetrate the pores will pass through the sytem unretained, whereas those that can penetrate all the pores will be retained. Those that can penetrate a fraction of the pores will be retained for an intermediate period of time between the two extremes. Thus the sample components elute from the separation column in order of *decreasing* molecular size.

A wide variety of materials is commercially available to perform GPC. For example, in the molecular weight range of 20,000 to 1,000,000 daltons, Superose 6 (Pharmacia), TSK 4000 (Toya Soda), Si 300 Polyol (Serva), and Zorbax GF-450 (USA) may be used, generally with buffers of about neutral pH containing salt to neutralize the electrostatic interactions, which may be present because of the small negative charge associated with unreacted silanol groups on the silica surface. *Note*: This is not required with Superose, which is agarose based.

Thus separation by GPC occurs between two clearly specified volumes, the void or interstitial volume V_0 (the volume outside the particles of packing) and V_t (the total volume contained within the column, which is the pore volume V_i plus the interstitial volume V_0). Because of this, GPC has often been called the perfect form of chromatography, in that elution times are clearly defined (providing adsorption and/or partition effects are absent).

The retention volume in GPC is controlled by the following equation:

$$V_R = V_0 + K_{GPC} V_i$$

Where K_{GPC} is the distribution coefficient of the solute. The other terms are as quoted above.

The retention characteristics are illustrated in Fig. 18, which shows the position of total exclusion V_0, and total permeation V_t. The plateau between these two points is where partial penetration of the pores occurs, and ideally, for maximum-separation capacity, this plateau should have as shallow a slope as possible, and the column efficiency should be as great as possible (see below).

9.1. Mobile Phase

In contrast to interactive methods of chromatography, the sole role of the mobile phase is to provide a suitable medium to dissolve the sample, to transport it through the chromatographic bed, and

Analytical LC of Proteins

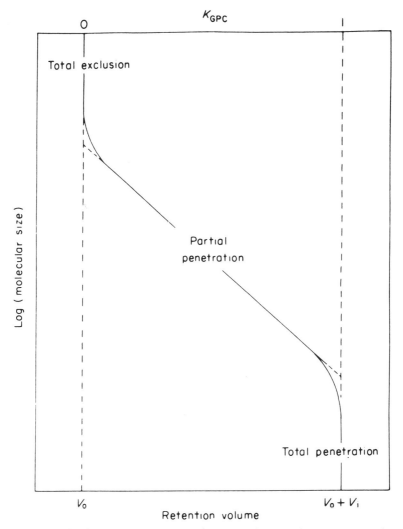

Fig. 18. GPC operating range, theoretical curve between two fixed points, total exclusion and total permeation.

to act as the "stationary phase" in the pores of the packing. Ideally the viscosity of the mobile phase should be low to permit fast diffusion of the solute molecules in and out of the pores of the packing. For biopolymers the mobile phase is usually a suitable aqueous buffer, normally in the pH range of 4.5 to 7.5. Care must be taken not to exceed this upper value because silica is soluble at pH values of greater than 7.5. Thus, the nature of the mobile phase is almost ideal for biopolymers. It is possible to modify the nature of the mobile phase using the following parameters:

Salt concentration or ionic strength of the solvent
pH value of the eluent
Addition of surfactants, preferably non-ionic, e.g., Triton X-100
Within limits, the temperature of the column

Care must be exercised in varying these parameters to avoid protein denaturation or precipitation. This is particularly important in preparative work (chapter 13).

9.2. Dispersion Mechanisms

In biopolymer separations, as distinct from artifical polydisperse polymers, the materials to be separated are present as discrete entities, and are subject to the same chromatographic dispersive effects as in the other forms of chromatography, i.e., the band width is a function of the residence time within the chromatographic bed and is described by the equation:

$$N = \{4V_r\}^2/w^2$$

where N is the plate number of the column and w is the bandwidth at the baseline intersection of the tangents at the point of inflexion of the elution peak.

The separation between two components is expressed in terms of the resolution, where:

$$R_S = 2(V_2 - V_1)/(w_1 + w_2)$$

To achieve good resolution, R_s must be greater than unity, i.e., $(V_2 - V_1) > w$.

In GPC, the value of K_{GPC} always ranges between 0 and 1, and because of this limited range, the number, n, of components in a sample that can be resolved is related to the column efficiency, N, by the following relationship suggested by Giddings (19):

$$n = 1 + 0.2N^{1/2}$$

Thus for $n = 21$, $N = 10,000$ plates, i.e., high column efficiencies are required for high-resolution separations.

A further factor controlling resolution is the pore size distribution of the packing, and it may be seen (Fig. 19) that separation power is inversely proportional to the slope of log (solute mw) vs V_R. Figure 20 shows the plot of elution volumes of various proteins on two Sephadex columns.

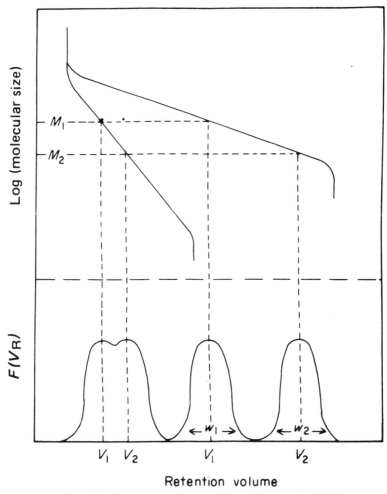

Fig. 19. Size separation and resolution in GPC.

Finally, it is clear that since separation by GPC depends upon the ability of the solute molecules to penetrate the pores of the packing, and because the diffusion rates of large molecules are low, it is necessary to operate GPC columns at relatively low flow rates (about 1 mL/min or less) to allow diffusion into and out of the pores to occur without causing excessive band broadening from nonequilibrium effects. High mobile phase flow rates can set up high shear forces within the column, and with some proteins it has been shown that these forces are sufficiently powerful to break the protein's peptide bond (20).

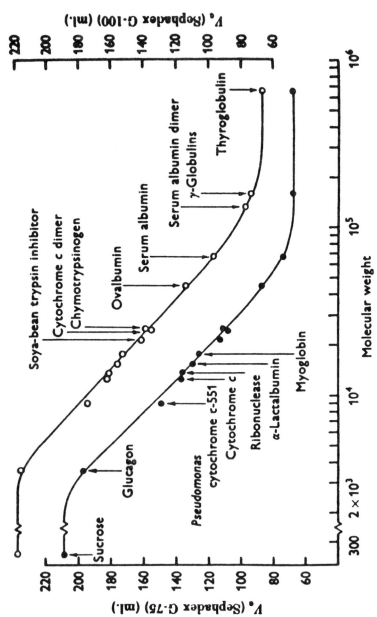

Fig. 20. Plot of elution volume vs molecular weight of proteins on two sephadex columns (from ref. 30).

The separation of some proteins by open-column and high-performance techniques are shown in Figs. 21, 22, and 23.

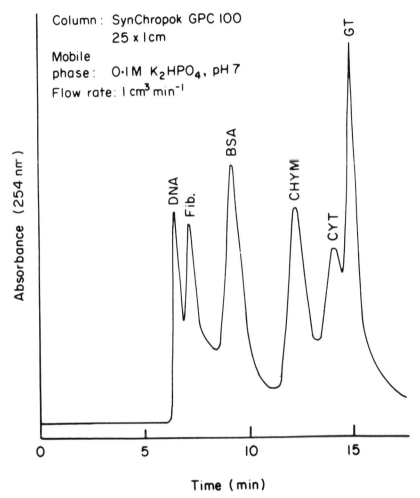

Fig. 21. Resolution of biopolymers by aqueous high performance GPC. Conditions Sychropak GPC 100 250 × 10 mm. Mobile phase 0.1M K_2HPO_4, 1 mL/min. Peak identity: DNA, deoxyribonucleic acid; Fib, fibrinogen; BSA, bovine serum albumin albumin; CHYM, chymotrypsinogen; CYT, cytochrome c; GT, glycyltyrosine (from ref. 27).

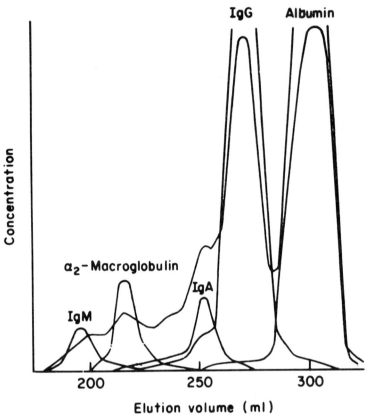

Fig. 22. Gel filtration of serum proteins on Sephacryl S-300 Superfine (courtesy of Pharmacia Fine Chemicals).

10. Field Flow Fractionation

Consider fluid flow through a capillary tube. Figure 24 illustrates the nature of this flow, showing it to be parabolic in character. The position of maximum flow is axial with the capillary, and progressively lower velocity streamlines occur as the cross-section of the tube is traversed, until, at the walls of the tube, the flow is effectively zero.

Note: This behavior illustrates the requirement in LC of minimizing the length and bore of connecting tubes between the injector/column and column/detector. Clearly band broadening can be brought

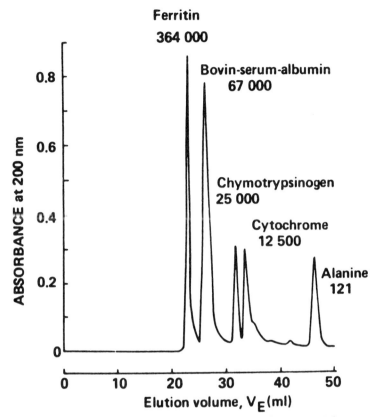

Fig. 23. Separation of proteins on Diol columns with aqueous phosphate buffer as eluent.

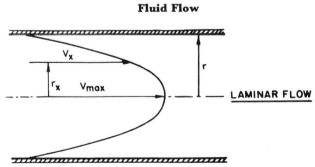

Fig. 24. Laminar flow in open capillaries.

about by this parabolic flow profile and the associated wash-out effect that occurs in injection valves.

Although the parabolic flow of fluids in narrow bore tubes has been recognized for many years, Giddings (21) first realized in 1966 that if molecules could be forced by an external field into the low-velocity flow streamlines, then, depending on how far they penetrated, it should be possible to achieve a separation between species of different molecular size and diffusivity.

This situation is illustrated in Fig. 25, which shows the basic concepts of the method. The sample is placed in a rectangular cross-section flow channel and the field, orientated perpendicular to the direction of fluid flow, causes the sample molecules, which are capable of interacting with the field, to migrate toward the walls of the channel. Diffusion prevents complete passage to the walls. Field flow fractionation has been called single-phase chromatography (if this statement has any meaning), and although the departure from conventional LC is great, the technique shares some of the underlying theoretical principles of LC. Because the flow channel is open and unobstructed by packing, the complex, geometry-dependent parameters of LC, such as pore size and particle size distribution, are absent. Thus, eddy diffusion is nonexistent. Further, since we are dealing with a one-phase system, such other parameters as partition coefficients and mass transfer between the stationary phase and mobile phases no longer exist. Hence those variables that bring about uncertainties in the chromatographic models do not appear in FFF. Thus the theory of FFF is far more tractable and indeed precise. A further advantage is that FFF may be supported by conventional LC equipment, i.e., mobile phase delivery systems, detectors, and so on.

10.1. Theory and Mechanism of FFF

Following the introduction of the sample into the separation channel (which is normally carried out in the absence of carrier flow), the external field is applied and begins to force the solute molecules toward the bottom wall of the channel. In the absence of any other effect, all the molecules would collect on this wall. However, diffusion counteracts the effect of the field, and shortly after injection a steady state is reached in which the various different solute types have concentrated into thin lines close to the wall. This concentration is exponential in nature and is expressed by:

Analytical LC of Proteins

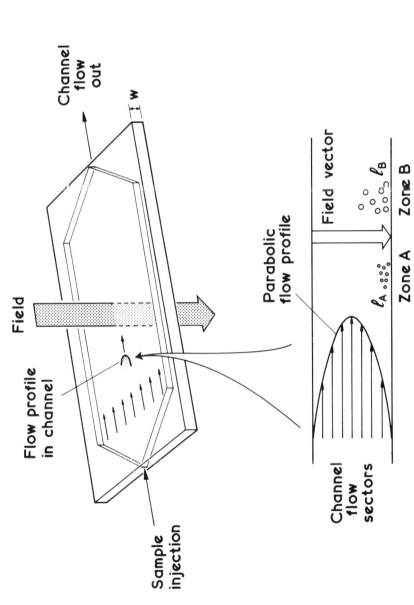

Fig. 25. A generalized FFF channel scheme showing the parabolic flow profile in detail (from ref. 22).

$$C = C_0 e^{[-x/l]}$$

where C = concentration of the solute at distance X from the lower wall, C_0 = concentration of the solute at the wall ($X = 0$), and l = approximately the layer thickness. The value of l depends upon two factors. (1) The strength of interaction of the solute molecules with the applied field. This causes the molecules to migrate with a velocity U toward the lower wall, and (2) the solute diffusivity, D, which acts to oppose the build-up of solute at the wall. Thus l contains both D and U, related by:

$$l = D/U$$

Thus since different solute types could be expected to have different diffusivities and different interaction strengths with the applied field, each different species would therefore be expected to have a unique l value.

Clearly the nature of the field is important and various types of field have been investigated that have different ranges of applicability; the nature of the solvent can also be varied to provide a suitable environment for a particular application (see later).

When the steady state has been achieved after the application of the field, the solvent flow is started and the solute molecules are transported downstream through the channel. Because the flow profile between the parallel plate is parabolic (Fig. 22), the greatest flow velocity is at the center of the channel. Thus, those solutes whose layers extend well into these rapidly moving streamlines will pass through the channel more quickly than those solutes compressed into the slower-moving streamlines. The output, monitored by a suitable detector, will show a series of peaks with the smaller molecules emerging before the larger.

10.2. Types of Field

Theoretically any field that can interact with the solute molecules can, in principle, be used in FFF. Four types of field have been investigated. The principle parameters that control separation are indicated in Table 5.

10.3. Some Separations by Various Modes of FFF

The principle of FFF is shown in Fig. 26, which indicates that enforced sedimentation (SFFF) is achieved by placing the flow chan-

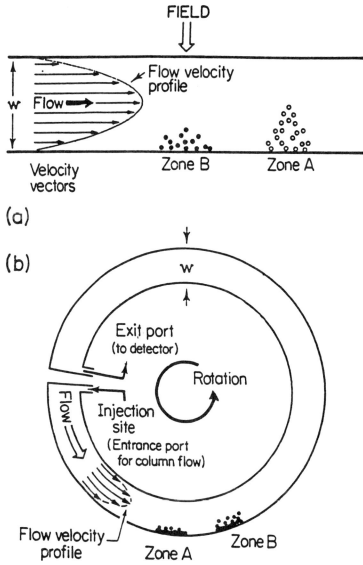

Fig. 26. Principle of sedimentation FFF (from ref. 23).

nel in a centrifuge basket. The flow channel is constructed of two polished stainless steel strips that sandwich another strip of stainless steel of the appropriate thickness in which the channel is cut. Full details of the construction of this and the other flow channels are

Table 5
Field Types and Controlling Parameters in FFF

FFF method	Principal parameter
Sedimentation	Molecular weight, M, Density
Electrical	Diff. coeff. D Mobility
Flow	Diff. coeff. D Stokes rad. a
Thermal	Thermal diff. coeff.

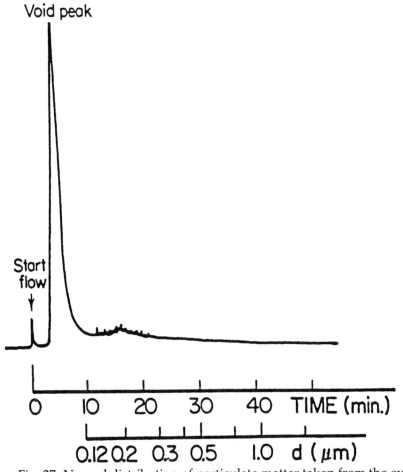

Fig. 27. Normal distribution of particulate matter taken from the eye of a 16-yr-old boy (SFFF) (from ref. 23).

Analytical LC of Proteins

given in ref. 22, which also provides relevant experimental details. An example of the use of SFFF is in the determination of cataractous lens tissue. Figure 27 shows the normal distribution of particulate matter taken from the eye of a 16-yr-old boy, whereas Fig. 28 shows the equivalent distribution taken from the eye of a 75-yr-old male.

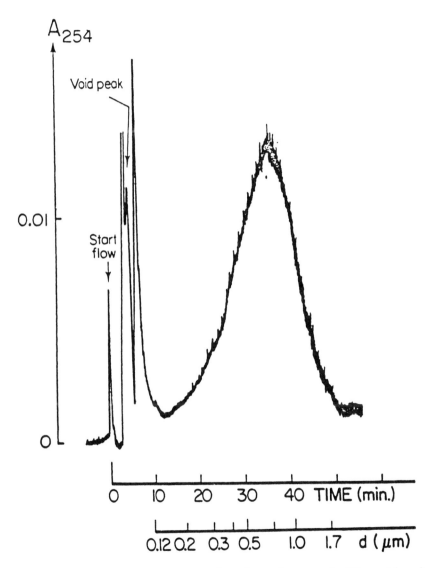

Fig. 28. Precipitated proteins taken from the eye of a 75-yr-old male (SFFF) (from ref. 23).

The difference is clear. *Note:* The particle size and distribution of the precipitated proteins may be determined by this method (23). Figure 29 shows the separation of some proteins using electrical FFF, (EFFF) Fig. 30. compares the protein constituents of rat and human plasmas by flow FFF, (FFFF) and Fig. 31 demonstrates the separation of two viruses by the same technique.

10.4. Advantages and Disadvantages of FFF

The advantages of FFF include the following:

1. The FFF method is a one-phase system. The carrier liquid can therefore be chosen for optimum interaction.
2. Equipment design is such that the minimum surface is exposed to the solute molecules, a major difference to LC, which operates with a high surface area exposed to the solute.
3. Programming is simple to perform. Simply change the applied field strength. This is capable of closer control than in LC.
4. Flow rate is easily adjusted, and the system back-pressure is low.
5. Separations are predictable providing λ is low.
6. It is possible to obtain reliable physicochemical data, which could be time consuming to obtain by other means.
7. Ready isolation of pure components is possible.
8. It can be used for the analysis of a variety of species not amenable to LC methodology.

Some possible disadvantages are:

1. Separation times are somewhat slow because the diffusion rates of large molecules are slow. More sophisticated designs of equipment may alleviate this problem, however.
2. The separation between the upper and lower walls of the flow channel is critical if predictable results are to be achieved.
3. The development of the system is relatively slow at the present time, principally because few people are using the system as yet. This situation will probably change in the relatively near future with the recent introduction of equipment for performing SFFF by DuPont.

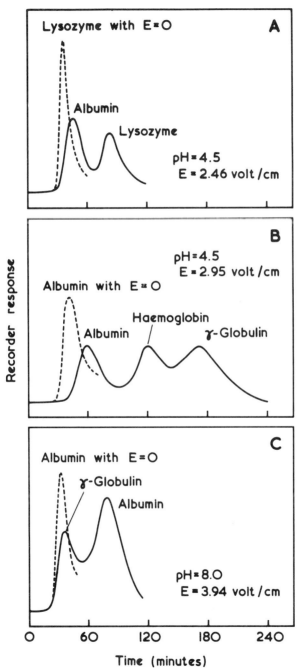

Fig. 29. Protein separation by electrical FFF (from ref. *31*).

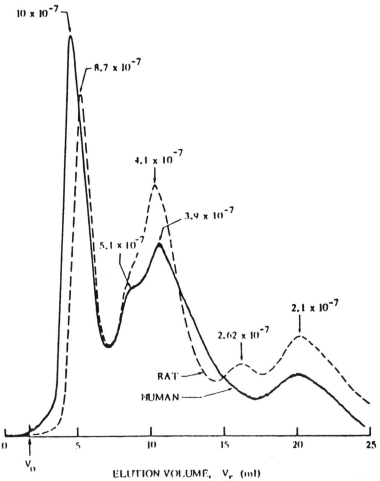

Fig. 30. Comparison of the proteins in human and rat plasmas by flow FFF (from ref. 32).

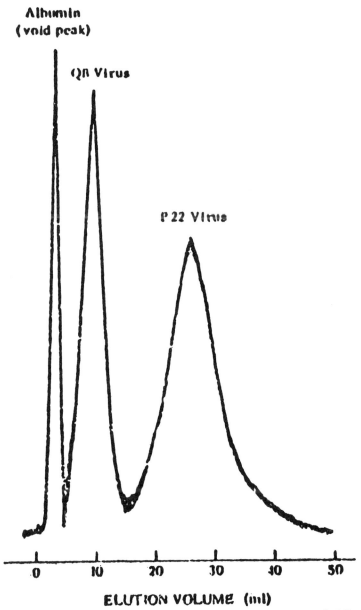

Fig. 31. Separation of viruses by flow FFF (from ref. 22).

References

1. R. P. W. Scott and P. Kucera, *Anal. Chem.* **45**, 749 (1973).
2. R. P. W. Scott, *Proceedings of Faraday Symposium of the Royal Society of Chemistry.* **15** (1980).
3. F. Hampson and C. F. Simpson, unpublished results.
4. K. Hostettmann, *J. Med. Plant Res.* **39** (1980), and K. Hostettmann, M. Hostettmann-Kaldas and K. Nakanishi, *J. Chromatog.* **170**, 355 (1979).
5. S. J. Singer and G. L. Nicholson, *Science* **175**, 720 (1972).
6. C. Horvath and W. Melander, *Am. Lab.* **10** (10), 17 (1978).
7. R. P. W. Scott and C. F. Simpson, *Proceedings of Faraday Symposium of the Royal Society of Chemistry* **15** (1980).
8. M. J. O'Hare and E. C. Nice, *J. Chromatogr.* **149**, 241 (1978).
9. I. Molnar, *Presentation at 8th International Symposium on Column Liquid Chromatography*, New York, no. 2a-B4 (1984).
10. M. T. W. Hearn, *J. Chromatogr.* **317**, (1984), and K. Benedek, S. Dong, and B. L. Karger, *J. Chromatogr.* **317**, 227 (1984).
11. M. T. W. Hearn, *Proceedings of the 9th International Symposium on Column Liquid Chromatography*, Edinburgh, 1985.
12. A. AE. Sluyterman and O. Elgersma, *J. Chromatogr.* **150**, 17 (1978).
13. A. AE. Sluyterman and J. Wijdenes, *J. Chromatogr.* **150**, 31 (1978).
14. A. AE. Sluyterman and J. Wijdenes, *J. Chromatogr.* **206**, 429–441 (1981).
15. T. W. Hutchens, R. D. Wiehle, N. A. Shahabi, and J. L. Wittliff, *J. Chromatogr.* **266**, 115 (1983).
16. T. W. Hutchens, W. E. Gibbons, and P. K. Besch, *J. Chromatogr.* **297**, 283 (1984).
17. H. Guilford, in *Practical High Performance Liquid Chromatography*, Heyden, London (1976).
18. Pharmacia Handbook, *Affinity Chromatography* Pharmacia, Sweden (1983).
19. J. C. Giddings, *Anal. Chem.* **39**, 1027 (1967).
20. M. T. W. Hearn, personal communication (1985).
21. J. C. Giddings, *Sep. Sci.* **1**, 123 (1966).
22. J. C. Giddings, M. N. Myers, K. D. Caldwell, and S. R. Fisher, in *Methods in Biochemical Analysis* (P. Glick, ed.) John Wiley, New York (1980).
23. K. D. Caldwell, B. J. Compton, J. C. Giddings, and R. J. Olson, in preparation.
24. R. P. W. Scott and C. F. Simpson, *J. Chromatogr.* **197**, 11 (1980).
25. H. J. Pownall, J. B. Massey, and A. M. Grotto Jr. (1981).
26. W. S. Hancock and J. T. Sparrow, in *A Laboratory Manual for the Separation of Biological Materials by HPLC* Marcel Dekker, New York (1981).
27. F. E. Regnier and K. M. Gooding, *Anal. Biochem.* **103**, 1 (1980).
28. G. W. Welling. HPLC of Proteins. Lecture at Amsterdam Summer School on HPLC (1985).

29. H. Guilford, in *Practical High Performance Liquid Chromatography* (C. F. Simpson, ed.) Heyden, London (1976).
30. P. Andrews, *Biochem. J.* **91**, 222 (1974).
31. K. D. Caldwell, L. F. Kesner, M. N. Meyers, and J. C. Giddings *Science* **176**, 296 (1972).
32. J. C. Giddings, F. J. Yang, and M. N. Meyers. *Anal. Biochem.* **81**, 395 (1977).
33. W. S. Hancock and J. T. Sparrow, in *High Performance Liquid Chromatography, Advances and Perspectives.* (C. Horvath, ed.) Academic, New York (1983).

Suggested Further Reading

Six international meetings have been held on protein separation and these have been published by Academic Press. These meetings were held in Washington, 1981; Baltimore, 1982; Monaco, 1983; Boston, 1984; Washington, 1985; and Baden-Baden, 1986 and represent the state of the art in this subject.

In addition the reader is recommended the book edited by C. Horvath entitled *High Performance Liquid Chromatography, Advances and Perspectives*, Academic Press, 1983, chapters by M. T. W. Hearn and W. Hancock.

Further information on affinity chromatography will be found in a book of that name by W. H. Scouton, published by John Wiley, 1981. Size exclusion is considered in detail by J. J. Kirkland, W. W. Yau, and D. D. Bly, in *Modern Size Exclusion Chromatography. Practice of Gel Permation and Gel Filtration Chromatography*, John Wiley, New York, 1979. General chromatography is given in *Techniques in Liquid Chromatograhy*, C. F. Simpson, ed., Wiley Heyden, 1982.

Chapter 12

Analytical Electrophoresis

Colin Simpson

1. Introduction

The first true use of electrophoresis for the separation of proteins developed from the pioneering work of Tiselius (1) in 1937, but the concepts and developments that led to the introduction of the technique occurred in the mid-nineteenth century with Faraday's laws of electrolysis. These laws predated the first accepted demonstration of the chromatographic technique by about 50 years. From Faraday's laws the concept of transport numbers was developed by Hittorf in 1850. In 1870, Kohlrausch demonstrated the measurement of conductivity using an ac Wheatstone bridge system with a telephone receiver as the null indicating device. From conductivities, the concept of ionic mobilities was developed. These three concepts are fundamental to the development of electrophoresis. The moving boundary method of Tiselius, however, although a superb separation method, was a very difficult technique to control and had the disadvantage that only the slowest and fastest migrating species could be obtained in pure form and then only in small quantities. It was therefore not suited for preparative use. Interestingly, it was Martin's development of partition and paper chromatography (2,3) that gave the stimulus to develop electrophoretic methods capable of separating components into zones using materials similar to those used in chromatography. Development was rapid, and there is now a powerful armoury of methods suitable for protein separation.

In this chapter, it will be assumed that the basic premises of electrophoresis (EP) are known and the established method of performing the technique are understood (*4,5*). Here the modern methods of performing electrophoresis will be considered. Zone electrophoresis (ZEP), including high voltage ZEP (HVZEP), isoelectric focusing (IEF), two-dimensional electrophoresis, isotachophoresis (ITP), and associated techniques, will be discussed.

2. Isoelectric Focusing

Isoelectric focusing (IEF) is a relatively new electrophoretic technique. Although the reports of its use date to the turn of the century, it is only in the last 20 years that the method as it is known today was demonstrated by Svensson (*6,7*) [based upon background information provided by Kolin (*8*)], who with Vesterberg developed the requisite materials (*9*) for forming the pH gradient. The technique aroused considerable interest and many innovative papers have resulted. It has also been the subject of three relatively recent books (*10–12*). Numerous reviews have appeared, as well. Here the recent advances in the technique will be reviewed.

Isoelectric focusing is the electrophoretic migration of ampholytes (i.e., species containing both positive and negative charges; proteins, for example) in a pH gradient. Those species possessing a net positive or negative charge will migrate toward the appropriate electrode until they reach a position in the pH gradient at which the pH equals the pI (isoelectric point) value of the ampholyte. At this point migration will cease. Should the molecules diffuse out of the zone in which pH = pI, they will automatically move back to the pI position. Thus at equilibrium, the dispersive forces are balanced by the electrophoretic migration forces, i.e., the ampholytes are focused.

2.1. pH Gradient

Svensson recognized (*13*) that at the electrodes there is a continuous production of hydrogen and hydroxyl ions. The anode naturally becomes more acidic and the cathode naturally becomes more alkaline with the passage of an electric current. This produces a steep pH rise at the cathode and a fall at the anode. If there is an amphoteric substance present with some buffering capacity at its pI, it will form a small pH "plateau" at the position of its maximum concentration. With several such substances, they will align themselves and produce a series of plateaus and a step pH gradient will tend to be generated. The more ampholytes that are present, the smoother the pH gradient will be, particularly if the individual ampholytes overlap to a certain extent. Materials that generate pH gradients in this way are called carrier ampholytes.

Clearly the success of any IEF experiment depends upon the establishment and maintenance of a stable pH gradient in an anticonvection gel. This in turn depends upon the nature of the materials used to produce the gradients. In general a linear gradient is satisfactory for most applications, and small deviations can be tolerated. Occasionally a flattened gradient may be used to increase the resolution between components. Of great importance, and not often considered, is that the pH gradient used should have a reasonably uniform conductance. Large variations in conductance can change the field strength and hence the migration rate of the ampholytes, which results in poor focusing. Variations in conductance can also give rise to temperature changes, which can lead to diffusional broadening of the zones. Conductance gaps, associated with gradients formed from mixtures of peptides, bring about a steep rise in temperature, if not complete breakdown in operation of the system. In addition, the ampholytes should have good buffering capacity at their pI value so that the integrity of the pH gradient is maintained even in the presence of relatively high concentrations of protein. For practical purposes, it is preferable for the ampholytes to have relatively low molecular weights, e.g., in the range of 300 to 1000, and also to have low absorbance in the UV for ease of scanning gels.

These requirements are adequately met by the commercial products available on the market, e.g., Ampholine (LKB), Servalyt (Serva), Pharmalyte (Pharmacia), and Biolyte (BioRad). All of these products produce good pH gradients. Ampholine is prepared by condensing unsaturated fatty acids with polyamines. Servalyt is similar, but contains, in addition, sulfonic and phosphoric acid groups. Pharmalyte is prepared by copolymerizing epichlorohydrin with amines, amino acids, and peptides, whereas BioLyte is identical to Servalyt. A comparison of these materials has been made by Radola (14) and Allen (15), who showed that the slope of the pH gradient and the conductivities varied among samples. Allen further showed that the best pH gradient was obtained from a combination of all the commercial products, which supports the view that the best gradients have the maximum number of ampholytes per pH unit.

2.2. pH Drift

One of the problems experienced with pH gradients in gels rests in the slow drift to the cathode of the pH gradient with extended focusing times. This results in a flattening of the gradient in the center of the gradient and a sharp rise in pH at the anode. The reason for the cathodic drift is still unknown in spite of numerous attempts to define it. Attempts to minimize drift by including sucrose, urea, methyl cellulose, and so on into the gel have not been wholly unsuccessful, presumably because the addition of these additives only results in an increase in viscosity, which will increase the focusing time through the lower mobility of the species being focused.

2.3. The Anticonvective Medium

As the title implies, the anticonvective medium should provide good anticonvective properties to support the concentrated focused zones and provide a stable pH gradient. For the latter requirement, it is necessary that the medium have zero, or near zero, electroendosmosis. Any electroendosmotic flow will move the whole pH gradient and focused zones. A further requirement for the medium is that it have minimal sieving effect on the protein mixture; otherwise, excessive time is spent in the migration of the ampholytes to their isoelectric points, which delays the attainment of the equilibrium.

Polyacrylamide gel (PAG) possesses virtually no electroendosmosis in the pH range of 3 to 10, and it has been used with success for a large number of separations. Unfortunately, it is unsuitable for the separation of the larger proteins, whose migration is hindered by the sieving effect. Gels prepared with a high percentage of cross-linking agent have a more open-pore structure, but are brittle and opaque, and do not adher well to glass plates unless they have been silanized (14).

Agarose possesses an excellent open-pore structure that allows for free migration of the ampholytes, even up to molecular weights of 5 Mdalton. Until recently, agarose possessed severe electroendosmosis, and was unsuitable for IEF. Rosen et al. (16), however, have produced an agarose with a very low electroendosmosis, suitable for focusing proteins with high molecular weight. For the low molecular weight proteins, however, PAG is still preferred.

Analytical Electrophoresis

Other materials have been used for IEF. For example, thin layers of Sephadex have been used very successfully for focusing; dispersion of the focused bands is low. Sephadex does not accept stains, however, so paper prints are generally taken and then developed; some zone broadening occurs at this stage.

2.4. Dependence of Zone Width on Operating Parameters

Svensson (13) developed the following equation to predict the width of a focused zone:

$$x_i = \pm \sqrt{\frac{D}{E - \left(\dfrac{du}{d(\text{pH})}\right)\left(\dfrac{d(\text{pH})}{dx}\right)}}$$

where x_i = the zone width taken at the point of inflexion of the focused zone, D = diffusion coefficient, E = electric field strength, $du/d(\text{pH})$ = rate of change of mobility with pH, and $d(\text{pH})/dx$ = pH gradient.

The equation shows that sharper zones with better resolution will be obtained by increasing the applied voltage. To halve a zone width, the applied potential must be increased by a factor of four, but the Joule heat developed is proportional to the square of the applied voltage. Thus halving a zone width brings about a 16-fold increase in the heat produced, and this places severe requirements upon the cooling of the system, because as the temperature of the gel/plate increases, the ampholyte diffusion rate also increases, and all the benefit gained in increasing the voltage to obtain narrow bands is lost.

Various means have been used to attempt to overcome this problem. One of the best in terms of heat removal is to use metal cooling blocks separated from the plate with a thin plastic film. This is potentially electrically hazardous, however. Allen (17) has used plates constructed from beryllium oxide, which is an excellent electrical insulator and has a high thermal conductivity. Using Peltier cooling units, Allen was able to operate such plates at high potentials, 300 V/cm, and was able to detect three zones per millimeter.

2.5. Ultra-Thin-Layer IEF

Probably the most successful way of minimizing the heating effect is by using ultra-thin-layer gels (18,19). This method has additional bonuses. The diffusion distances are reduced, so that the bands are sharper. Staining and destaining are much more rapid than with conventional gels, so the protein bands are more rapidly fixed on switching off the power, again minimizing zone diffusion. Last, because the depth of the gel is only about 100 µm, the pH gradient is likely to be more homogeneous, which will promote more narrow zones.

One of the problems associated with the use of thin-layer gels is that the gels detach from the surface of the supporting plate during the staining and destaining procedures, thus imposing considerable strain on the gel during the subsequent visualization steps. This problem is exacerbated as the thickness of the gel is reduced. Radola's technique (19) prevents this from occurring and provides a method in which batches of plates may be prepared and stored for subsequent use.

Glass plates, 1 mm thick, are thoroughly degreased and the surface silanols reacted with methacryloxypropyltrimethoxysilane (Polyfix 1000) in ethanol/water (1:1), followed by air drying. Polyacrylamide gel is polymerized onto this surface, which participates in the reaction, thus chemically binding the gel to the surface. Radola established that 5% T and 3% C gels prepared from acrylamide and N,N'-methylene-bisacrylamide were the best formulations to use. Ampholine or Servalyt (3%) was included as the carrier ampholyte. Polymerization was carried out using the optimum amount of ammonium persulfate as initiator (determined by prior experiment). Prepared gels could be stored for several months, providing the surface was covered by a hydrophilized polyester film.

The technique of throwing the gels to provide the ultra-thin layer required is relatively simple using the flap technique (20) (Fig. 1). This consists of two 4-mm thick glass plates. One plate serves as the support for the silanized glass plates, and the thickness of the gel is determined by placing a thin strip of self-adhesive tape of known thickness along both edges of the silanized plate. The other thick glass plate is pretreated by rolling on to it a hydrophilized polyester film coated with a thin film of water. The calculated amount of polymerization mixture is applied to one end of the silanized plate, and the top plate is flapped down on to it. This causes the polymerization mixture to spread evenly along the

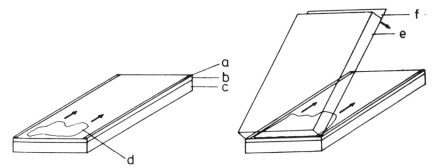

Fig. 1. Radola's flap technique for ultrathin layer gels (from ref. *19*). See text for details.

plate, with the slight excess exuding at the ends. After polymerization, the top plate is removed (the cover film protects the surface of the gel during this step) and the gel is ready for use after 2 h. This technique may also be used for the preparation of gels on silanized polyester films. It is important to use the thick glass plates in this method to ensure that the gel film is even. The working plates tend to bend during the coating procedure, resulting in an uneven gel layer. Miniature plates prepared on microscope slides may be prepared in the same way. *Note*: BioRad markets a system very similar to that described above, and LKB markets a sliding spreader for ultra-thin gels.

The performance of gels prepared in the way is excellent. Figure 2 shows the separation and reproducibility obtained on marker proteins and a crude fungal extract, and Table 1 compares the performance of ultra-thin IEF with conventional methods.

It is clear that there are considerable advantages in using the ultra-thin-layer method of IEF. The time of analysis is small, staining and destaining (see later) is markedly faster, and the consumption of carrier ampholytes is reduced. The resolution between zones is high. It is possible to distinguish three zones per millimeter. This is largely because of the excellent heat dissipation in the ultra-thin layer, which allows for higher field strengths (100–500 V/cm), thus minimizing the diffusion of the zones. There is, however, still a problem of pH gradient drift, but even this effect is reduced because of the shorter focusing times.

2.6. The Immobiline System

Probably the most significant development in the last few years has been the development of the Immobiline gels for IEF (*21*), which

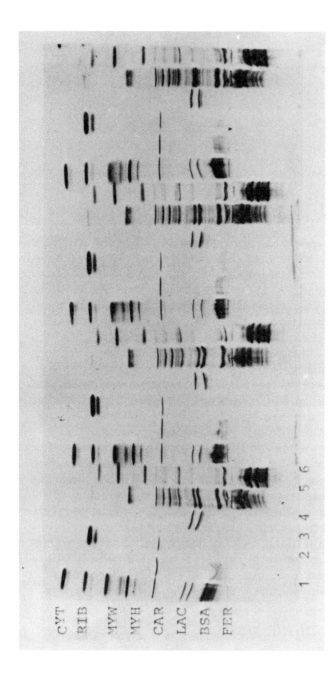

Fig. 2. Ultrathin layer separation of some fungal proteins and standards (from ref. 19. Gel thickness 100 μm on silanized polyester film. Standards, tracks 1–4, samples track 5,6. CYT, cytochrome C; pI 10.1; RIB, ribonuclease, pI 9.3; MYW, sperm whale myoglobin, pI 8.3 (major component); CAR, carbonic anhydrase, pI 5.9; LAC, beta-lactoglobulin, pI 5.14; BSA, bovine serum albumin, pI 4.9; FER, ferritin, pI 4.2–4.5.

Table 1
Comparison of Different Methods of IEF[a]

Method	Volume per sample, mL	Load protein, mg	Focusing time, min	Evaluation time, min
Minature ultra-thin layer IEF, 50–100 µm, 3 cm	0.005–0.01	0.003–0.001	10–20	15–30
Ultra-thin layer IEF, 50–100 µm, 12 cm	0.06–0.12	0.001–0.01	60–150	15–30
Thin-layer IEF, 1 mm, 12 cm	1.2	0.002–0.02	60–150	60–1200

[a]From ref. 20.

has completely solved the problem of pH gradient cathodic drift.

Immobilines are marketed by LKB (22). The system consists of a series of specially synthesized buffering acrylamide derivatives with the general structure:

$$CH_2 = CH - \underset{\underset{\displaystyle O}{\|}}{C} - \underset{\underset{\displaystyle H}{|}}{N} - R$$

where R contains either a carboxylic acid or a tertiary amino group. The derivatives copolymerize with acrylamide and bis to produce an immobilized pH gradient. Each of the derivatives has a defined and known pK value. The pH gradient is formed by mixing two buffered solutions: (1) basic (light), and (2) acidic (dense). In a gradient mixer the basic solution is drawn into the acidic solution, which is simultaneously withdrawn to give a linear concentration gradient. Provided the Immobiline concentrations are chosen according to the recipies given in ref. 22, the resulting pH gradient will have the form corresponding to a titration curve that is symmetrical about a pK corresponding to the midpoint of the gradient. Full details of the recipes required are given in ref. 22. Figure 3a shows the effect of performing IEF with various gradient slopes on the separation of ovalbumin. It will be seen that very narrow pH ranges may be used and it is possible to focus proteins with pI values differing by as little as 0.001 pH units (21). Figure 3b compares a separation of PEA proteins using (1) Ampholines and (2) Immobilines.

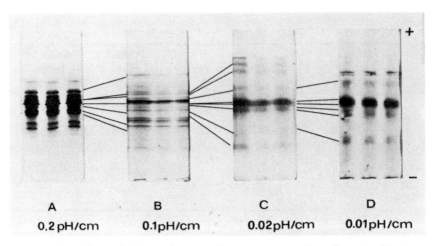

Fig. 3a. Effect of pH gradient on the separation of ovalbumin. Various Immobiline Systems (Courtesy of LKB). The electrofocusing was carried out on 0.5-mm polyacrylamide gels using pH gradients of different slopes. Gradient A was prepared from Ampholine carrier ampholytes, whereas B, C, and D were prepared from Immobilines.

The advantages of the Immobiline system may be summarized as follows:

1. Cathodic pH drift is eliminated.
2. Resolution between zones may be enhanced because of the increased control of the pH gradients.
3. The gradients have linear conductances, buffering capacities, and ionic strength, and good control is possible with these quantities.
4. The pH gradient slope may be chosen to suit a particular separation.
5. The gradients produced are very reproducible.
6. The gradient is insensitive to minor disturbances.

3. Zone Electrophoresis

Advances in zone electrophoresis (ZEP) in gels have not been as dramatic, nor as intensely investigated as IEF. The principle improvements in separation have come about through the application of ultra-thin-layer techniques, as indicated above.

However, the development of high-voltage capillary zone electrophoresis (HVCZE), which was discussed briefly in chapter 10

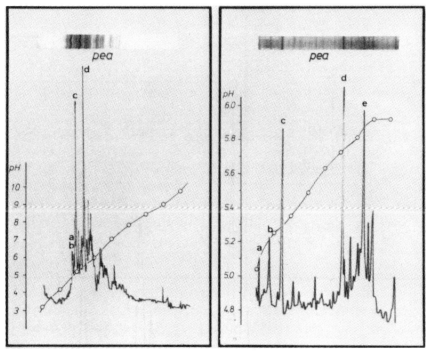

Fig. 3b. Example of the increased resolution obtained using Immobilines. Analytical isoelectric focusing of pea proteins: comparison of the increased resolution obtained by the use of Immobilines vs Ampholine carrier ampholytes. Left: Ampholine carrier ampholytes pH 3.5–9.5, right Immobiline system, pH 5.2–5.8. Note the increased resolution obtainable in the Immobiline system. Densitometer scans were obtained using a LKB UltroScan Laser Densitometer (Courtesy of Dr. A. Gorg et al., Technical University, Freising-Weihenstephan, FRG).

as applied to the analysis of amino acids and small peptides, has also been used for protein separations, although only limited work has as yet been performed. One of the problems associated with this application resides in the adsorption of the proteins on the surface of the quartz capillary. The surface-to-volume ratio is high in this technique (*see* chapter 11). Again the problem is to generate a "soft" surface that will not adsorb or denature the protein (although denaturation is not really a problem, since the method cannot be used for preparative purposes).

Jorgenson and Walbroehl (23) have demonstrated some elegant separations of protein standards (Fig. 4) on a silica capillary that had been modified by chemically binding glycol groups to its inner

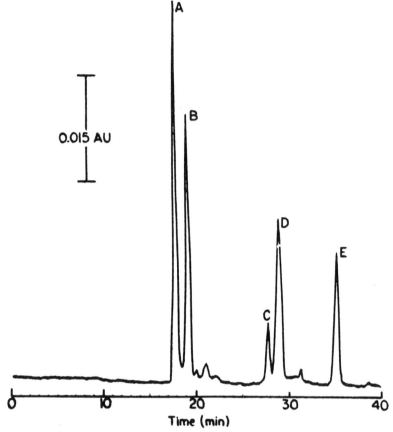

Fig. 4. Separation of some protein standards by HVCZE (from ref. 23). The electropherogram was carried out in a 75 cm × 75 μm surface modified capillary filled with 0.025M phosphate buffer pH 6.8, applied voltage 15 kV. Sample, 0.2% w/v of each protein. Peak identities: A, egg white lysozyme; B, horse heart cytochrome c; C, bovine pancreatic ribonuclease a; D, bovine pancreatic alpha-chymotrypsinogen; E, equine myoglobin.

surface (24). Detection of the separated zones was by UV monitoring at 229 nm, using the capillary as the detector cell, i.e., on column detection. In a recent paper, Lauer and McMonigill (63) have demonstrated protein separations using untreated fused silica capillaries by including additives in the migration buffer that dynamically modify the capillary wall. Typical additives used are zwitterionic compounds (cyclohexylamino) propane sulfonic acid (pK_a, 10.40), glycylglycine (pK_a, 8.40), and N-(2-acetamido)-2-aminoethansulfonic acid (pK_a, 6.88) (see ref. 63).

A further example of the use of capillaries for ZEP is provided by Hjerten (25), who used capillaries of about 0.15 mm id filled with a molecular sieving polyacrylamide gel (T = 6%; C = 3%). Figure 5 shows the separation of some oligimers of bovine serum albumin (up to $n = 5$) in 45 min. It has been impossible to separate $n = 4$ from $n = 5$ by molecular seiving chromatography using cross-linked agarose or TSK SW gel (26,27).

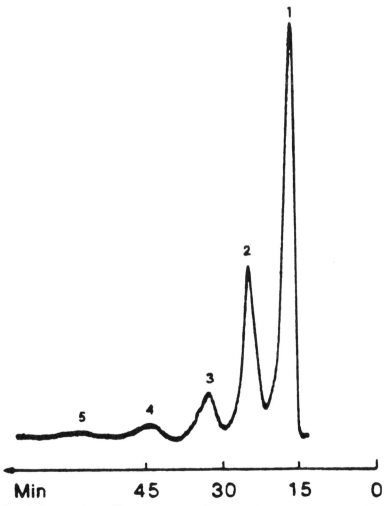

Fig. 5. Separation of bovine serum albumin oligimers using capillaries filled with PAG (from ref. 25). Sample size 0.1 µg. Peak identities: 1, monomer; 2, dimer; 3, trimer; 4, tetramer; 5, pentamer.

Clearly, this new technique, which is capable of generating millions of theoretical plates, may well become the method of choice for analytical protein separation in the future. At the present time, however, the method is restricted by the lack of a suitably high-sensitivity detection system. Proteins are not very easy to derivatize and, when successful, there is the possibility of multiple derivatives being formed, making quantitative work difficult to justify. The on-column detector reported by Jorgenson (25) goes some way toward solving this problem, but a high-sensitivity detector remains to be constructed.

4. Isotachophoresis

A brief background to isotachophoresis (ITP) is given in Chapter 10, in which it is shown that ITP can be used with success for the analysis of amino acids. Indeed ITP has principally been used for the analysis of small molecules, and Everaerts (28) has suggested that in order to predict the feasibility of the method for a given analysis, the ratio of the molecular weight to effective charge should not exceed 3000. This naturally limits the scope of the method to relatively small molecules. Proteins have been separated by ITP, however, and the problem is not that proteins cannot migrate isotachophoretically, but that the number of components present in a simple-cell extract or serum sample can range from hundreds to thousands of different proteins, all of which are unique but may possess only very small differences in mobility or be present in very small amounts, thus making the isotachopherogram difficult, if not impossible, to interpret. Nevertheless there are instrumental problems; for example a PTFE capillary tube may not necessarily be the best material for protein separation in that it is not such an inactive material and it is capable of adsorption. It may well be that specially treated silica surfaces similar to those used in HVCZE (see above) would be the preferred material. Using treated silicas, however, could introduce electroendosmosis into the system, with adverse affects on the separation obtained. Also, detection of the separated zones may be difficult to observe because of the fine distinctions between the zone boundaries at low concentrations.

This problem of interpretation has been addressed by Holloway et al. (29), who studied the separation of polyclonal human serum immunoglobulin G (IgG). Three series of experiments were carried out: (1) increasing concentrations of IgG, (2) inclusion of amino acid spacers, and (3) inclusion of Ampholines of different pH ranges. The results obtained showed that from (1) the IgG runs as an apparently homogeneous zone, but the steady rise of the thermometric trace indicates that the zone is not homogeneous. From (2) there was obtained a series of patterns to which no interpretation could be given, other than if n spacers were included in the sample, $n + 1$ zones appeared. This is clearly ridiculous and is an artificial situation probably not restricted to this class of protein. With (3) it was possible to see differences in the patterns obtained, but the interpretation was difficult and caution must be observed in reporting this type of result.

Keeping these cautionary remarks in mind, it is nevertheless possible to achieve useful separations by ITP. An interesting application is given by Bours (*30*), who studied the distribution of rat lens crystallins with age. Figure 6 shows the distribution of the crystallins present in rats at different ages. The reproducibility between the separations is excellent and it is possible to distinguish between the relative amounts of the separated crystallins.

Many other examples of protein separation by ITP have been reported in the literature, but in many ways it would appear that the resolving power of ITP, although high, is insufficient for protein separation in biological fluids in its own right. The combination of ITP with IEF (*31*), however, could provide excellent two-dimensional separations.

5. Two-Dimensional Separations

Two-dimensional (2-D) separations have been used since Gross demonstrated the use of the method for the separation of amino acids (*see* chapter 10). The method was subsequently extended by Smithies and Poulik (*32*) to a variety of different systems. Dale and Latner (*33*) first separated proteins by the use of IEF and PAGE in the first and second dimensions, respectively, in which the proteins are characterized by their p*I* and effective mobilities, which is a function of their charge and molecular size. The polyacrylamide gel, because of its limited pore size structure, acts as a molecular seive. This effect can be a disadvantage in that the separation of some proteins may be obscured by their charge and by their molecular size being in opposition, but if a PAGE gradient is constructed over the range of 4.5 to 26%, the proteins will migrate electrophoretically according to their molecular size or Stokes radius to

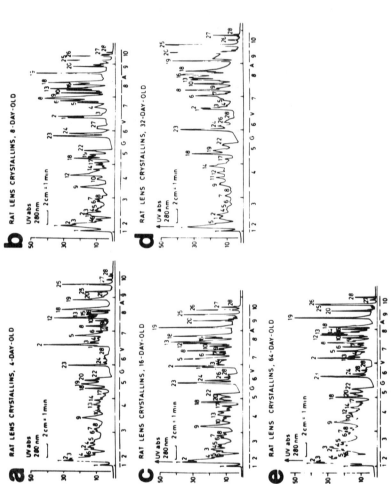

Fig. 6. Isotachophoretic separation of rat lens crystallins of various ages (from ref. 30). Fingerprints of the Sprague Dawley rat lens crystallins separated by isotachophoresis. a, 4-d-old; b, 8-d-old; c, 16-d-old; d, 32-d-old; e, 64-d-old. The numerals 1–28 indicate the single crystallins separated. The numerals under the abscissa refer to (1) pre-alpha-crystallin, (2) alpha-crystallins, (3) beta-I-crystallins, (4) beta-II-crystallins, (5) beta-III-crystallins, (6) beta-IV-crystallins, (7) gamma-I-crystallins, (8) gamma-II-crystallins, (9) gamma-III-crystallins, and (10)

a position of quasiequilibrium. Hence an estimate of their molecular size may be made. A further improvement in technique that eliminates charge differences is through the use of sodium dodecyl sulfate (SDS). The SDS binds to the protein to form a complex in which the charge-to-mass ratio is constant. Separation is thus primarily caused by differences in molecular size. The first true use of 2-D separations for the analysis of proteins, however, is credited to O'Farrell (34), who demonstrated the separation of some 1100 proteins from *Escherichia coli*, a separation that shows the complexity of the protein composition of cells. O'Farrell used IEF in the first dimension and SDS-PAGE in the second; the quality of the separation achieved is shown in Fig. 7.

Anderson and Anderson (35) subsequently modified O'Farrell's technique to enable multiple runs to be undertaken under identical conditions. Their method, labeled ISO-DALT, is now a widely used standard method of separation of acidic and neutral proteins. The method requires modification, however, if basic proteins are to be analyzed. This arises because of the instability of the basic end of the pH gradient in the first-dimension gel. The basic proteins that enter the gel are not well resolved. This problem can be surmounted by not allowing equilibrium to become established in the first-dimension IEF gel by using short electrophoresis times followed by a reversal of the direction of focusing so that the basic proteins lead in the separation. This process was called nonequilibrium pH gradient electrophoresis (NEPHGE) by O'Farrell et al. (36), who demonstrated that this modification had similar resolving power to the ISO-DALT method. Indeed, NEPHGE has been used as the first-dimension separation in the ISO-DALT method, and to distinguish between the two methods, this is known as BASO-DALT. The two methods are used to separate acidic and neutral proteins and basic proteins, respectively.

Since the introduction of these powerful methods of separation, a variety of combinations of first and second separation stages have been attempted, as indicated in Table 2.

Of the systems indicated above, it will be seen that all of the two-dimensional methods incorporate IEF either as the first- or second-dimension separation stage, usually the first. Further, the large majority depend upon the molecular weight of the species for the second characteristic.

In a recent meeting (45) on advances in electrophoresis, some 30 papers were presented, the majority of which were concerned with the ISO-DALT or BASO-DALT techniques. This indicates the interest in, and utility of, the two methods. Some of the problems

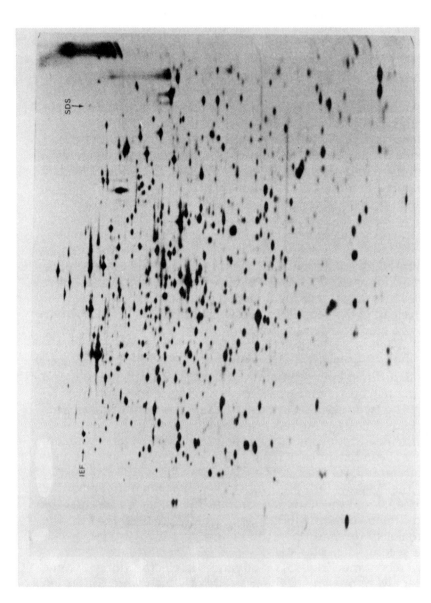

Fig. 7. Two-dimensional separation of proteins derived from *E. coli* (from ref. *34*). *E. coli* was labeled with ^{14}C amino acids in vivo. The cells were lysed by sonication, treated with DNase and RNase, dissolved in 9.5M urea, 2% Nonidet P-40, 2% Ampholine pH 5–7, and 5% 2-mercapto-ethanol. The soluble native proteins (containing 180,000 cpm per 25 µL) were subjected to IEF in PAG in the first dimension and to SDS gradient PAGE

Table 2
Examples of Methods of 2-D Separations

1st dimension	2nd dimension	1st separation	2nd separation	Reference
IEF	Gel EP	pI	Charge	(33) (37)
IEF	Gradient gel EP	pI	Charge/mass ratio	(38) (39)
IEF	SDS gel EP	pI	Molecular weight	(40) (41)
IEF	SDS gradient gel EP	pI	Molecular weight/charge	(34)
SDS gradient PAGE	IEF	Molecular weight	pI	(42)
ITP	IEF	Mobility	pI	(43)
IEF	Immunology	pI	Antibody/and so on	(44)
NEPHGE	SDS gradient PAGE	pI (basic)	Molecular weight	(36)

encountered in two-dimensional techniques are discussed in refs. 19 and 46. The use of ultra-thin-layer techniques for separation in both dimensions is becoming more frequent with the improvement in resolution that the technique offers. For example, Gorg et al. (47) demonstrate some elegant separations using ultra-thin-layer techniques with either the laying-on or laying-in techniques (Fig. 8). It will be interesting to see the further improvement in protein mapping (as the 2-D method is sometimes called) on using Immobilines for the first-dimension separation. This may well eliminate the problem of instability of the pH gradient at the basic region and remove the necessity of using BASO-DALT.

6. Visualization and Detection

Detection methods for the instrumental methods of electrophoresis, i.e., ITP and HVZEP, use on-column UV or fluorescence photometric detector or conductomeric, potential drop, or electrochemical in-column detectors. These systems have the advantage that they continuously monitor the column eluent that contains the separated zones. Thus a record of the separation similar to those obtained from chromatographic separations is produced.

With the other forms of electrophoresis in gels, it is necessary to use stains to visualize the position of the protein zones. A wide variety of stains have been recommended over the years; indeed, the choice of stain is often a personal matter. One stain in common use is Coomassie Brilliant Blue, which has high sensitivity. There is controversy, however, as to the relative merits of R250 and G250. Righetti and Chillemi (48) favor the use of colloidal dispersion of G250 in 12% trichloracetic acid and $0.5M$ H_2SO_4, whereas Bibring and Baxendale (49) prefer R250 in 15% trichloracetic acid. By these means the gel matrix remains unstained. An interesting new stain that is said not to stain carrier ampholytes or gels is Fast Green (50), which gives a longer linear calibration curve than Coomassie Blue, but is not as sensitive. For rapid work in which the sensitivity is not important, however, Fast Green could be of great use.

The most sensitive method of visualization is undoubtedly the silver stain, and a number of different recipes with varying degrees of complexity and cost have been proposed. In a recent comparison (51) between six different workers' staining techniques, the following conclusions were drawn.

The techniques, given in decreasing order of sensitivity are: Sammons > Merill (1) > Wray > Switzer > Merill (2) > Oakley.

Analytical Electrophoresis

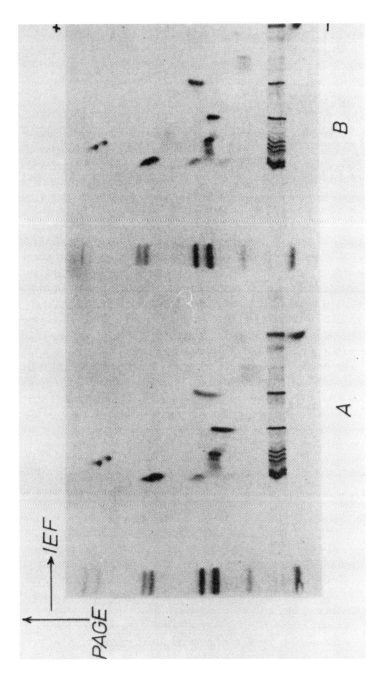

Fig. 8. Ultrathin layer two-dimensional protein separation (from ref. 47). Two methods of achieving the second dimension are demonstrated. (A) Laying-on technique; gel strip with the focused proteins is laid on the second dimension gel and (B) laying-in technique; the gel strip with focused proteins is laid into a moulded trough in the second dimension gel. Zone identities from left to right: human serum albumin, beta-lactoglobulin, human transferrin, carbonic anhydrase, horse myoglobin, sperm whale myoglobin, ribonuclease.

The Switzer method was the most expensive, and the Sammons and Merill (2) methods were the cheapest. Oakley, Wray, and Merill's two methods are simple to perform, whereas those of Switzer and Sammons were considered complicated, although the latter method was the most reproducible because it followed a strict timing schedule.

Peats (52) describes a silver staining procedure that is 100 times more sensitive than Coomassie Blue, whereas Marshall's method (53) appears to be useful for 2-D staining and has a sensitivity of 0.01 ng/mm² of bovine serum albumin. A rapid staining method that only take 10 min to complete has been reported by Lasne et al. (54).

Clearly the use of these stains provides qualitative information, for quantitative work densitometry provides an output similar to a chromatogram in which the peak height is proportional to the zone intensity and hence the amount of that component. Alternatively, the output from the densitometer may be analyzed by computer (55,56), which can pinpoint a zone in a 2-D protein map to within a micrometer and the intensity to within 0.001 AU.

Probably the most sensitive method of trace protein visualization is by autoradiography, which will detect zones that stains will not help to visualize. Weinstein (57) has calculated that the method is at least 150 times more sensitive than the common protein stains, but considerably more time-consuming. Indeed O'Farrell's 2-D separation of *E. coli* proteins required 825 h of exposure, a major disadvantage of this method. For labeling, ^{14}C-amino acids, ^{32}P-inorganic phosphate, ^{35}S-methionine, or ^{3}H-amino acids are commonly used.

7. Blotting

One of the main reasons for the long visualization times of proteins in gels arises from the diffusion time of the reagent in the gel matrix. In addition, correspondingly long washing times are required, and, most important, some of the resolution obtained will be lost through diffusion of the separated components in the gel. This problem was recognized by Southern (58), who developed a technique (Southern blot) for the electrophoretic transfer of focused zones of DNA onto the surface of a nitrocellulose membrane, where it is immobilized. This technique was extended by Stark et al. (59) to include the transfer of proteins from polyacrylamide gel to cellulose paper, where they are immobilized with bound diazonium groups (Northern blot). Towbin et al. (60) concurrently developed protocols for the quantitative transfer of proteins (Western blot) and their immobilization on DEAE cellulose paper. Numerous papers have been published on the technique, and the advantages of blotting are summarized here:

1. The immobilization of the protein on the surface of a membrane renders it more accessible for staining or immune responses.
2. The transfer and binding results in no loss in resolution of the separated zones and biological activity is usually retained.
3. The membranes are considerably more robust than the gels.
4. The blotting procedure may be performed several times on a single gel so that multiple analyses with selective stains can increase the information available from a single EP run.
5. Blotting may be carried out on any electrophoretic separation in gels, and the quality of the results obtained can be enhanced, particularly in relatively thick gels. There is probably little advantage in using the blotting procedure for ultra-thin gels because their staining/destaining time is so fast. Only with multiple determinations would there be any advantage, but even here blotting will be of limited use because of the very small sample size used in ultra-thin methods.

Two relatively recent reviews of this subject are given in refs. 61 and 62.

References

1. A. Tiselius, *Trans. Faraday Soc.* **33**, 524 (1937).
2. A. J. P. Martin and R. L. M. Synge, *Biochem. J.* **35**, 1358 (1941).
3. R. Consdon, A. H. Gordon, and A. J. P. Martin, *Biochem. J.* **38**, 224 (1944).
4. I. Smith, ed., *Chromatographic and Electrophoretic Techniques* vol. 2, *Zone Electrophoresis* 4th ed., Heinemann, London (1976).
5. O. Gaal, G. A. Medgyesi, and L. Vereczkey, *Electrophoresis in the Separation of Biological Macromolecules* Wiley-Interscience, New York (1980).
6. H. Svensson, *Acta. Chem. Scand.* **15**, 325 (1961).
7. H. Svensson, *Acta. Chem. Scand.* **16**, 456 (1962).
8. A. Kolin, in *Methods in Biochemical Analysis* vol. 6 (D. Glick, ed.) Interscience, New York (1958), p. 259.

9. O. Vesterberg, *Acta Chem. Scand.* **23**, 2653 (1969).
10. N. Catsimpoolas, ed., *Isoelectric Focussing* Academic, New York (1976).
11. N. Catsimpoolas and J. W. Drysdale, eds., *Biological and Biomedical Applications of Isoelectric Focussing* Plenum, New York, London (1977).
12. P. G. Righetti and J. W. Drysdale, in *Laboratory Techniques in Biochemistry and Molecular Biology* (T. S. Work and E. Work, eds.) Elsevier, Amsterdam (1976).
13. H. Svensson, *Acta Chem. Scand.* **15**, 325 (1961).
14. B. J. Radola, *Electrophoresis* **1**, 43 (1980).
15. R. C. Allen, *Electrophoresis* **1**, 32 (1980).
16. A. Rosen, K. Ek, and P. Aman, *J. Immunol. Meth.* **28**, (1979).
17. R. C. Allen, P. M. Oulla, P. Arnaud, and J. S. Baumstark, in *Electrofocussing and Isotachophoresis* (B. J. Radola and D. Graesslin, eds.) W. de Gruyter, Berlin (1977), pp. 255.
18. A. Gorg, W. Postel, and R. Westmeier, *Anal. Biochem.* **89**, 60 (1979).
19. B. J. Radola, in *Electrophoresis '79* (B. J. Radola, ed.) W. de Gruyter, Berlin (1980), p. 79.
20. B. J. Radola, in *Electrophoretic Techniques* (C. F. Simpson and M. Whittaker, eds.) Academic, London (1983).
21. B. Bjelliquist, K. Ek, P. G. Righetti, E. Gianazzi, E. Gory, R. Westermeister, and W. Postel, *J. Biochem. Biophys. Methods* **6**, 317 (1982).
22. V. Gasparic, B. Bjellqvist, and A. Rosengren, Various patents, see LKB application note 321.
23. Y. Walbroehl and J. W. Jorgenson, *J. Chromatog.* **315**, 135 (1984).
24. J. W. Jorgenson and K. D. Lukacs, *Science* **222**, 266 (1983).
25. S. Hjerten, *J. Chromatog.* **270**, 1 (1983).
26. S. Hjerten, *Acta Chem. Scand.* **B36**, 203 (1982).
27. M. E. Himmel and P. G. Squire, *Int. J. Peptide Protein Res.* **17**, 365 (1981).
28. F. Everaerts and Th. P. E. M. Verheggen, in *Electrophoretic Techniques* (C. F. Simpson and M. Whittaker, eds.) Academic, London (1983), pp. 149-196.
29. C. J. Holloway, R. V. Battersby, B. Bussenschutt, G. Bulge, and J. Lustorff, in *Analytical and Preparative Isotachophoresis* (C. J. Holloway, ed.) Walter de Gruyter, Berlin (1984), pp. 271-289.
30. J. Bours, in *Analytical and Preparative Isotachophoresis* (C. J. Holloway, ed.) Walter de Gruyter, Berlin (1984), pp. 315-324.
31. C. H. Brogren, in *Electrofocussing and Isotachophoresis* (B. J. Radola and D. Grasslin, eds.) W. de Gruyter, Berlin (1977), pp. 549-558.
32. O. Smithies and M. D. Poulik, *Nature* **177**, 1033 (1956).
33. G. Dale and A. L. Latner, *Clin. Chem. Acta* **24**, 61 (1969).
34. P. H. O'Farrell, *J. Biol. Chem.* **250**, 4007 (1975).
35. N. G. Anderson and N. L. Anderson, *Anal. Biochem.* **85**, 331 (1978) [*see also* M. J. Dunn and A. H. M. Burghes, *Electrophoresis* **4**, 973 (1983), for a recent review with 215 references].

36. P. Z. O'Farrell, H. M. Goodman, and P. H. O'Farrell, *Cell* **12**, 1133 (1977).
37. C. Wrigley, *Biochem. Genetics* **4**, 509 (1970).
38. K. G. Kenrick and J. Margolis, *Anal. Biochem.* **33**, 204 (1970).
39. D. H. Leaback, in *Isoelectric Focussing* (J. P. Arbuthnott and J. A. Beeley, eds.) Butterworths, London (1975), pp. 201.
40. H. Stegmann, H. Francksen, and V. Macko, *Z. Naturforsch* **28**, 722 (1973).
41. T. Barrett and H. J. Gould, *Eur. J. Biochem.* **93**, 329 (1979).
42. G. P. Tuszynski, C. A. Buck, and L. Warren, *Anal. Biochem.* **93**, 329 (1979).
43. C. H. Brogren, in *Electrophoresis and Isotachophoresis* (B. J. Radola and D. Grasslin, ed.) W. de Gruyter, Berlin (1977), pp. 549.
44. P. D. Eckersall and J. A. Beeley, *Electrophoresis* **1**, 62 (1980).
45. 2-D Electrophoresis, *Clinical Chemistry*, **30** No. 12 (1985).
46. J. S. Fawcett, in *Electrophoretic Techniques* (C. F. Simpson and M. Whittaker, eds.) Academic, London (1983), pp. 57–80.
47. A. Gorg, W. Postel, and R. Westermeier, in *Electrophoresis '79* (B. J. Radola, ed.) W. de Gruyter, Berlin (1980), p. 67.
48. P. E. Righetti and F. Chellemi, *J. Chromatog.* **157**, 253 (1978).
49. T. Bibring and J. Baxendale, *Anal. Biochem.* **85**, 1 (1978).
50. R. E. Allen, K. C. Marsak, and P. K. McAllister, *Anal. Biochem.* **104**, 494 (1980).
51. H. D. Kay, *Electrophoresis* **2**, 304 (1981).
52. S. Peats, *Analyt. Biochem.* **140**, 178 (1984).
53. T. Marshall, *Analyt. Biochem.* **136**, 340 (1984).
54. F. Lasne, O. Benzerra, and Y. Lasne, *Analyt. Biochem.* **132**, 338 (1983).
55. J. Bossinger, M. J. Miller, K. P. Vo, P. Geudushik, and N. H. Xung, *J. Biol. chem.* **254**, 7960 (1975).
56. E. Kitatoe, M. Mayhara, N. Kiraska, H. Ueta, and K. Utumsi, *Anal. Biochem.* **134**, 295 (1984).
57. J. D. Weinstein, T. G. Obrig, and J. B. Weiggle, *J. Chromatog.* **157**, 198 (1978).
58. E. M. Southern, *J. Mol. Biol.* **98**, 503 (1975).
59. J. Renart, J. Reiser, and G. R. Stark, *Proc. Natl. Acad. Sci. USA* **76**, 4350 (1979).
60. H. Towbin, T. Staehelin, and J. Gordon, *Proc. Natl. Acad. Sci. USA* **76**, 4350 (1979).
61. J. M. Gershoni and G. E. Palade, *Anal. Biochem.* **131**, 1 (1983).
62. J. Symington, in *Two-Dimensional Gel Electrophoresis of Proteins: Methods and Applications* (J. E. Celis and R. Bravo, eds.) Academic, New York (1983).
63. H.H. Laver and D. McManigill, *Anal. Chem.* **58**, 166 (1986).

Chapter 13
Preparative Liquid Chromatography

Colin Simpson

1. Introduction

Preparative liquid chromatography (PLC) potentially can be undertaken using all of the techniques described in chapter 11, and the basic requirement for efficient preparation is to obtain the maximum throughput in unit time. This requires optimization of the methods employed, which is the subject of this chapter. However, there is an additional constraint in preparative protein separations; that is, the prepared proteins should retain their biological activity, and this imposes severe restraints upon the operating system. Three effects can be listed that potentially can affect protein denaturization: (1) the "hardness" of the surface, (2) temperature effects, and (3) shear forces.

1.1. Surface "Hardness"

In chapter 11 it was shown that proteins can be separated using all of the techniques of liquid chromatography. Here it is of no significance if denaturation occurs apart from the (real) possibility of a protein unfolding, thus giving rise to different areas of hydrophobicity with concomitant different retention times and hence presenting a fallacious analysis. In preparative work this is unacceptable (although protein refolding can occur), and hence the systems used are mainly restricted to ion exchange and related methods and size exclusion, which present relatively soft surfaces for interaction. (Affinity chromatography is also used, but this does not necessarily require column systems.)

1.2. Temperature Effects

The passage of the mobile phase through the chromatographic bed gives rise to an increase in temperature of the bed through frictional heating, and this problem is exacerbated as the bore of the preparative column is increased through the decrease in the surface area to volume ratio and the poor thermal conductivity of the packing and mobile phase. Fortunately in protein separations, through their low diffusion rates, low mobile phase flow rates are used so this problem is not too severe, but it must be borne in mind. An associated problem lies in the narrow bore of the delivery line from the column end to the sample collector. Here the total column flow, which can be as high as 500 mL/min in high capacity preparative columns, can give rise to high temperature rises, albeit the residence time of the sample is small in such a tube. Nevertheless problems of denaturation can occur here.

1.3. Shear Forces

Shear forces can also bring about denaturation as indicated in Chapter 11 in the case of size exclusion chromatography particularly, and again, the delivery tube to the sample collector can contribute here.

The term "preparative chromatography" can be used to cover a wide range of sample loads, from the nanogram level to tens of grams (or more) being separated at any one time. The size of the prepared sample usually depends upon the subsequent application. Thus the sample loads may be processed using simple analytical scale columns and equipment, intermediate scale columns with standard equipment, to specially designed systems capable of handling high loads at high volumetric throughputs.

There are four interrelated attributes of the chromatographic process that effectively control what is possible to achieve with the technique; namely, resolution, speed, scope, and load. In PLC, the requirement is to increase the load while maintaining resolution between sample components (although some loss is bound to occur). Thus, speed and scope will be impaired. Unfortunately, guidelines for the optimization of preparative work are far less well established than in the analytical field.

Preparative Liquid Chromatography

The aim of any preparative separation (to maximize the sample throughput and to minimize the use of high-purity solvents) necessitates that the sample band eluting from the column be as concentrated as possible. This in turn means that the sample bands must be overloaded, which can have important consequences on the chromatography involved. With protein separations there is an additional constraint, that of maintaining biological activity, which can limit the nature of the phase systems it is possible to use.

2. Rationale for Sample Overloading (1)

Initially it is worthwhile considering the maximum sample load that can be applied to the column without overloading. This can be derived from the plate theory of chromatography (2) and may be shown to be directly proportional to the plate volume and $N^{1/2}$. From the plate theory:

$$M = AN^{1/2}(v_m + Ka_s)$$

where M = limiting mass of sample, A = constant, N = column efficiency, v_m = volume of mobile phase/plate, a_s = surface area/plate, and K = distribution coefficient.

Now the retention volume

$$V_R = N(v_m + Ka_s)$$

Thus,

$$M = AV_R/N^{1/2}$$

If the solute is well-retained, so that $v_m << Ka_s$, then

$$V_R \to NKa_s \to VdKA_s \to \pi r^2 ldKA_s$$

where V = volume of the column, d = packing density of the packing, g/mL, l = length of the column, mm, r = radius of the column, mm, and A_s = surface area of the packing in the column, m²/g.

Hence

$$M = A\pi r^2 lKA_s/N^{1/2}$$

If it is assumed that N is a fixed value, $(N)_c$, that just provides the necessary resolution between the solute of interest at the maximum sample load. Then

$$(N)_c = l/(H)_c$$

where $(H)_c$ (HETP), is a constant at a given linear mobile phase velocity.

For well-retained peaks, mass transfer is the controlling factor in the HETP equation. Thus

$$H = \beta dp^2/l$$

where β = constant and dp = the particle size.

Thus the final equation that describes the maximum load is:

$$M = A\pi r^2 l K d A_s (\beta dp^2/l)^{1/2} c$$

The significance of these terms is as follows:

2.1. Column Radius, r

The maximum sample load will increase with the square of the column radius. This also means that the plate volume will increase leading to a greater volume of packing per plate. This will not increase retention times, providing a proportional increase in volumetric mobile phase flow is made to provide the same linear mobile phase velocity. Therefore solvent usage remains the same, and solvent consumption per unit mass of sample also remains constant.

2.2. Distribution Coefficient, K

As K increases, an increase in retention volume is obtained that gives an increase in loading capacity, but at the expense of increased solvent consumption. Solvent economy is unchanged.

2.3. Packing Density, d

Packing density has no effect, provided the columns are well packed.

2.4. Surface Area, A_s

Loading capacity increases with the surface area of the packing, but this brings about a decrease in the number of components that may be separated under isocratic conditions in a finite time, i.e., the scope of the separation is diminished, together with the speed of the separation.

2.5. Column Length (l) and Particle Size, dp

From the equation above, these two parameters are related. The ratio dp^2/l must remain constant to maintain the necessary minimum column efficiency to achieve the required resolution. Thus loading capacity increases with $l^{1/2}$ (providing kinetic contributions are solely responsible for band broadening), but dp must be increased in proportion to maintain the constant ratio for minimum resolution. *Note*: As l increases at a constant particle size, the pressure drop will also increase, which may prove to be outside the operating range of the pump. Thus dp must also be increased to improve column permeability.

These conclusions are summarized in Table 1.

Table 1
Relationship Between Column Length, Particle Size, Relative Permeability, and Relative Loading Capacity for Columns With Constant Cross-Sectional Area and With the Same Efficiency

Column length, mm	dp, μm	Relative load capacity	Relative permeability
25	10	1.0	1.0
50	14	1.4	2.0
75	17	1.7	3.0
100	20	2.0	4.0

The data in Table 1 show that in order to improve sample throughput, it is necessary to overload the column. The question therefore immediately arises, what is the effect of column overload on the separation obtained, and what criteria must be observed?

3. Column Overload

There are three ways in which a column may be overloaded.

1. With sparingly soluble materials, overload may occur by the sheer volume of sample solution that must be applied to the column to achieve the required mass load.
2. With very soluble materials, overload occurs because of the mass of sample applied.

3. The use of a different solvent to provide a more concentrated solution for application to the column.

Point 3 may be dealt with immediately. The use of a different solvent to the mobile phase used to develop the separation can disturb the delicate equilibrium that is present in the column. If the sample solvent is stronger than the mobile phase, this may completely destroy the separation by the solvent adsorbing on the stationary phase and hence deactivating it. This would thus present a different phase system to the solute molecules, which may elute unresolved. It may be advantageous for the solvent to be weaker than the mobile phase, when the sample will be concentrated at the column head, thus minimizing the applied sample band (and possibly mass overloading the column). This is acceptable. However, it is always preferable to apply the sample in the mobile phase and accept a somewhat lower sample throughput.

In point 1 (column overload from feed volume), the maximum feed volume that may be applied to a column is a function of α, the separation ratio between the peaks of interest. From the plate theory, it may be shown (1) that the maximum feed volume V_L may be calculated from the expression

$$V_L = V_0[(k'_B - k'_A) - 2(2 + k'_A + k'_B)/N^{1/2}]$$

where N = column efficiency.

Utilizing this equation with typical column dimensions, e.g., 3600 mm long × 25 mm id with a dead volume of 900 cm^3, possessing 8000 theoretical plates, for two components with k' values of 8.7 and 10.9, the maximum sample feed volume that may be injected is 1545 cm^3, which may appear to be surprisingly large, but in fact provides complete separation between the components of interest (e.g., Fig. 1).

Figure 2 shows the sample volumes it is possible to inject at various α values and k' values.

Point 2 (column overload from sample mass) is far more difficult to evaluate quantitatively. Little work has been performed to assess the effect of mass overload. There are two points that should be considered here.

1. The normal kinetic band broadening effects will be modified by the sample overloading for a finite distance down the column during the elution of the components.
2. The effect of a high sample concentration at the head of the column, which can change the normal elution properties

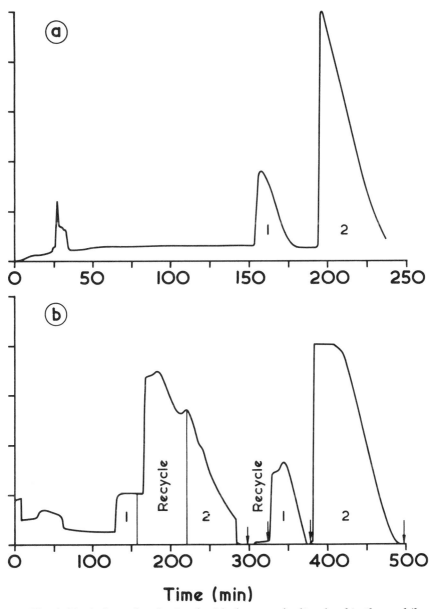

Fig. 1. Typical results obtained with the sample dissolved in the mobile phase. (a) 500 mL of 0.2% solution injected, baseline separation obtained. (b) 4 dm^3 of 0.15% solution injected, overloads the system. Central position of the first set recycled gives separation (from ref. 6).

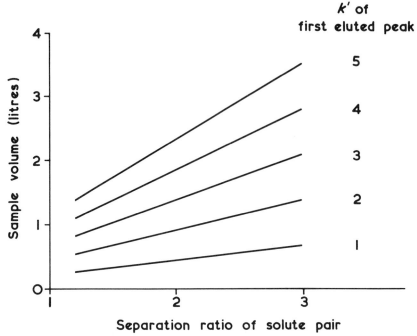

Fig. 2. Relationship between the maximum sample volume permissible at various separation ratios and k' values (from ref. 1).

of the sample components. This leads to what has been termed thermodynamic band broadening.

If we consider 1, kinetic effects produce the normal band broadening processes that occur within the column, brought about by the various dispersive mechanisms that are functions of the nature of the stationary phase, the flow streamlines, which the solute molecules sample during their passage through the column, and the so-called resistance to mass transfer of the solute between the stationary and mobile phases. Kinetic band broadening is proportional to the square root of the residence time of the solute in the column, i.e., proportional to $N^{1/2}$. This is predictable. When the column is overloaded, the normal band broadening mechanisms are substantially modified.

Considering 2, when a large concentration of sample is placed at the head of the column, the stationary phase present is effectively saturated with the sample, so that the migration of the sample constituents takes place between adsorbed sample components and potentially saturated mobile phase. Thus the normal chromatographic migration laws are not applicable, and band broadening is directly proportional to the distance migrated, at least until the sample concentration diminishes to substantially "normal" levels, i.e., where the stationary phase is not overloaded when normal band broadening processes again occur. It has been shown that this linear dependence is not wholly justified (2), but is at least partly true. The net result of this is that retention times of the components will be significantly decreased. A further consequence of this effect is that the sample peaks will possess substantial tails that can overlap other later developing components and, in general, the band widths will become significantly larger.

One way to overcome this problem is to increase the column length, providing the pressure drop is within the capability of the solvent delivery system. Should it not be so, an alternative way to artificially increase the column length is to recycle the sample through the same column.

4. Recycle Technique

In the recycle mode, the sample constituents (after their first passage through the column) are pumped back to the inlet of the column for a second pass, and so on until the separation between components is such that pure constituents can be isolated (Fig. 3).

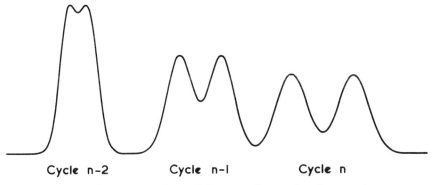

Fig. 3. Representation of the recycle mode of operation.

This procedure cannot be carried out indefinitely because of the "race track effect," that is, the faster migrating component will "lap" the slower migrating component, and separation will be lost.

If the resolution between components on a single pass is given by

$$R_1 = 2(V_{R2} - V_{R1})/(w_1 + w_2)$$

then after n passes, the resolution is increased according to

$$R_n = (V_{R2} - V_{R1})/2[n^{1/2}w_1^2 + (n-1)w_{1A}^2 + n^{1/2}w_2^2 + (n-1)w_{2A}^2]$$

where w_{1A} and w_{2A} is the band broadening in the detector and extracolumn transfer lines.

It may be shown (4) that if

$$\lambda = w_A^2/w_C^2$$

(where w_c^2 is the volume broadening assuming equal variances for each peak), that the condition to have $R_n > R_s$, is when

$$\lambda < n$$

and obviously the smaller λ, the better the increase in separation with recycle operation. A complete analysis of this situation is beyond the scope of this article. For a complete understanding of the maximum number of passes possible before overlap reoccurs, see ref. 4.

A modification to this technique (called shave/recycle) involves the leading and trailing edges of the partially separated components being collected while the unresolved section is recycled (Fig. 4).

Here, the direct and shave/recycle methods are compared for the preparation of 3 g of a reasonably complex mixture. There are two points to notice. In the shave/recycle method, the total sample is processed in one injection; the total preparation time is 16 min and the solvent consumption is 6.3 L. In the direct method, three injections have to be made to process the sample; the total time is 48 min, and the solvent consumption is 24 L. The advantages of the shave/recycle method are obvious.

5. Equipment for Preparative Liquid Chromatography

5.1. The Column

5.1.1. Nanogram to Microgram Range

Within this range, normal analytical scale columns may be used (4.6 mm id). The only precautions that need to be observed are

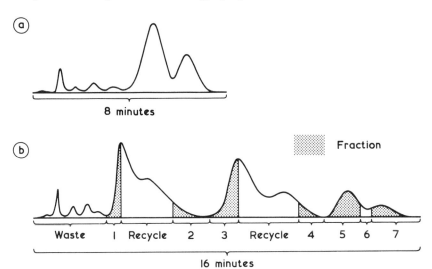

Fig. 4. Typical chromatogram of the shave recycle technique for sample recovery. (a) Lightly loaded; R_s = 1.25, 0.5 g sample load, 500 cm³/min run six times to process 3.0 g. Throughput 0.06 g/min, separation time 48 min, solvent consumption 24 l. (b) Heavily loaded, R_s = 0.7 on the first pass, 3.0 g load, 500 cm³/min. Throughput 0.19 g/min, separation time 16 min, solvent consumption, 6.3 dm³ (courtesy of Waters Associates).

(1) to ensure that the sample is applied over the entire cross-section of the column to minimize mass overloading at the column head, which (at this level of preparation) has only a limited effect on the peak shape and retention time, and (2) to limit the length of delivery tube from the detector outlet so that when the sample is detected, the time lag for collecting the sample is minimized. To calibrate this time delay (which is really only important when closely eluting components are to be collected), perform a blank experiment with a suitable dyestuff.

5.1.2. Microgram to Tens of Milligrams

Columns to cover this level of preparation have internal diameters in the range of 6.0 to 25.4 mm, and are chosen according to the sample mass to be prepared (to minimize solvent consumption). These columns perform well analytically and possess efficiencies usually greater than 50,000 plates/m, i.e., similar to standard analytical columns; here the principal requirement is to ensure that the volumetric flow rate is in proportion to the square of the radius for optimum performance. Thus, if the mobile phase flow for a 4.6 mm id column is 2.0 cm^3/min, then for a 25.4 mm id column, the flow should be 30 times greater. Clearly this requires pumps capable of delivering this flow rate, a point to be considered when choosing the column system. Other conditions are similar to those quoted above.

5.1.3. Hundreds of Milligrams to Tens of Grams

Two pieces of equipment capable of undertaking the range of sample load indicated are available commercially. These are the Water's Associates Prep 500 and the Jobin Yvon preparative chromatograph.

In the Water's system, stationary phase cartridges (with flexible walls) filled with the appropriate stationary phase (it may be necessary to pack one's own cartridges) are placed inside a close-fitting stainless steel container, where end-seals are made. Pressure is applied to the annulus between the cartridge and the container walls, thus compressing the stationary phase contained therein. In this way, any inhomogeneity of the packing is eliminated and maximum column efficiency is obtained. This radial compression is beneficial in two ways. (1) It produces a compact stationary phase bed with a packing density associated with the applied annular pressure, and (2) most importantly, because the cartridge walls are flexible and relatively soft, the excess pressure used in compressing the bed results in the packing at the walls actually being partially buried in the wall of the cartridge (Fig. 5). This eliminates the usual somewhat faster flow of mobile phase close to the walls

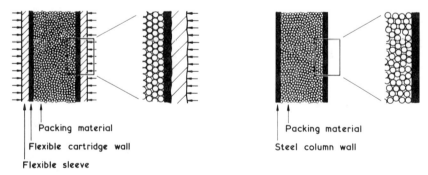

Fig. 5. Comparison between the state of the packing adjacent to the walls in (a) Radial Compression System™, and (b) rigid steel walls in a conventional column (courtesy of Waters Associates).

of the column that is responsible for a loss in column efficiency (the infinite column diameter effect) (5) in the analytical or preparative mode. This is because of the inhomogeneous packing density across the diameter of the column, particularly close to the walls. This beneficial effect has its shortcomings, however, that the net effect of radially compressing the cartridge will inevitably result (because of the construction of the cartridge) in a "waisting" of the cartridge with a concomitant variation in the mobile phase flow along the axis of the column, thus potentially impairing the best separation obtainable. Nevertheless the equipment performs well, but the large diameter of the column (3 in.) requires a high mobile phase flow-rate (approximately 500 cm^3/min). Therefore the separation requires optimization before applying to the preparative column. *Note*: The example of shave/collect given above was obtained on a Water's Prep 500. (Fig. 4).

The Jobin Yvon preparative chromatograph (Fig. 6) operates on the principle of axial compression, in which the stationary phase is placed into the column blank, and a piston compresses the packing axially to promote a well-packed bed against inflexible walls. This system has the advantage that for a given application, the appropriate bed length may be chosen to obtain maximum throughput with the sample. It must always be remembered that the maximum load that can be applied to a column is proportional to the separation factor between components of interest. An advantage of this equipment lies in being able to pack the column with the stationary phase of choice, and this can, and ideally should, be the same as the material used to develop the separation. Different sizes of column blank and piston are available.

Alternatively, it may be preferable to construct equipment for specific purposes. Figure 7 shows a typical preparative system capable of undertaking a wide variety of different applications, both directly, under the recycle mode of operation and under column switching conditions. It may be argued that all the time that the sample is not being collected is wasted time. The use of column switching techniques can be useful here. Systems of this nature are necessarily reasonably complex, to provide flexibility in operation. One point worth noting in the figure is the presence of two detectors (items 15 and 17). There is nothing more wasteful than performing a collection when the detector response changes direction in the middle of a peak! The sample has already been contaminated. If a detector is placed in the column system so that 90% of the column has been traversed by the sample, then the separa-

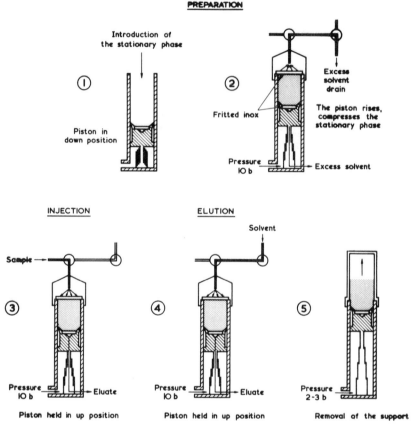

Fig. 6. Phases in the setting up of the axial compression system (courtesy of Jobin Yvon).

tion will be 90% complete, and sufficiently so to allow prediction of when a collection should be stopped to prevent cross-component contamination. If preparative work is a major part of one's work, it would be worth considering building a preparative chromatograph incorporating this modification (*see* also ref. 6).

5.2. Sample Application

Normal analytical methods of sample application are of no use in the preparative mode. In PLC it is required to place the sample in as narrow an *even* band as possible, but covering the entire column cross-section. Point injection will lead to sample overloading

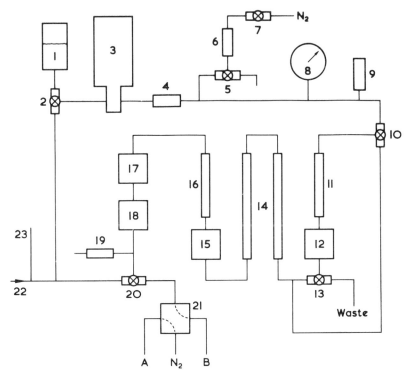

Fig. 7. Preparative LC recycle system: (1) sample reservoir, (2) pump inlet select valve, (3) pump, (4) filter, (5) pulse dampner valve, (6) pulse dampner, (7) nitrogen inlet valve, (8) pressure gage, (9) 1500 psi relief valve, (10) precolumn select/bypass valve, (11) precolumn, (12) precolumn detector, (13) precolumn transfer valve, (14) main columns (up to six), (15) preview detector, (16) final column, (17) exit detector, (18) RI detector, (19) 50 psi relief valve, (20) recycle valve, (21) fraction collect valve, (22) eluent inlet, and (23) eluent debubbler (adapted from ref. 6).

at the column head with all its attendant problems. If the sample is applied to cover the entire cross-section, the column, albeit overloaded, will provide a better situation for sample development. In addition the total sample load that can be processed in a single pass will be increased! Various methods have been developed to effect sample introduction, but one of the simplest is that recently described (3), in which the sample is introduced into a short precolumn packed with ballotini beads and fitted with a bottom vent to eliminate back pressure in the injector. In effect the sample displaces mobile phase from the precolumn, and when the desired volume has been added, the vent is closed and the mobile phase flow diverted to the head of the precolumn. The ballotini disperses the sample evenly across the bore of the precolumn, and hence into the head of the separation column (*see* also ref. 6).

5.3. Column End Fittings

The use of wide-bore columns necessitates a special design of column end-fitting to ensure that the whole of the bore of the column at the outlet is swept with mobile phase evenly into the outlet collection tubes. Figure 8 illustrates the problem and its solution.

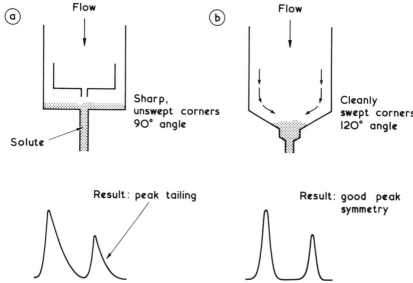

Fig. 8. Flow in column end fittings showing the necessity to use tapered fittings.

6. Technique in Preparative Chromatography

6.1. Developing a Separation

In cases in which separations are to be performed on a greater scale than can be conveniently undertaken on analytical equipment, it is advisable to develop the separation by an analytical method first. The main reason to do so is that if large-scale apparatus is used, the solvent consumption, and more importantly, the sample usage, is high and losses in valuable material will occur. Separation

Preparative Liquid Chromatography

should therefore be carried out initially on suitable equipment that will allow for direct scale-up with minimum wastage in material. Ideally the analytical preruns should be carried out on the same stationary phase as it is proposed to undertake the preparative separation.

6.1.1. Thin Layer Chromatography (TLC)

Thin layer chromatography may be used very efficiently for providing a cheap, simple-to-use, multi-sample monitoring system for providing guidelines on the course of a separation one wishes to upgrade to a preparative level. Although it may not have the same resolving power of analytical column chromatography, it can nevertheless indicate the course of a separation. Indeed, in the form of HPTLC, the resolving power can equal the performance of analytical LC, albeit with markedly reduced sample loads, which reduces its worth as a micropreparative system. Both silica gel and reversed-phase plates are available for use, but it must be remembered that the surface area of the materials used in TLC are about twice that of the column packings. Hence retention will be about doubled. Thus it is advisable that the TLC separation be developed so that $R_f < 0.3$, where

$$R_f = \frac{\text{distance of origin from the spot center}}{\text{distance of origin from solvent front}}$$

$$= U_{band}/U$$

$$= 1/(1 + k')$$

where U_{band} and U are the migration velocities of the sample and solvent front, respectively.

Hence,

$$k' = (1 - R_f)/R_f$$

i.e., when $R_f = 0.3$, $k' = 2.33$. Hence α can be obtained from

$$\alpha = k'_2/k'_1, \text{ where } \alpha \text{ is the separation factor.}$$

One point to remember is when TLC is being performed using mixed solvents for the mobile phase; almost inevitably the two solvents used will have different volatilities. Thus the partial vapor pressures of the solvent components in the vapor phase above the bulk solvent in the developing tank will be different to the bulk composition in the liquid phase. In TLC the plate is allowed to achieve equilibrium with the solvent vapors in the tank. Thus chromatography is carried out between the bulk solvent and the adsorbed solvent vapors on the surface of the plate, i.e., a form of liquid/liquid/adsorption chromatography is being undertaken that will be somewhat different from that carried out with the same solvent system in a closed column system. Thus great care has to be employed in transferring mixed solvent mobile phases to a column system, although a good separation may have been obtained on the thin layer plate. With single solvent mobile phases, this problem does not arise and scale-up may be carried out with confidence, bearing in mind the restriction in R_f.

6.1.2. Analytical Chromatography

Scaling-up from analytical LC is a relatively simple matter, providing the same packing is used in both the analytical and preparative columns. The principle requirement is to maximize the α value between the components of interest because it is this value, together with the resolution it is possible to achieve, that dictates the maximum sample load that can be processed per pass. It must be remembered that k' will decrease with sample size, particularly if overload conditions are used, so that very low values of k' are not desirable. Similarly, very high values of k' are equally undesirable because of the high solvent consumption that will be experienced. A useful range of k' to aim for is 1 to 5 in the analytical runs, with maximum resolution between the components of interest. This will give good sample loading with reasonable solvent consumption.

A useful design (6) that combines an analytical and preparative column in the same system, so that they are both subject to the same mobile phase changes, is illustrated in Fig. 9. Here the two columns have the appropriate proportion of mobile phase flow passing through them so that if a solvent gradient must be employed, although this is not desirable, but often necessary in protein separations, then the packings are influenced to the same extent. Clearly this is wasteful of solvent, but is highly necessary if successful scale-up is to be achieved. The system illustrated can be easily constructed by the addition of a simple six-port valve to divert the analytical and preparative flows through the detector. *Note*: The inclusion of a suitable splitter associated with the preparative column so that only a fraction of the total column flow is diverted to the detector.

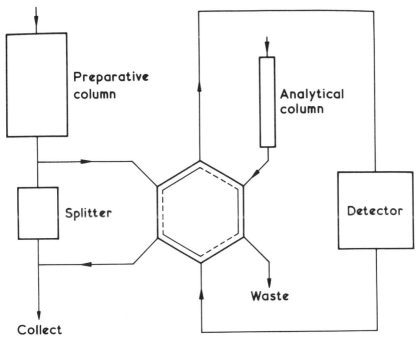

Fig. 9. Combined analytical/preparative chromatographs showing the flow diagram (from ref. 6). *Note*: Water's Associates now markets a chromatograph of this nature.

Care must be exercised to ensure that the extracolumn transfer lines from the preparative column are balanced, i.e., the detector and splitter lines; otherwise cross-contamination can occur. This system has the additional advantage in providing an immediate way of estimating the purity of the collected samples.

References

1. R. P. W. Scott and P. Kucera, *J. Chromatogr.* **119**, 467 (1976).
2. R. P. W. Scott in *Contemporary Liquid Chromatography*. John Wiley, New York (1976).
3. Hazel Pyper, Doctoral thesis, University of Edinburgh, Scotland (1984).
4. M. Martin, F. Verillon, C. Eon, and G. Guiochon. *J. Chromatogr.* **125**, 17 (1976).
5. J. J. de Stefano and H. C. Beachell, *J. Chromatogr. Sci.* **8**, 434 (1970).
6. C. E. Reese, in *Techniques in Liquid Chromatography* (C. F. Simpson, ed.) Wiley-Heyden, John Wiley (1982).

Chapter 14

Preparative Scale Electrophoresis

Colin Simpson

1. Scale

The design of the equipment used for preparative scale electrophoresis (EP) depends principally on the amount of purified protein required for a particular application. Should it be required to raise antibodies, for example, then the amount required may be low and capable of being obtained by isolating the protein from analytical equipment. This level of preparation will be discussed only briefly. When larger quantities are required, limited scale-up of analytical equipment may suffice, providing the problem of heat dissipation, which is an intrinsic byproduct of electrophoresis, can be surmounted. Other factors are the complexity of the sample, and the separation between the component of interest and the impurities. In general, the more complex the mixture, the smaller (for the most part) the size of sample possible to process in a given run. For large-scale preparation, i.e., grams and larger quantities, simple scale-up is impossible, and over the last decade several elegant solutions to this problem have been proposed. These have necessitated the design of novel equipment that has successfully eliminated the three major problems in electrophoresis; namely, Joule heat, convection, and hydrodynamics.

2. Microscale

2.1. Conventional Analytical Methods (Zone EP and Isoelectric Focusing)

The techniques discussed in chapter 12 may be used to prepare samples in the micro-range. Samples may be applied in a trough cut over nearly the whole width of the gel (edge effects preclude using the whole gel width), and the plate is electrophoresed in the usual way. The principal modification in technique lies in not using conventional staining methods to recognize the separated zones. Two methods may be employed. (1) A strip of the gel may be removed, stained, and destained in the normal manner and the highlighted zones used to indicate the equivalent positions in the main body of the gel. This method assumes a constant gel thickness and a homogeneous distribution of buffer or pH gradient across the width and length of the gel. Clearly this may not always be so, and contamination between closely spaced zones could occur. (2) In the contact strip method, a piece of filter paper with the same dimensions as the gel is carefully laid on the surface (taking care to ensure good contact). It is left there for a few minutes to allow a fraction of the separated zones to diffuse into the paper. On removal, the paper may be stained and destained using a general stain to visualize the zones when, by placing the gel plate on top of the paper, the separated zones in the gel may be accurately removed, crushed, and the samples eluted with a suitable buffer. Alternatively, the isolated gel strip may be frozen and, after thawing, centrifuged at $40,000g$ for 15 min, resuspended and centrifuged again. Using this technique, recoveries approaching 100% may be obtained. Clearly, the second method provides the best way of determining the position of the zones. It is also possible to cut the paper into strips and stain successive strips with selective stains to highlight components of interest.

2.1.1. Isotachophoresis (ITP)

2.1.1.1. Capillary Isotachophoresis

Arlinger [1] and Kobayashi [2] have adapted capillary ITP for micropreparative purposes. Their approaches are discussed below.

2.1.1.1.1. Arlinger's Method [1]

The modifications necessary to adapt the L.K.B. Tachophor to microsample collection are illustrated in Fig. 1. The figure compares

Preparative Scale Electrophoresis

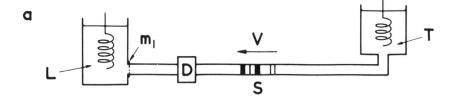

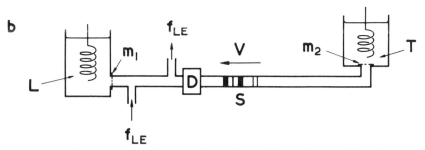

Fig. 1. Comparison of (a) analytical capillary ITP, and (b) the preparative modifications to the basic equipment. L = LE, T = TE, D = UV detector, S = sample zones, V = migration velocity, m_1, m_2 = semipermeable membranes, f_{LE} = flow of leading electrolyte to elute sample zones (adapted from ref. 1).

the analytical with the modified version. Two modifications are apparent. (1) A membrane, m_2 is included immediately after the TE reservoir. This is incorporated to prevent counter-flow through the capillary, and (2) inlet and outlet T pieces are situated immediately after the detector, through which a low flow of leading electrolyte (LE) is passed to sweep separated zones out of the capillary for collection. The sample volumes can be as low as 20 nL. Hence only a very low flow of LE is required; otherwise considerable dilution will occur. This is normally set at about 10% faster than the migration velocity of the zones. Clearly this is a very low sample volume to collect by conventional means. Arlinger solved this problem by collecting the eluting zones on a moving cellulose acetate strip synchronized with the recorder chart speed, so that the position of any zone could be correlated with its appearance on the recorder paper. The appropriate section of cellulose acetate strip can be removed and the protein isolated. There is a time delay between

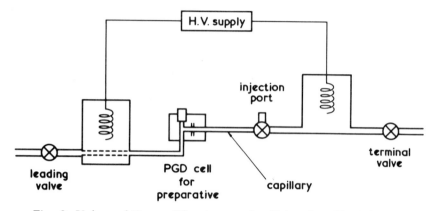

Fig. 2. Kobayashi's modification to the Shimadzu IP-18 for microsampling, with replacement of the potential drop detector with a modified one (adapted from ref. 2).

when the detector registers the presence of a component and its passing out of the T on the surface of the strip. This time constant, at a given LE flow, can be established by electrophoresing a suitable dyestuff, e.g., indigo tetrasulfate, under identical operating conditions as the sample. Arlinger also showed that the proteins could be removed from the surface of the strip with good recovery, i.e., proteins were not irreversibly adsorbed, at least for the compounds tested.

2.1.1.1.2. Kobayashi's Method (2)

The modifications to the Shimadzu system IP-1B for microsampling are shown in Fig. 2. These consist of replacing the potential drop detector with one incorporating a sampling port. Operation of the system is identical to that of the analytical methodology, but when the sample passes out of the detector and reaches a point P (Fig. 3), the driving potential is switched off and the separated zone removed with a syringe. Again, the time lag for a sample to pass from the detecting element to point P must be established by prior calibration under identical operating conditions using a dyestuff. In contrast to Arlinger's method, in which the sample is collected under dynamic conditions, i.e., the power is on, in this method the power is switched off. This can bring about mixing at the zone boundaries if the sampling time is excessive, but Kobayashi has shown that the power can be off for at least 10 min before zone mixing begins (*see* Fig. 4).

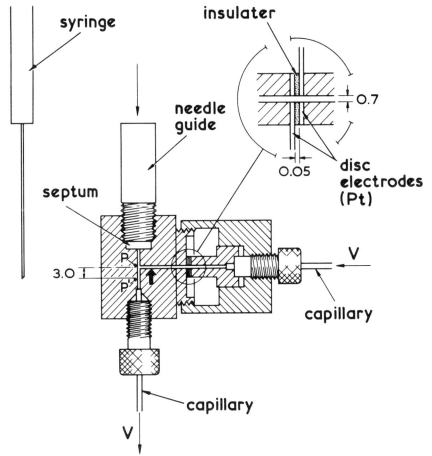

Fig. 3. Detail of the modified detector cell (adapted from ref. 2).

3. Semimicro Scale

Semimicro scales are used for the preparation of up to several hundred milligrams of sample.

3.1. Flat-Bed Preparative Isotachophoresis

Svendsen and Rose (3) first applied isotachophoresis (ITP) for preparative work in gels using a flat-bed technique. Although well documented, this method is not used as frequently as it could be.

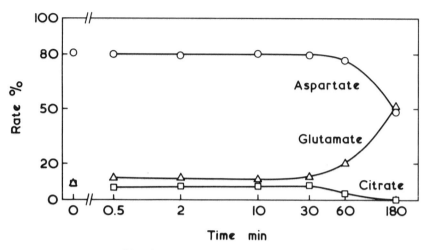

Fig. 4. Time profile of the sample zones showing the effect of sample diffusion with time after switching of the applied potential (adapted from ref. 2).

In practice, the apparatus requirements are similar to those used for zone EP. As indicated in chapter 12, the method uses leading and terminating electrolytes and the sample is placed at the junction of these materials. The anticonvective gel (polyacrylamide gel, PAG) is prepared in a known concentration of leading electrolyte and layered in the usual way in a suitable gel tray, taking care to ensure the gel surface is completely flat. The prepared gel is placed in suitable electrophoresis equipment, well thermostated, with the anode compartment filled with leading electrolyte (LE), the cathode compartment filled with terminating electrolyte (TE), and both connected to the gel with impregnated strips of filter paper. A short prerun is performed at high-current density to move the junction between the LE and TE onto the plate. A drop of bromothymol blue is added to the cathode end of the gel to demark the junction. After about 5 cm migration, the current is switched off and the sample is applied at the junction between the electrolytes. Samples to be collected may contain small quantities of spacer ampholytes (see later) in TE to indicate the boundaries between separated zones. The constant current power supply is switched on and the separation allowed to develop. At the end of the run, the separated zones are detected using the contact strip method or, if fluorescent ampholytes have been used, by irradiation with UV light. Zones are then removed and the protein isolated, as indicated above. The main disadvantage to this method is the dissipation of the Joule heat generated. This limits the level of the applied voltage and the irregular temperature across the gel brought about by uneven cooling, which results in variations in migration velocities of the separating species.

3.2. Column Isotachophoresis

A disadvantage to flat-bed ITP lies in the equipment dimensions, which restrict the length of the gel that may be used. An alternative format is to use a column to contain the gel when the controlling parameter is the voltage drop across the gel (and the heat generated). Care should be exercised in the preparation of the gel to ensure that the surface is absolutely flat. As in the flat-bed method, the gel precursor is prepared in a known concentration of leading electrolyte, which is carefully poured into the column to obviate the formation of bubbles. This is then topped with a one-tenth dilution of LE to ensure an absolutely flat surface. Photopolymerization is used rather than the more usual catalytic process. A further point to note is that the column material is preferably constructed of plastic. Polyacrylamide gel adhers to glass surfaces. When a sample is applied to the gel, its concentration (as used in preparative work) may cause swelling, leading to deformation of the surface and detriment to the symmetry of the applied zone.

The sample is applied to the gel dissolved in TE, to which marker carrier ampholytes may be added if desired. Application of the electric current causes the sample and carrier ampholytes to migrate into the gel and commence to disengage to form the steady-state stack. The stacked components migrate through the gel and elute out of the gel into a small conical zone that is swept by LE into sample collection tubes in a fraction collector. Alternatively, the eluate may be passed through a flow cell and the components detected and collected. The presence of fluorescent ampholytes can be of assistance in detecting the boundaries between zones, particularly if the sample components have only weak absorbance.

The principal restriction to this method lies in the problem of heat dissipation. It is difficult to remove heat from the center of the gel column, making it impossible to operate the system at the higher potentials that would be desirable for high-resolution separations (*see* chapter 12).

3.3. Annular Isotachophoresis

The novel system reported by Hampson and Martin (4) eliminates the problems associated with heat dissipation. The rationale for this method is worth considering in some detail. Unfortunately the equipment is not commercially available, but is relatively simple to construct.

As discussed above, the principal problem in electrophoresis lies in the heat generated on passage of an electric current through the system. Additional problems when dealing with large sample sizes arise from the differences in conductivity of the successive zones, which hence produce different potential gradients across them. This effect also changes the rate of electroendosmosis (which can never be completely eliminated) within the various zones and causes swelling and contraction of the gel. The obvious way to prevent deformation of the gel (which would occur in the column method) is to support it on only one side, thus allowing for free variation in gel thickness as electroendosmosis rates differ. In turn this necessitates immersing the gel in a nonconducting liquid, which will perform two functions: (1) prevent evaporation from the surface of the gel, and (2) assist in removing heat generated in the system.

The apparatus shown in Fig. 5 shows the practical solution to these problems. It consists essentially of a cold finger on which the gel is deposited by the simple expedient of plunging the cold finger held at 10°C into molten agarose (1% wt/vol in water) held at 70°C. This is performed in the apparatus shown in Fig. 6, which consists of a vertical tube joined at the top and the bottom to a second tube containing a magnetic stirrer at its base. The stirrer forms a crude centrifugal pump that circulates the molten agarose around the system. The cold finger is lowered concentrically into the molten agarose and refrigerated water passed through the finger causing the agarose to be deposited on the walls, which should be roughened to provide a key. Coating is stopped by ending chilled water circulation, withdrawing the tube, and plunging it into a bath of water at 55°C to remove any uncongealed agarose. The prepared gel is soaked in LE and is ready for use. It is necessary to limit the heating of the agarose; 2 h appears to be the maximum time the agarose can be held in the molten state. Longer periods lead to a surface reminiscent of "orange peel," which is unsatisfactory for preparative work. Providing care is taken, highly reproducible gels may be prepared by this technique.

Preparative Scale Electrophoresis

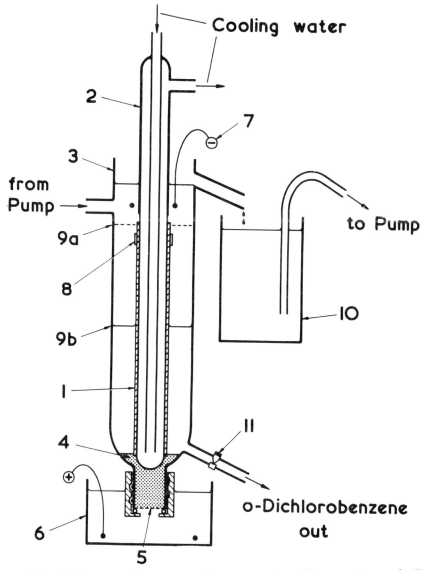

Fig. 5. Diagram of apparatus in cross-section. (1) separating gel, (2) cylindrical glass cooler and support, (3) glass container, (4) agarose gel plug saturated with LE, (5) electrically conducting (cellophane) membrane, (6) bottom electrode compartment, (7) top electrode, (8) filter paper impregnated with sample at the beginning of the separation, (9a) terminator o-dichlorobenzene interface at the beginning of the separation, (9b) interface partway through the separation, (10) terminating electrode reservoir, (11) electronically controlled magnetic valve (adapted from ref. 5).

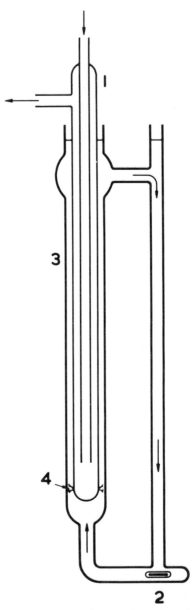

Fig. 6. Diagram of the device for coating the glass tubes with thin uniform films of agarose. (1) glass tube to be coated with agarose, (2) magnetic stirrer, (3) container in which circulation occurs, (4) brass collar with three small projections to center cooling tube (adapted from ref. 5).

The prepared gel on its support is placed in the separation chamber containing the nonconducting liquid (this should be significantly denser than water). In Hampson's work this was o-dichlorobenzene, but perfluorokerosene would be preferred because it is completely inert. The bottom end of the finger dips into a plug of agarose saturated with LE. The top of the agarose coating is arranged so that it is above the nonconducting liquid and in contact with TE.

In this configuration, electrophoresis is carried out in a downward direction, so that any heat generated rises; further it is a feature of this system that the level of the insulating liquid may be lowered automatically as the separation progresses. The extent of the gel that is subject to low conductance is thus kept to a minimum, minimizing the heat effect in the region of the TE. This system shows a remarkable absence of syneresis induced by electroendosmosis, and the zones are symmetrically arranged in the agarose annulus. According to Hampson, even if the sample is asymmetrically applied to the gel, it rapidly redistributes around the annulus.

The sample may contain fluorescent spacer ampholytes to demark the separated zones. Indeed, Hampson produced these substances for this purpose. Figure 7 shows a typical separation obtained between two widely different concentrations of BSA Cohn Fraction V (65 mg) and bovine hemoglobin (1.5 mg). It is clear that as a result of the separation, the concentration levels in the gel are equivalent, which is a characteristic of separations carried out by isotachophoresis. The separated zones may be isolated as indicated above. Full details of this technique are given in ref. 5.

The principal disadvantage of the methods outlined above is that they are batch processes, and although they are satisfactory for the isolation of limited quantities of protein, if larger amounts are required the task can become onerous through repetition. To alleviate this situation, a continuous method of electrophoresis has recently been described by Trop et al. (6). The technique is described below.

4. Continuous Preparative Electrophoresis

4.1. Flat-Bed—Trop's Method

In batch processes, the sample constituents separate according to their mobilities along a single pathway. Hence there is no way

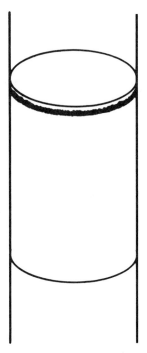

Fig. 7. Displacement electrophoresis of a mixture of BSA Cohn Fraction V, 65 mg and bovine hemoglobin 1.5 mg in 1 cm³ water. LE Tris phosphate pH 8.1, 0.0076M Tris, TE 3-aminobutyric acid. Loading time 25 min, running time 10 min (adapted from ref. 5).

that a second sample can be applied to the gel before the first sample has completed elution development; otherwise the slower migrating components of the first sample will be overtaken by the faster migrating species in the second. Trop et al. have solved this problem by applying an intermittent second potential to point electrodes orthogonal to the main driving potential so that the sample constituents are diverted from their linear pathway and hence traverse individual pathways through the main body of the gel, thus allowing the application of a continuous slow stream of sample to be applied to the plate. Figure 8 shows the positions of the main and secondary electrodes, and it will be seen that the secondary field exerts a nonhomogeneous field to the gel. Thus the faster migrating species will experience a stronger diverting field as they pass down the plate, removing them from the linear pathway normally followed. This system has been investigated by computer simulation, and Fig. 9 shows the effect of this method of develop-

Preparative Scale Electrophoresis

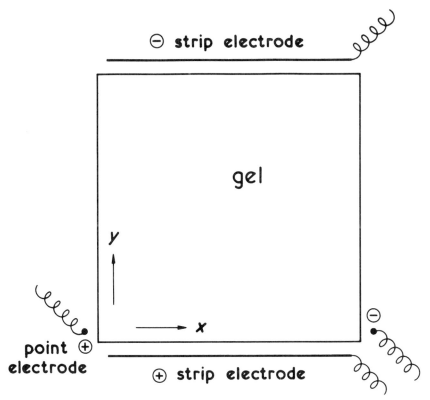

Fig. 8. Diagram of the electrode positions for continuous zone electrophoresis (adapted from ref. 6).

ment, whereas Fig. 10 shows a practical example of a separation obtained on a series of food dyes.

Clearly this method requires more evaluation before it achieves routine status, but there can be little doubt that this method should prove invaluable in the near future.

The problem of continuous separation has also been studied by other workers using a free-flowing technique. Here the sample is allowed to flow down between two plates in a suitable buffer and a potential is applied orthogonal to the direction of flow. The sample constituents are deflected from their linear pathway by the applied field and are collected at the bottom of the flow channel. Alternatively sheets of thick filter paper are impregnated in buffer, and the sample allowed to percolate down through the paper. A potential is applied across the paper thickness and the sample components migrate through the sheets until they emerge at the

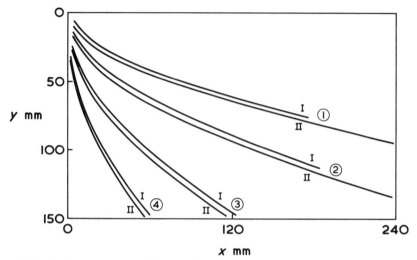

Fig. 9. Computer simulation of the migration pathways under various applied voltages and time sequences (adapted from ref. 6).

bottom of the paper and drip into suitable containers. These methods are perfectly satisfactory for relatively simple mixtures, but for more complex samples, in which resolution is of paramount importance, they do not possess sufficient separating power.

Various methods have been used to prevent mixing, e.g., density gradient stabilization (7,8). This method, although simple to set up, requires excellent flow control and considerable amount of stabilizer to provide the density gradient, e.g., sucrose, and these are the major disadvantages in the technique (Fig. 11).

Enhanced viscosity has been used to stabilize laminar flow in an annulus by Dobry and Finn (9) to obtain separations with some degree of success. They showed that for protein separations, however, the required viscosity would have to be as high as 10^{-2} Pa s, which could give rise to operating problems.

Hannig (10) developed a system similar to that outlined above using simple buffers (Fig. 12). This was quite successful, but extreme care is required to maintain an accurate flow and control the temperature of the system.

4.2. Free-Flow Elecrophoresis — Wagner and Mang

Wagner and Mang (11, 11a,b), following other workers, notably Hannig, have studied free-flowing methods of electrophoresis and

Preparative Scale Electrophoresis

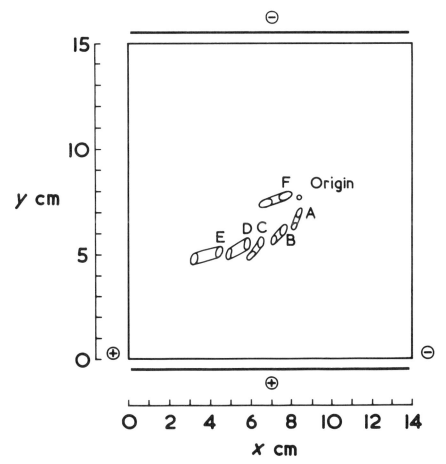

Fig. 10. Separation obtained for various food dyes (adapted from ref. 6).

described a variety of ways to use the technique. Their equipment is marketed by Bender and Hobein, and includes facilities for performing (1) free-flow isoelectric focusing, (2) free-flow isotachophoresis, (3) free-flow field step electrophoresis, and (4) free-field zone electrophoresis. These methods are now briefly considered.

The separation chamber consists of a thick glass front plate and a thick stainless steel back plate. These sandwich a thin glass, or glass/plastic, plate with a spacer to separate the two glass plates. The sample and buffer are admitted to the top of the gap between the plates through a series of entrance ports. A potential is applied

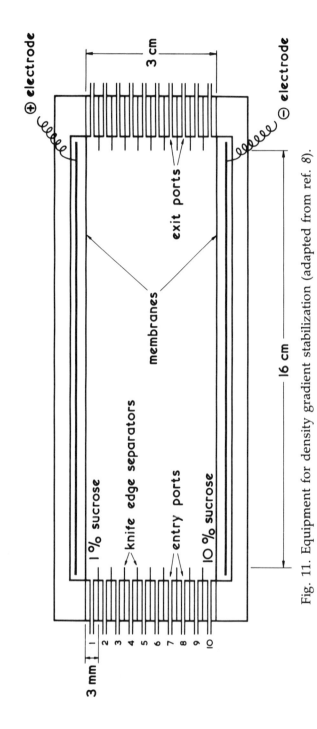

Fig. 11. Equipment for density gradient stabilization (adapted from ref. 8).

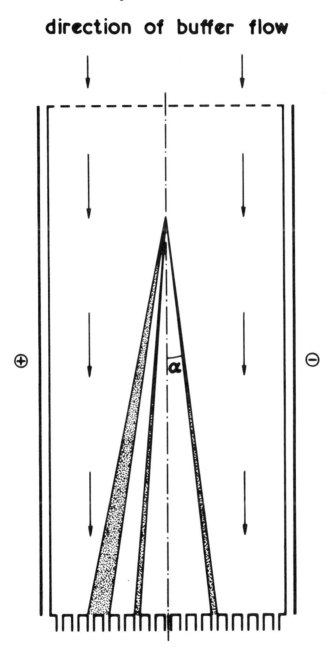

Fig. 12. Hannig system for normal fluid viscosity (adapted from ref. 10).

at right angles to the direction of liquid flow and the sample constituents are diverted from their linear flow by an amount proportional to their electrophoretic mobility. At the bottom of the separation chamber are 90 exit ports (maximum) leading to 90 individual containers for sample collection. The separation between the outlet ports is 1.1 mm and the system can accommodate flows of up to 980 mL/h (depending on the size of the separation chamber).

The four modes of operation are shown in Figs. 13–16. In mode (1) the sample is applied in a solution of ampholytes. Under the

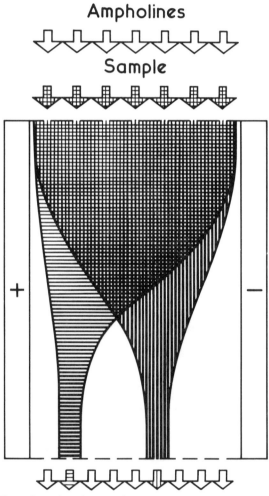

Fig. 13. Free-flow isoelectric focusing, after Wagner et al. (adapted from Bender and Hobein information leaflet.)

influence of the electric field, a pH gradient is rapidly established, and the sample components migrate in the gradient to the position of their p*I*, when they are carried down through the separation chamber to be collected. Field strengths may vary by up to 300 V/cm and residence times range between 20 and 60 min. In mode (2), three different solutions are applied to the chamber. The LE is allowed to enter through the majority of the inlet ports, followed by the sample, and finally the TE (Fig. 14). Thus the requirements for ITP are upheld. Under the influence of the applied field,

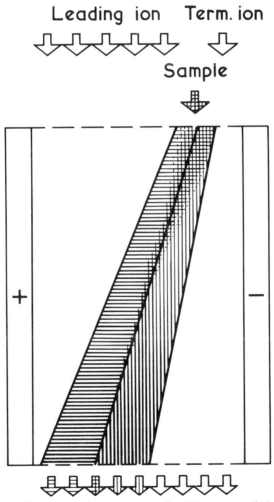

Fig. 14. Free-flow isotachophoresis, after Wagner et al. (adapted from Bender and Hobein information leaflet).

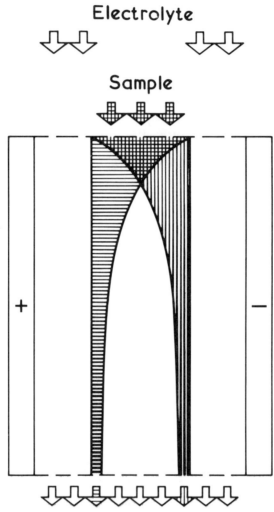

Fig. 15. Free-flow step electrophoresis, after Wagner et al. (adapted from Bender and Hobein information leaflet).

the constituents of the cell form a stacked configuration (albeit horizontal), and are carried down the cell to the collection ports. The field strength may be up to 150 V/cm, and separation times range between 10 and 40 min. Under mode (3), a low-conductivity sample solution is flanked by two electrolyte solutions of high ionic strength. Under the influence of the applied field, the tendency is to equate the field strength across the cell. Hence the sample

Preparative Scale Electrophoresis

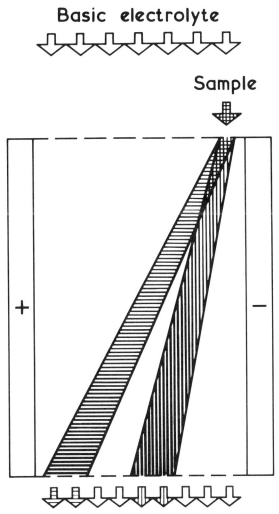

Fig. 16. Free-flow zone electrophoresis, after Wagner et al. (adapted from Bender and Hobein information leaflet).

ions will migrate, according to their charge, to positions next to the boundaries of the sample and the electrolyte solutions. Thus the sample is focused at these boundaries and hence collected. Field strengths are in the range of 200 to 800 V/cm and transit times are typically in the range of 40 to 120 s. In mode (4), an appropriate buffer is admitted to the whole width of the cell, sample is introduced into the laminar flow, and, because of the difference in mobilities, the constituents will migrate to different extents under the influence of the applied field and hence will flow through the cell to different collection ports.

Provision is made to provide adequate cooling of the cell, and importantly, the cell dimensions may be varied to accommodate different sample flows, and hence may be used for small-scale to quite large-scale separations. The equipment may be used also for analytical purposes, and a detecting system is available to provide a trace of the separations achieved based upon differences in optical density of the various flow streamlines. Thus information on the component concentration and migration distance may be recorded.

5. Large-Scale Preparative Electrophoresis

The methods outlined above, with the possible exception of Wanger's method, provide means for preparation of up to the gram scale, and indeed this level may be sufficient for most purposes. Two methods have been developed for large-scale separations: one, by Martin and Hampson is still in the development stage, the other, developed at Harwell, UK, is marketed by CJB Developments Ltd. These methods are outlined below.

5.1. Preparative Isoelectric Focusing

Martin and Hampson (12,13) studied the problem of producing stable pH gradients using simple buffers (the cost of ampholytes would be prohibitive on a large scale). The essence of their method depends on the production of amphoteric membranes to stabilize the pH gradient and to remove the Joule heat developed by employing external cooled reservoirs, pumping the various buffers through the various cells of the equipment arranged in series and electrically connected to each other by the semipermeable isoelectric membranes (Fig. 17), which will produce a stepwise gradient. The principal requirement for the membranes is that they should not transfer electrolyte from cell to cell by electroendosmosis. By careful design in producing membranes of appropriate pI, it was shown to be possible to promote stable pH gradients over extended periods of

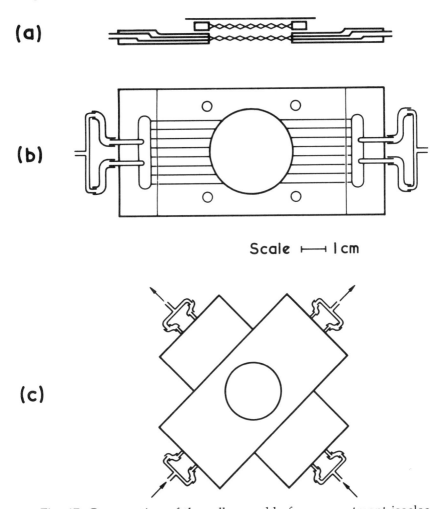

Fig. 17. Construction of the cell assembly for compartment isoelectric focusing. (a) section through single compartment, (b) separator, (c) method of assembly of a number of cells (adapted from refs. *12,13*).

time, and to prepare samples of protein that are focused in a given cell of appropriate pH for the pI of the protein. Although the experiments performed were on a relatively small scale, there seems to be no limit to the size of the cells that can be used, so that it is potentially possible to process a kilogram at a time. Full details of this novel method are given in ref. *13*.

5.2. Large-Scale Free-Flow Electrophoresis

5.2.1. The Biostream Separator

The Biostream Separator (14) is the trade name given by CJB Developments Ltd to the Harwell-designed (15,15a,b) free-flow preparative electrophoretic separator.

In this system the three main problems in large-scale production EP mentioned in the introduction, i.e., Joule heat, hydrodynamics, and convection, have been successfully eliminated. The system is based upon carrying out EP in an annulus between two concentric cylinders in which a laminar fluid flow is maintained by rotating the outer wall of the annulus around a stator. The sample is introduced into the bottom of the system through a thin circular slit and is carried up the annulus in a thin film to the top. A potential is applied between the inner and outer walls of the annulus via semipermeable membranes separating concentrated electrolyte solutions that continuously remove electrolysis products and gases from the electrode compartments. The pH values of the buffers used in the equipment are normally arranged so that the constituents of the sample will all be in the same charged state (usually negative), so that individual components will migrate away from the stator (usually held negative) across the annulus toward the anode. The distance migrated is a function of the mobilities of the species being separated and the upward flow velocity. Thus, separated components emerge from the top of the equipment in a series of concentric rings, where they are collected by a novel collection stack capable of collecting some 30 components into individual collection tubes.

Figure 18 illustrates the principle of the apparatus, indicating the various flow streams in the equipment. Typical dimensions are:

Electrode section length	30.5 cm
Stator diameter	8.0 cm
Annulus width	0.5 cm

Although Joule heat is nevertheless produced in this system, smooth laminar flow is maintained by the angular velocity gradient in the annulus, even though density changes occur in the carrier buffer by the heating effect. This may be as large as 18°C, but all solutions should be cooled to about 0–2°C prior to entry into the annulus. This temperature rise has not proven to be detrimental

Preparative Scale Electrophoresis

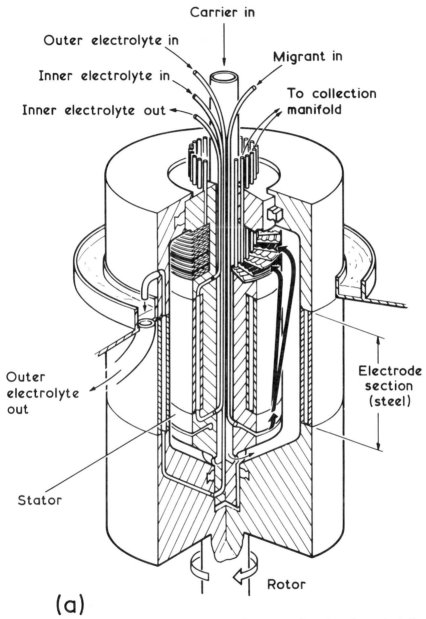

(a)

Fig. 18a. A cutaway of the commercial system showing the principle construction details.

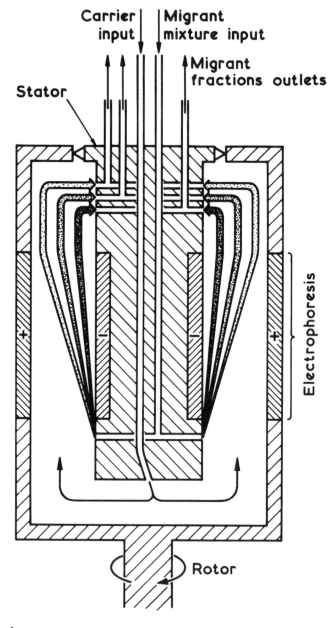

Fig. 18b. Diagram of the principle flow streams (from ref. *14*).

Preparative Scale Electrophoresis 411

in the large number of applications studied in which typical inlet flow rates of 1–20 mL/min have been used with residence times of 15–60 s in the equipment.

Clearly, different dimensions to those given above will affect the time of separation and the volume processed in unit time proportionally.

5.2.1.1. Factors Affecting Zone Disperison

There are two principal factors that affect the successful operation of this free-flow method: the hydrodynamic considerations that control the laminar flow in the annulus, and electrophoretic dispersion associated with the actual separation step.

5.2.1.2. Hydrodynamic Factors

Laminar flow is maintained in the annulus by rotational stabilization of the carrier buffer. There is a maximum rotation speed of 390 rpm for the dimensions given above; this is associated with the rotational Reynold's number and is a function of the ratio of the annular gap and the radius of the outer wall. Associated with this is the migrant zone width at the point of injection at the base of the annulus, which can affect the resolution between the electrophoretically separated zones. This has been shown to be a function of the ratio of the migrant flow to carrier flow and should be small. Finally, the fraction collector system depends upon the rotational speed. With no rotation the sample will be removed symmetrically in all directions, whereas under rotation the sample is removed axially, thus providing the means to withdraw separated components from a given flow streamline.

5.2.1.3. Electrophoretic Factors

The flow profile in the annulus is parabolic in nature, the flow being effectively zero at the containing walls and at a maximum in the center of the annular region. Hence the residence time of any ion in the electric field depends on its radial position in the annulus. This provides a self-sharpening effect to zones that only pass to the center of the annulus, i.e., the leading edge of an initial zone will enter a faster flow streamline than the trailing edge. Thus, it will spend a shorter time in the annulus and be electrophoresed over a shorter radial distance than the trailing edge, which is subject to the field for a longer period, thus allowing it to catch up. The reverse situation occurs after the midpoint of the annulus when zone broadening will occur. Thus the resolution between zones, which is a function of dispersive forces and the separation between the zone maxima, rises to a maximum up to the center of the annulus, falling off as the zone approaches the rotor.

5.2.1.4. Characteristics of the Biostream Separator

Because the separator operates under a continuous mode, the system is characterized by:

(1) High throughput: Up to 2.4 L/h migrant flow is possible at concentrations of up to 50 g/L protein.
(2) High yields: Virtually quantitative recoveries of native protein and enzymes have been regularly obtained.
(3) Wide fractionation range: The equipment is capable of treating samples composed of small molecules (e.g., dyes) to whole cells (e.g., erythrocytes).
(4) Wide pH range: (3–11): *Note:* Restricted in ionic strength.
(5) Good resolution.
(6) Simultaneous multicomponent output. All fractions emerge in less than 1 min; therefore residence time will have little effect on sample components.
(7) Dilution: Because the migrant is injected into the carrier, which is at a much greater flow (up to 2 L/min), dilution occurs, necessitating sample recovery by other techniques, e.g., ultrafiltration, dialysis, reverse osmosis.

References

1. L. Arlinger, *J. Chromatog.* **119**, 9 (1976).
2. S. Kobayashi, Y. Shiogai, and J. Akiyama, in *Second International Symposium on Isotachophoresis* Sept. 9–11, 1980. Analytical Chemistry Symposium Series. vol. 6, Elsevier, Amsterdam, p. 47.
3. P. J. Svendsen and C. Rose, *Sci. Tools* **17**, 13 (1970).
4. F. Hampson and A. J. P. Martin, *J. Chromatog.* **174**, 61 (1979).
5. F. Hampson, in *Electrophoretic Techniques* (C. F. Simpson and M. Whittaker, eds.) Academic, New York (1983), pp. 215–230.
6. M. Trop, A. Kushelevsky, and C. Frenkel, *J. Chromatog. Sci.*, in press.
7. J. S. Fawcett, *Ann. NY Acad. Sci.* **209**, 112 (1973).
8. M. Bier, in *Proceedings of the International Workshop on Technology for Protein Separation and Improvement in Blood Plasma Fractionation* (H. E. Sandberg, ed.) DHEW publication no. NIH 78-1422, Washington, DC (1977), p. 456.
9. R. Dobry and R. K. Finn, *Science* **127**, 697 (1958).
10. K. Hannig, in *Modern Separation Methods of Macromolecules and Particles* (Th. Gerritsen, ed.) John Wiley, New York (1969), p. 45.

11. H. Wagner and V. Mang, in *Second International Symposium on Isotachophoresis* Sept. 9–11, 1980, Analytical and Preparative Isotachophoresis, Walter de Gruyter, Berlin (1984).
11a. H. Wagner, D. Newport, and K. Schlick, *J. Chromatog.* **115**, 357 (1957).
11b. H. Wagner and W. Speer, *J. Chromatog.* **157**, 259 (1978).
12. A. J. P. Martin and F. Hampson, *J. Chromatog.* **159**, 101 (1978).
13. F. Hampson, in *Electrophoretic Methods* (C. F. Simpson and M. Whittaker, eds.) Academic, New York (1983), pp. 231–251.
14. CJB Developments Ltd., Portsmouth, Hampshire, UK.
15. J. St. L. Philpot, in *Methodological Developments in Biochemistry* vol. 2, *Preparative Techniques* (E. Reid, ed.) Longmans, UK (1973), p. 81.
15a. A. R. Thomson, P. Mattock, and G. F. Aitchison, in *Proceedings of the International Workshop on Technology for Protein Separation and Improvement of Blood Plasma Fractionation* (H. E. Sandberg, ed.) DHEW publication no. NIH 78-1422, Washington, DC (1977), p. 441.
15b. P. Mattock, G. F. Aitchison, and A. R. Thomson, *Separation and Purification Methods* (1980), in press.

Chapter 15

Use and Production of Polyclonal and Monoclonal Antibodies

A. W. Schram

1. Introduction

To obtain information on the properties and structure of proteins, the use of specific probes is of great advantage in many cases. These probes may be, for example, substrate analogs (in case of enzymes) that firmly bind to the enzyme, or even suicide substrates that bind and inactivate the enzyme. Another group of very specific probes is represented by antibodies that specifically recognize structural parts of the protein. Over the last 15 years, the use of antibodies in detecting, characterizing, and purifying proteins has become a common method.

The basic principle of the production and use of antibodies is that upon introduction of a foreign protein into an individual, the organism responds by producing antibodies to the foreign protein. The antibodies specifically recognize and bind the protein. This binding is an essential step in the elimination of the protein from the organism; antibodies therefore play an essential role in the defense of the organism against, for example, bacterial infections.

In this chapter information is provided on the production and detection of antibodies. The first part deals with the production of polyclonal antibodies and the second part with monoclonal antibodies.

2. Polyclonal Antibodies

2.1. Immunization Procedure

In most cases the choice of an animal in raising an antiserum depends on the recognition by the experimental animal of the antigen as a foreign molecule. If considerable homology exists, the use of another animal should be considered. Although the large amount of serum that can be obtained from goats, sheeps, pigs, and so on is an advantage, the necessary facilities to keep these animals make it in many cases attractive to choose smaller animals, such as rabbits, rats, mice, and so on.

The immunization procedure itself generally includes a first subcutaneous (sc) and intramuscular (im) injection of 50–500 μg antigen into, for example, a rabbit. To obtain an adequate immunoresponse, it is advisable to suspend the antigen (present in, for example, phosphate buffered saline) in an equivalent volume of Freund's complete adjuvant. One or two weeks after this initial immunization, a couple of booster injections with similar amounts of protein, but using Freund's incomplete adjuvant and a time interval of approximately 10 d, should be sufficient to obtain an antiserum. After a few booster injections, a small sample of blood should be taken to test the serum for the presence of antibodies against the antigen (*see* section 2.2). If after a few booster injections no further increase in concentration of specific antibodies occurs, the animal should be bled or samples of blood should be taken. It should be noted that, even after a long period of rest, a simple booster injection could be sufficient to trigger the immunosystem of the animal and to obtain a second batch of antiserum of good quality.

The blood samples obtained from the animal should be left at room temperature for a short period to allow coagulation to occur. After coagulation, the fibrin cloth and the blood cells are removed by centifugation. To inactivate the complement system, the sample is heated for 30 min at 56 °C and subsequently dialyzed against phosphate buffered saline (PBS), divided into aliquots, and stored at −20 °C. It should be noted that repeated freezing and thawing of immunoglobulins gives rise to loss of reactivity of the serum. It is therefore recommended that the frozen aliquots thawed for use should not be frozen again, but stored at 4 °C in the presence of 0.01% (wt/vol) sodium azide (to prevent bacterial growth).

2.2. Detection of Specific Antibodies

2.2.1. Precipitation of Antigen

Antibodies are able to specifically bind to certain regions (epitopes) of the antigen, thus complexing the antigen in a network that is easily precipitated upon centrifugation. By monitoring the removal of antigen from the supernatant or, alternatively, the appearance of antigen in the pellet, one would be able to detect the presence of antibodies and to define the dilution of antiserum that is just sufficient to obtain maximal precipitation of the antigen (the titer of the antiserum). It should be noted that inhibition of enzyme activity upon adding antibodies (in case the antigen is an enzyme) could be a method of screening, but eliminates antibodies that recognize domains of the protein that do not contribute to the catalytic site of the enzyme.

In some cases the antibody obtained is unable to complex with the antigen to form a network. To obtain precipitation of immunocomplexes in this case, several methods are available.

1. After the incubation of antigen with antiserum (1 h at 20 °C; 20 h at 0 °C) add a second antibody that specifically recognizes the first antibody. This second antibody will induce the formation of a network that is precipitable.
2. After incubation of the first antibody with antigen, add immobilized protein A or immobilized second antibody. Protein A is a protein present in, and purified from, *Staphylococcus aureus*. It exhibits a high affinity for several classes of immunoglobulins and is very useful in precipitating immunocomplexes when attached to a solid matrix (for example sepharose, agarose, or other polymers).

2.2.2. Enzyme-Linked Immunoadsorbent Assay (ELISA)

In order to screen many batches of serum using low amounts of antigen, an ELISA could be useful. In this procedure the antigen is adsorbed to the wells of a microtiter plate. The 96-well plates are prepared from polystyrene and have a high binding capacity for proteins. After adsorption of the protein, the wells are incubated with serial dilutions of antiserum, followed by an incubation with a second antibody that specifically recognizes the first antibody. The second antibody is conjugated to an enzyme, which gives a colored or fluorescent reaction product (β-galactosidase, peroxidase, phosphatase), or to a radioactive isotope, ^{125}I. The enzyme activity or radioactivity present in the well is thus a direct reflection of the presence of second and, therefore, first antibody in the well.

It is essential that between every incubation the wells are carefully washed with buffer (PBS) containing detergent [0.02% Tween 20 (vol/vol)]. Furthermore, consideration should be given to incubating the wells after incubation with antigen with 1% gelatin (wt/vol) or 1% bovine serum albumin (wt/vol). This blocking procedure reduces aspecific adsorption of immunoglobulins to the wells.

2.2.3. Immunoblotting (Western Blotting)

Using ELISA, information is obtained on the presence or absence of antibodies recognizing the antigen. In many cases, however, it is essential to establish whether the antiserum is monospecific, i.e., only recognizes the antigen or is reactive to more components present in the antigen preparation (for example impurities). The immunoblotting technique provides a way to visualize the number of proteins recognized by the antiserum. The first step in the procedure is the separation of proteins in the antigen containing preparation using polyacrylamide gel electrophoresis (PAGE) in the absence or presence of sodium dodecyl sulfate (SDS). Another possibility to separate the proteins is to perform flat-bed isoelectricfocusing. After completion of the separation, the proteins are electrophoretically transferred to a sheet of nitrocellulose. After this blotting procedure, the nitrocellulose sheet (which is a replica of the gel) is incubated first with antiserum, followed by an incubation with second antiserum, to which an enzyme or an isotope is conjugated (see section 2.2.2). As an alternative for the second antibody, protein A conjugated to ^{125}I could be used. The protein bands are visualized by incubating the sheet with a substrate that is enzymatically converted to a colored, insoluble product or by performing fluorography. If (prestained) molecular weight standards are included in the procedure (SDS-PAGE), an estimation of the relative molecular weights of the proteins recognized by the antiserum can be obtained.

2.2.4. Enzyme-Binding Assay

ELISA makes use of protein that is absorbed to the plastic of a microtiter plate or to another solid support. Sometimes, however, it is essential to test whether the antiserum binds the antigen when native or, in case of an enzyme, in its active configuration.

The enzyme-binding test provides a way to investigate this. In this procedure the second antibody (not conjugated) is adsorbed or immobilized to a solid support (microtiter plate, sepharose beads). The second step is the incubation of the immobilized second antibody with the antiserum of interest. The antibodies, if present, will bind to the second antibody. The third incubation is done with the native antigen or antigen-containing solution. After this last incubation, a specific substrate for the enzyme of interest is added to the immuno-immobilized enzyme and the reaction is allowed to occur. If enzyme has been bound (without interference of the catalytic site), conversion of substrate should be detectable.

3. Monoclonal Antibodies

3.1. Principle of the Procedure

In general, proteins exhibit different antigenic sites. If a polyclonal antiserum is raised, an antibody for each antigenic site on the molecule is obtained. In some cases it is desirable to obtain an antibody directed against one of the antigenic sites (determinants). This is of particular interest in those cases in which one wants to discriminate between different molecular or configurational forms (aggregates) of one protein or between a mutant protein and the wild type protein.

In trying to obtain a monospecific polyclonal antiserum, impurity of the antigen could be a problem. If a homogeneous antigen preparation is not available, the conventional procedure for producing an antiserum is unsatisfactory. Moreover, the reproducibility within the different batches of polyclonal sera is sometimes a problem.

With regard to the problems described above, the monoclonal antibody technology is of great importance. Each mammalian lymphocyte has the ability to produce a monospecific antibody. The problem is, however, that plasma cells that secrete the antibody are differentiated lymphocytes with a short lifetime. They cannot normally grow in culture. To immortalize the lymphocyte, it is essential to fuse the cell with an immortal cell that does not produce antibodies (or produces irrelevant antibodies). The basic principle of the hybridoma technology is (1) that such a fusion can be performed relatively easily using polyethyleneglycol and myeloma cells as fusion partners and (2) that the myeloma cells are vulnerable to the cell culture conditions, so they can not survive unless they have participated in the fusion. One way of achieving this is to use a parent tumor line that lacks either the enzyme thymidine kinase or hypoxanthine phosphoribosyl transferase. These are enzymes involved in nucleic acid synthesis and are essential to cells growing in the presence of aminopterin. This compound blocks the main pathway of nucleotide synthesis. After the fusion, the cells are therefore grown in medium containing hypoxanthine, thymidine, and the inhibitor aminopterin (HAT medium).

3.2. In Vivo Immunization and Fusion

The first step in the production of monoclonal antibodies is to immunize a mouse or rat with the antigen of interest. The choice of the animal is limited because of the number of myeloma cell lines available. The immunization is carried out essentially as described in section 2.1. However, a few days before removal of the spleen and fusion of the splenocytes with the myeloma, a hyperimmunization is advisable. This can be performed by iv injection of the protein (no adjuvant present!) in the tail of the animal. After removal of the spleen, the splenocytes are teased from the capsule, suspended in medium, and washed carefully. The splenocytes are subsequently mixed with myelomas, and polyethylene glycol is added. After incubation for a short period, the cells are diluted in medium, centrifuged, and resuspended in the wells of a microtiter plate.

3.3. Cloning Procedure

The hybrid cells are grown for 7–14 d in medium containing amino acids, HAT, vitamins, glucose, buffer, antibiotics, and fetal calf serum. After this initial culturing of the hybridomas, several clones should be visible through the light microscope. After this period it is essential to assay the media, in which the hybridomas have been growing, for the presence of antibodies. This can be done easily using, for example, an ELISA (section 2.2.2) or an enzyme-binding assay (section 2.2.4). Those clones that produce antibodies are harvested, diluted in medium, and divided into the wells of a new sterile microtiter plate. The recommended cloning procedure is to seed hybridomas at approximately 1 cell/well and to grow them in the presence of feeder cells (for example, spleen cells of a nonimmunized animal). One of the functions of the feeder cells, which are essential at low hybridoma concentrations, is to produce growth factors such as lymphokines.

This cloning procedure must be repeated several times and suitable clones should be selected (*see* sections 2.2 and 3.4) for expansion and eventually for large-scale production of antibodies (*see* section 3.5).

3.4. Screening of the Clones

A very important decision to be made during production of monoclonal antibodies is the choice of the assay to be used for screening the hybridomas. The specificity of the monoclonal antibodies could be such that only denatured or native protein is recognized. By using a positive reaction in an ELISA as a sole criterion to subclone wells, the result could be that no clones are obtained that are able to recognize the native protein. It is therefore essential to decide beforehand about the use of the antibodies. If it is essential to have antibodies that recognize reduced antigen, β-mercaptoethanol should be included in the screening incubations. If the antibody is to be used in detecting native enzyme, an enzyme-binding assay (section 2.2.4) should be performed. Considerations of this type will help to limit the amount of useless work, i.e., screening and culturing of hybridomas producing useless antibodies.

3.5. Large-Scale Production of Monoclonal Antibodies

It is not very difficult to expand the scale of production of monoclonal antibodies. The cells can be transferred from microtiter plates to even 175 cm^2 flasks and grown at a cell density of 10^6 cells/mL medium. As an average, the cells produce up to 100 μg/mL of antibody. It is, however, important to check the characteristics of the antibody regularly and to reclone the cells at regular intervals. If a larger scale of production is necessary, the technology today allows one to grow the cells in bulk quantities to yield gram amounts of antibody. Expansion of hybridomas in vivo is also possible. The strategy is to introduce 10^6 hybridoma cells into the intraperitoneal (ip) space of mice. It is advisable to pretreat the mice with pristane, which encourages hybridoma growth. After sufficient growth of the hybridomas, the ascites cells and fluid can be removed from the mice. The fluid represents a rich source of antibodies, but it should be noted that the ascites fluid also contains serum proteins of the mouse. The ascites cells recover well from freezing and can be frozen again in the same way as tissue culture cells and reintroduced into animals without difficulty.

3.6. In Vitro Immunization

In order to efficiently generate hybridomas that secrete antibodies against specific antigens, a source of activated, antigen-specific B-lymphocytes is needed. The usual approach employed for the production of murine- or rat-derived hybridomas is in vivo immunization followed by fusion of the animals' spleen cells (*see* section 3.2). However, in some situations in vitro rather than in vivo immunization is more desirable. For instance:

1. When a very small amount of the antigen of interest is available.
2. When the antigen is evolutionarily highly conserved.
3. When human-derived monoclonal antibodies are desired and there is no source of previously antigen-sensitized human lymphocytes.
4. When antigen is highly toxic in vivo, but not in vitro.
5. When empirical evidence indicates that no satisfactory hybridomas can be obtained using conventional in vivo immunization.

In the in vitro procedure, spleen cells from a nonimmunized mouse are incubated in vitro for a few hours to a few days with antigen. After incubation the spleen cells are fused with myelomas and the procedure continues as described in section 3.2.

To obtain a sufficient immunoresponse, the composition of the medium is of particular interest. It is advisable to use serum-free medium during in vitro immunization and thymocyte-conditioned medium during the procedure. Extreme antigen concentrations will result in no positive hybrids. A titration of soluble antigen in the nanogram and microgram range is thus a wise choice (see below). The in vitro immunization procedure has a couple of major advances.

1. Conservative proteins seem to be antigenic in this system.
2. There is a response at low doses of antigen (>50 ng).

3. The whole procedure, including fusion, takes hours to days instead of weeks.
4. It seems that the percentage of positive clones reacting positively to the antigen is increased.
5. It is possible to divide the splenocyte suspension before immunization and to compare different immunization procedures (antigen presentation, concentration).

The main disadvantage of the procedure is that a large proportion of the antibodies obtained are of the IgM, and not IgG, class. Whether this can be manipulated by changing immunization conditions is unknown.

As an alternative to the above procedure, it is possible to immunize a mouse in vivo and to stimulate the spleen cells before fusion in vivo. This has the advantage of a higher percentage of IgG and an increased percentage of positive clones. It also allows one to make antibodies against poorly antigenic proteins.

3.7. Epitope Analysis

A question one raises about a panel of monoclonal antibodies is whether they are directed to the same or to different epitopes of the protein. The basic principle of the test used is to investigate a possible competition between different antibodies for binding to the protein. This can be done in various ways. One strategy is to adsorb the protein to the well of a microtiter plate and to compete between a radioactively labeled and a second unlabeled monoclonal antibody. In case of a common epitope, competition occurs and the amount of label bound to the well depends on the absence/presence of the other unlabeled monoclonal antibody.

Another important parameter is the affinity of the antibody to the epitope recognized. To obtain information on the affinity, different amounts of radioactively labeled antigen and a constant amount of antibody are allowed to react. After reaction the amount of free and bound antigen is determined and plotted according to the SIPS equation:

$$\log[b/Ab \text{ tot} - b] = a \log K + a \log c$$

where Ab = antibody and tot = total.

If $\log [b/(Ab \text{ tot} - b)]$ is plotted against $\log c$, $K = 1/c$ when the ordinate is zero. The equilibrium constant is a measure of the affinity of the antibody for the epitope.

4. Characterization and Purification of Antibodies

The antibody class is most readily determined by the use of class-specific antibodies in an ELISA. It is also common to analyze in vivo labeled antibody on SDS-PAGE and isoelectrofocusing. This is particularly helpful in the determination of the relative proportions of the antibody chains and as a proof that the antibody is, for example, really monoclonal. One of the characteristics of an antibody is also the affinity for protein A. Because of the wide use of protein A (see this section and 2.2.3), it is an important parameter to investigate. It should be noted that not all immunoglobulins are bound to protein A.

Antibodies can be purified using different procedures. A crude preparation (and a concentration) can be obtained using $(NH_4)_2SO_4$ precipitation. Chromatography on DEAE cellulose or other ion exchangers allows for large-scale purification methods. High-performance liquid chromatography using hydroxyapatite columns leads to a preparation of higher quality, but requires an investment in equipment. More refined methods are the purification using immobilized protein A (not applicable to every antibody) or using affinity chromatography. In the last case the antigen is covalently bound to a solid matrix and the antibody-containing serum or medium is allowed to incubate with the affinity support. By lowering the pH to 2–3, the bound antibody can be recovered from the matrix giving rise to a monospecific, immunopurified antibody. In many cases it depends on the use of the antibody which method of purification should be applied.

5. Application of Antibodies

Antibodies can be used in a variety of applications. They are especially suitable for detecting the presence and quantity of proteins and to analyze a sample on the presence of different molecular forms of the same proteins, mutated proteins, and so on. For this purpose, the methods described in this chapter for detecting antibodies can be applied. Besides these methods, other techniques such as radioimmunoassays, rocket electrophoresis, and so on are available. It is without the scope of this chapter to treat these methods extensively. The reader is referred to the list of references and to the many laboratory handbooks available that give detailed description of the individual techniques. Antibodies are also of great importance in histochemical studies in which the (sub)cellular localization of proteins in tissue is investigated. Also of great potential is the use of immobilized antibodies as affinity ligands in purifying proteins. Very elegant one-step procedures to extract a protein using this methodology have been described.

The introduction of the in vitro immunization technology without doubt increases the possibility of using antibodies for these and other purposes. It should in this respect be noted that, at present, using minor amounts of (even conservative) proteins or peptides derived from these proteins, antibodies can be obtained.

Suggested Reading

Literature of particular importance and recent literature is listed below. The complete literature from before 1984 and many protocols and procedures are summarized in references cited with an asterisk.

R. A. Barneveld, F. P. W. Tegelaers, E. I. Ginns, P. Visser, E. A. Laanen, R. O. Brady, H. Galjaard, J. A. Barranger, A. J. J. Reuser, and J. M. Tager (1985) Monoclonal antibodies against β-glucocerebrosidase. *Eur. J. Biochem.* **134**, 585–589.

C. A. Borrebaeck (1983) In vitro immunization for the production of antigen-specific lymphocyte hybridomas. *Scand. J. Immunol.* **18**, 9–12.

*A. M. Campbell (1984) Monoclonal Antibody Technology: The Production and Characterization of Rodent and Human Hybridomas, in *Laboratory Techniques in Biochemistry and Molecular Biology*, Elsevier, Amsterdam.

H. Hengartner, A. L. Luzzati, and M. Schreier (1978) Fusion of in vitro immobilized lymphoid cells with X63Ag8. *Curr. Topics Microbiol. Immunol.* **81**, 92–99.

J. Hilkens, J. M. Tager, F. Buijs, E. M. Brouwer-Kelder, G. M. Van Thienen, F. P. W. Tegelaers, and J. Hilgers (1981) Monoclonal antibodies against human acid α-glucosidase. *Biochim. Biophys. Acta* **678**, 7–11.

K. Inaba, R. M. Steinman, W. C. Van Voorhis, and S. Muramatsu (1983) Dendritic cells are critical accessory cells for thymus-dependent antibody responses in mouse and man. *Proc. Natl. Acad. Sci. USA* **80**, 6041–6045.

G. Kohler and C. Milstein (1975) Continuous cultures of fused cells secreting antibody of predefined specificity. *Nature* **256**, 495.

M. M. S. Lo, T. Y. Tsong, M. K. Conrad, S. M. Strittmatter, L. D. Hester, and S. H. Snijder (1984) Monoclonal antibody production by receptor mediated electrically induced cell fusion. *Nature* **310**, 792–794.

R. A. Luben and M. A. Mohler (1980) In vitro immunization as an adjunct to the production of hybridomas producing antibodies against the lymphokine osteoclast activating factor. *Mol. Immunol.* **17**, 635–639.

R. A. Luben, P. Brazeau, and R. Guillemin (1982) Monoclonal antibodies to hypothalamic growth hormone releasing factor with picomoles of antigen. *Science* **218**, 887–889.

K. Lundgren, M. Wahlgren, M. Troye-Blomberg, K. Berzins, H. Perlmann, and P. Perlmann (1983) Monoclonal anti-parasite and anti-RBC antibodies produced by stable EBV-transformed B-cell lines from malaria patients. *J. Immunol.* **131**, 2000–2003.

R. L. Pardue, R. C. Brady, G. W. Perry, and J. R. Dedman (1983) Production of monoclonal antibodies against calmodulin by in vitro immunization of spleen cells. *J. Cell Biol.* **96**, 1149–1154.

C. L. Reading (1982) Theory and methods for immunization in culture and monoclonal antibody production. *J. Immunol. Meth.* **53**, 261–291.

C. Stahli, V. Staehlin, J. Miggiano, J. Schmidt, and P. Haring (1980) High frequencies of antigen-specific hybridomas: Dependence on immunization parameters and prediction by spleen cell analysis. *J. Immunol. Meth.* **32**, 297–304.

H. Towbin, T. Staehelin, and J. Gordon (1979) Electrophoretic transfer of proteins from polyacrylamind gels to nitrocellulose sheets: Procedures and some applications. *Proc. Natl. Acad. Sci. USA* **76**, 4350.

J. Van Ness, U. K. Laemmli, and D. E. Pettijohn (1984) Immunization in vitro and production of monoclonal antibodies specific to insoluble and weakly immunogenic proteins. *Proc. Natl. Acad. Sci. USA* **81**, 7897–7901.

*D. M. Weir (1978) *Handbook of Experimental Immunology*, Blackwell, Oxford.

Chapter 16

Large-Scale Methods for Protein Separation and Isolation

Peter J. Lillford

1. Aims of Large-Scale Processes

Unlike laboratory processing, large-scale methods are primarily concerned with efficiency and economics of scale. In handling biological materials there is constant "negotiation" between the engineer and chemist. The former is primarily concerned with efficient heat and mass transfer—because of its efficient utilization of energy, whereas the latter is concerned with maintenance or selective modification of the protein conformational state. As we will see, both criteria are not always optimally achieved. Furthermore, the transient and nonuniform conditions in large-scale plants have not been mathematically modeled, so that most progress still occurs by a combination of past experience and empiricism. Nonetheless, these derived guidelines are of value and lead to avoidance of the most obvious mistakes.

One concept on which both engineer and chemist can agree is that the yield at each isolation step must be as high as possible and the number of isolation steps minimized. It is simple to identify that even if each process step results in 90% recovery then for n steps

$$\text{Yield (\%)} = 0.9^n \times 100$$
$$= 59\% \text{ for } n = 5$$

We will now examine some of the unit operations common to large-scale isolation. The majority of the information derives from downstream treatment of fermentation processes for the isolation of enzyme, and a more extensive background can be found in *Fermentation and Enzyme Technology* (1).

2. Description of Raw Materials

Except for liquid feedstocks (e.g., milk, blood), or the recovery of extracellular enzymes from fermentation broths, the first step in protein isolation requires disruption of the structure in which the proteins are compartmentalized. Unfortunately most disruption processes release not only protein but other undesirable fractions that at best contaminate and at worst cross-react with the required product. On a large scale, the *selective* efficiency of disruption may dominate the whole process efficiency. Ironically, since the emphasis on yield is less important in laboratory practice, little information is available to the large-scale operator. Figure 1 shows methods that have been used (2); the choice of optimum method obviously relates to the mechanical properties of the raw material. Very little quantified information exists, but for hard, brittle structures, such as seeds, fracture dominates the empirical methods, and mechanical disruption by shear during cracking, flaking, or grinding is a prerequisite. Further disruption by liquid shear is then employed in aqueous extraction (*see* chapter 18). Table 1 summarizes the response of softer structures to disruption (3), and examples of nonmechanical methods are given.

For commodity food proteins, selective disruption and separation are sufficient to provide a significant fractionation of proteins, e.g.:

1. Mechanical means of bone and flesh separation for small fish produce a high-protein mince for restructuring.
2. New methods for meat deboning simultaneously remove large amounts of connective tissue (elastin and collagen), leaving an upgraded lean meat (myofibrillar and sarcoplasmic protein) fraction.
3. Air classification of crushed and flaked seeds provide enriched protein fractions (*see* chapter 18).
4. Washing and kneading of wheat dough releases starch to produce the "vital gluten" fraction.

Large-Scale Methods

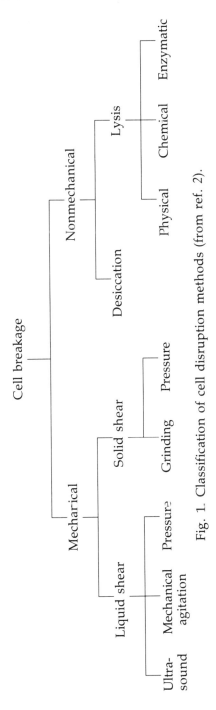

Fig. 1. Classification of cell disruption methods (from ref. 2).

Table 1
Susceptability of Various Cells in Suspension to Disintegration

	Sonic	Agitation	Liquid pressing	Freeze pressing
Animal cells/vegetative cells	7	7	7	7
Gram− ve bacilli and cocci	6	5	6	6
Gram+ ve bacilli	5		5	4
Yeast	3.5	3	4	2.5
Spores	2		2	1
Mycelium	1	6		5
Seeds	0	0	0	0

Scale of probability 0 → 7

Nonmechanical methods

Chemical plasmolysis	Yeast mixed with chloroform. Filtration removes most soluble protein, but leaves active enzyme with the debris.
Desiccation	Yeast slurry, spray dried at 10–18% solids at low outlet temperatures (54–95°C), allows for selective extraction of enzymes (e.g., β-galactosidase).
Enzymatic action	Lysis of *Micrococcus lysodeikticus* by lysozyme permits extraction and recovery of catalase.

3. Extraction

All extraction (and subsequent precipitation) methods, whether for total protein or selective fractions, rely on the differential solubility of proteins in aqueous solution (*see* chapter 3). In general, adjustment of the ionic and pH environment is used to obtain minimal electrostatic interaction. Studies of protein solubility usually precede development of a large-scale process, and Fig. 2 shows the solubility profile/extraction efficiency typical for legume storage proteins (soya, groundnut, pea, and so on).

In this case, the protein is stored by the seed in a highly concentrated form in the protein body. Low ionic strength produces a disruptive "osmotic shock"; reducing agents facilitate the rate and extent of extraction, presumably by cleavage of labile disulfides formed during the storage process; the pH is adjusted away from the isoelectric point, and provided pH extrema are avoided, the proteins can be maintained in a "native" state. The published information on protein solubilities is not always relevant since it refers to high solvent:protein ratios. Large-scale operation under these conditions results in large volumes of material in subsequent handl-

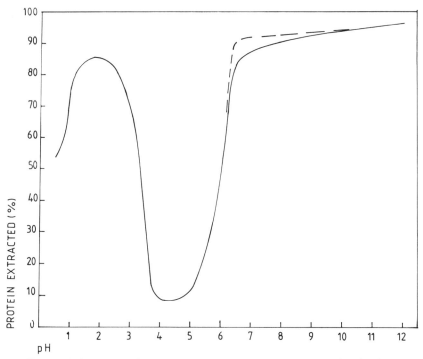

Fig. 2. Schematic of a typical protein extractability profile for legume proteins. The dotted line shows the enhanced extractability in the presence of reducing agent.

ing, which can have a major effect on equipment costing. It is always advantageous to operate at much lower solvent:protein ratios, but two factors that are not encountered in the laboratory then inhibit yield. First, solubility of proteins may be concentration-dependent, and, second, the kinetics of dissolution will be significantly altered both by reduced molecular diffusion and, more importantly, by reduced mass transfer during mixing. The latter can be compensated for by greater mixing shear rates (4). This does not necessarily lead to structural damage. It is the opinion of the authors that, although increased mixing shear may possibly denature proteins of high axial ratios, much of the reported damage to globular proteins, such as enzymes, results not from the liquid shear but from the concurrent entrainment of air, which produces significant interfacial area at which the protein can irreversibly unfold.

In general, for enzyme isolation, in which biological function must be maintained, the extraction conditions must be mild and override simple criteria of extraction efficiency.

4. Precipitation and Fractionation

4.1. Precipitation Methods

4.1.1. "Salting Out"

Ionic strength manipulation provides for protein precipitation on the basis of solubility as a function of salt concentration. This "salting in" is followed either by ionic strength reduction or by further increase in ionic strength to "salting-out" levels (see chapter 3).

Salting out is usually performed with ammonium sulfate: other salts may also be used, but univalent salts are not particularly effective. Figure 3 gives an empirical relationship for salting out, where

$$\log S = \beta - KC$$

where S is solubility, C is salt concentration (depends on T and pH), K is a constant, and β is dependent on temperature, pH, and the nature of the protein, but independent of the nature of the salt. K depends on the salt and protein, but is independent of either pH or temperature.

In practice, ammonium sulfate may be added as a solid, or it may be added as a saturated solution to give a "% saturation" in the protein dispersion (5). On any scale of operation, colloidal precipitation is not instantaneous and the mechanical procedure for contacting salt modifies the apparent yield (6) (Fig. 4).

Hoffmeister (lyotropic) effects are normally observed for the salting-out phenomenon. As such, hydrophobic effects are important and therefore temperature can be an important variable—increasing temperatures may decrease solubility—and increasing ionic strength enhances the protein–protein hydrophobic interactions (see chapter 5).

4.1.2. Isoelectric Precipitation

The different proportions of acidic and basic amino acid residues in a protein result in the existence of a pH at which its *net* charge

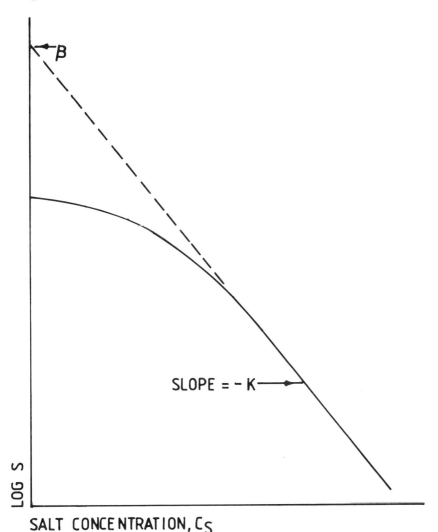

Fig. 3. Typical salting-out curve for soluble proteins.

(not surface charge) will be zero. At this pH, protein solubility is minimal, and by definition only albumins are soluble at low ionic strength. This results in a widely used and relatively simple protein-recovery method. Unfortunately, whereas many seed and vegetable proteins are globulins and can be recovered by this means, most enzymes are albumins and can be recovered only by simultaneous temperature adjustment. This results in losses of activity and limits the applicability of the method.

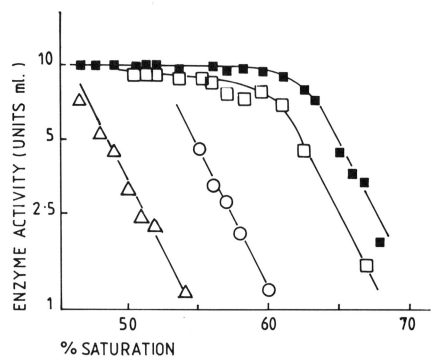

Fig. 4. The effect of contacting procedure and time on the salting out of fumarase by ammonium sulfate. △, solid, batch, 12 h contact; ○, saturated solution, batch, 12 h contact; □, saturated solution, continuous, 12 h contact; ■, saturated solution, continuous, 1.6 min contact. Initial protein concentration was 40 mg/mL at pH 5.9 (reproduced from ref. 5).

For globulins the method is widely used (chapter 18), but, compared with laboratory practice, large-scale operation requires the use of concentrated acids or alkalis—to reduce handling volumes—with the concurrent danger of denaturation by local, transient pH extremes. Some work has been reported on the *rates* of acid denaturation that show that the results are protein specific (Fig. 5) (6).

For proteins with labile sulfhydryl groups, the presence of reducing agents apparently acts in a protective manner at low pH, presumably by decelerating aggregation (*see* Table 2).

The acid anion also apparently influences the degree of denaturation, presumably by shifting the point of pH denaturation (*see* Table 3).

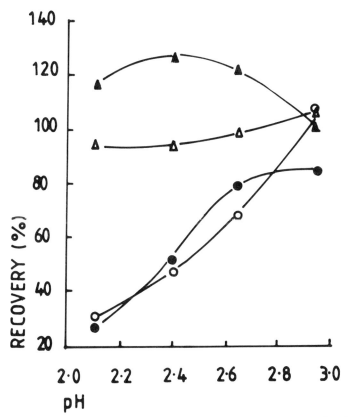

Fig. 5. The effect of exposure to low pH for 5 min prior to back titration upon soya conglycinin ($\triangle$, $\blacktriangle$) and glycinin ($\bigcirc$, $\bullet$). Open symbols represent recoveries measured by gel electrophoresis and closed symbols, by differential scanning calorimetry (from ref. 6).

4.1.3. Thermal Coagulation

Thermal denaturation of a protein frequently results in insolubilization (at pH values near to the isoelectric point); this is followed by coagulation and precipitation. The product is necessarily denatured, but for some food and nutritional uses, this may not matter. The formation and properties of the precipitate are dependent on the net charge (pH) and the presence of added or incipient flocculating agents in the extract. In leaf protein recovery for food and feed, the latter can depend on the maturity of the crop (7).

Table 2
Effect of Very Low pH on the Recovery of Unchanged Soya Proteins
During Precipitation and Redissolution as Measured in
Polyacrylamide Gel Electrophoresis

	Percentage recovery of unchanged protein			
	Lowest pH 2		Lowest pH 4.7	
	Glycinin	Conglycinin	Glycinin	Conglycinin
Without mercaptoethanol	52	48	86	67
With mercaptoethanol	78	93	92	85

Table 3
Percentage Recoveries of Unchanged Soya Proteins
After Acid Precipitation by Different Acids

	% Recovery			
	Conglycinin		Glycinin	
Acid	PAGE	DSC[a]	PAGE	DSC
Sulfuric (10N)	109	112	96	67
Nitric (7N)	87	131	69	53
o-Phosphoric (9N)	80	150	25	19
Hydrochloric (3N)	77	133	26	0
Trichloroacetic (10N)	32	0	7	0

[a]DSC, differential scanning calorimetry.

4.1.4. Addition of Organic Solvents

Protein solubility is modified by the addition of water-miscible organic solvents in a manner similar to salting out. An exceptional example of its use is the Cohn method for blood serum fractionation (8). The outline of the method and the fractional recovery are given in Fig. 6 and Table 4.

The transfer of this laboratory exercise to a large-scale method is an excellent case study for would-be biochemical engineers. The particular problems originally encountered were (9):

1. The temperature change produced by the heat of mixing of alcohol and water shifts the solubility of components.
2. pH control and flow metering is a general problem also encountered in isoelectric precipitation.
3. Good mixing, without excessive turbulence, which causes foaming and denaturation.
4. Coagulation, to allow efficient dewatering and purification of fractions.

Large-Scale Methods

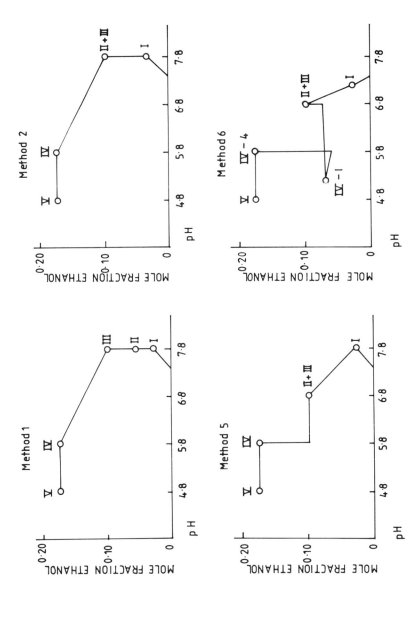

Fig. 6. Ethanol concentration and pH for separation of blood plasma fractions by various methods (from ref. 8).

Table 4
Distribution of Plasma Proteins into Fractions by Various Methods Estimated by Electrophoretic Analysis and a Nitrogen Factor of 6.25 for All Proteins

Fraction	Method	Albumin	α-Globulin	Cholesterol	β-Globulin	γ-Globulin	Fibrinogen
Plasma		33.2	8.4	1.6	7.8	6.6	4.3
I	1	0.3	0	0	0.2	0.7	3.0
I	2	1.0	0.3	—	0.4	0.2	2.4
I	5	0.2	0.2	0.02	0.8	0.5	2.6
I	6	0.3	0.3	0.01	0.6	0.3	2.3
II	1	1.5	0	0.2	1.3	3.7	0.3
III	1	1.5	0.1	0.3	2.6	2.6	0.2
II + III	2	0.6	0.9	—	5.9	4.7	1.4
II + III	5	0.7	1.8	1.1	6.2	6.0	1.6
II + III	6	0.6	0.9	1.3	6.7	5.7	1.6
IV	1	5.6	4.2	0.7	3.0	0.4	0.2
IV	2	4.9	4.9	—	3.7	0.6	0
IV	5	1.0	5.4	0.4	3.1	0.2	0
IV-1	6	—	3.9	0.4	3.1	0.2	0
IV-4	6	0.9	2.7	0.04	2.2	0	0
V	1	26.0	0.3	0	0.4	0	0
V	2	27.0	0.6	—	0.3	0	0
V	5	29.0	0.6	<0.01	0	0	0
V	6	28.4	1.2	<0.01	0.3	0	0
VI	1	2.2	0.1	0.02	0	0	0
VI	2	0.5	0.1	—	0	0	0
VI	5	0.3	0.3	—	0	0	0
VI	6	0.7	0.2	—	0	0	0

Large-Scale Methods

The process is optimized for the recovery of albumin. Despite these process difficulties, the advent of improved separation methods, and the increasing value of the minor components, this original process still remains dominant.

4.1.5. Use of Exclusion Polymers

Exclusion polymer precipitation consists of the addition of high molecular weight neutral polymers (dextroses, polyethylene glycol, and so on) to protein dispersions resulting in separation of a protein-rich phase by mutual exclusion. Ingham and coworkers have published useful literature in this area (10), and for human albumin, the linear dependence of log S on polyethylene glycol concentration suggests that this precipitation method should be treated conceptually along with traditional salting out. Indeed, coacervation (phase separation of lyophilic colloids in the presence of salts and nonelectrolytes), salting out, and organic solvent precipitation may all be considered within the same scientific framework, i.e., a competition with the protein for the water.

4.2. Formation of Precipitates

In the previous section, reference has been made to differences in the *kinetics* of precipitation. In laboratory and batch processing the time scales are not particularly significant, but in continuous, large-scale operation, *rates* of formation of precipitates become dominant.

The scientific basis of kinetic studies uses the theories of colloidal flocculation and both nucleation and growth rates of aggregates must be considered.

4.2.1. Perikinetic Aggregation

Under conditions leading to self-association, molecules collide and aggregate by random motion or thermal agitation. Growth continues to a size at which fluid motion becomes important in promoting further collision or aggregate disruption. This first growth process is termed "perikinetic" and applies to a size range of up to 0.1 to 10 μm, respectively, for high- and low-shear fields. The time dependence of the aggregate molecular weight ($M_w(t)$) is given by:

$$M_w(t) = M_w(o)\,(1 + 8\pi\,Dd\,m_o t)$$

where m_o is the molar concentration of the aggregate species, D is the particle diffusivity, and d is the particle diameter. A more detailed review can be found elsewhere (11).

4.2.2. Orthokinetic Aggregation

For large particles, fluid motion promotes collision and hence the aggregates grow, i.e., the total number of particles decreases. The rate of decrease of particle number N is then given by (12).

$$\frac{dN}{dt} = \frac{2}{3} G d^3 N^2$$

where G is the shear rate, d the particle diameter, and N the collision effectiveness factor.

Note that this equation predicts very rapid growth and a simple dependence on shear rate. A point is reached, however, at which collision favors break-up. Simple theories relate the maximum diameter (d_{max}) in a stirred vessel to the impeller speed (N_s), the power dissipation (E) or the mean velocity gradient (G) by

$$d_{max} \, N_s^{-3j} \, E^{-j} \, G^{[-2j]}$$

where j is a measured parameter, $0.1 < j < 0.8$ (13,14).

For a more detailed treatment of colloidal events in precipitation and centrifugation, see ref. 15. It is obvious, however, that most large-scale developments still derive from empirical observation.

5. Solid/Liquid Separation

Solid/liquid separation is necessary for the clarification of extracts and the recovery of precipitates. Usually several passes through a single device or multiple-separation processes are used.

5.1. Filtration

Upscaling laboratory methods produces the drum vacuum filter and the plate and frame filter. Drum filters are often used for the removal of cell debris. Filter aids (e.g., Kieselguhr) can be used for operational efficiency, but this frequently results in absorption of the dissolved protein or contamination of the downstream liquid with abrasive fines.

Large-Scale Methods

The filter press is cheap, but slow and cumbersome, and can only be used in batch operation. A recent article (16) identifies a number of commercial systems.

Sartorius disk and cartridge membranes with 10,000 L/min flow rates are made from plastic polymers with pore sizes down to 0.01 μm. Sinclair rotary filters, operating under gravity, with 1–3 rpm rotation, are self-cleaning. Flow rates of 26–1170 L/min for water and separation of particles down to 75 μm are quoted. A Chemapex Funda-Filter is a "horizontal leaf pressure filter" that may be used in combination with a specially designed centrifuge. This system is completely closed and allows for operations under sterile conditions.

Stockdale rotary drum vacuum filters are particularly aimed at biotechnology processes (complete discharge of solids and high standards of sterility). An Edwards and Jones Ltd. filter press is aimed at protein processes and allows for recovery of either solid or liquid phases.

5.2. Centrifugation

5.2.1. Basket Centrifuges

Filtration can also be achieved with the aid of centrifuges, in which the centrifugal field acts as the pressure. These basket centrifuges are the engineers' equivalent of the domestic spin drier. They are essentially batch devices and are limited by the mechanical properties of the suspended solids. Protein precipitates are essentially soft solids because of their associated moisture. In a basket centrifuge, an increased centrifugal pressure is likely to cause blinding of the filters, by extrusion of the precipitate into the filter pores, and will not necessarily increase the dewatering. By comparison, attention to the precipitate structure by aging (17), ultrasonic conditioning (18), or even freezing (19) has a greater effect on efficiency than the application of higher pressures. In general, the precipitate for efficient dewatering by filtration has an open structure, strong enough to withstand the applied pressure and allowing for efficient drainage of the occluded liquid. This is rarely achievable for the precipitates of globular proteins; hence the method is intrinsically limited.

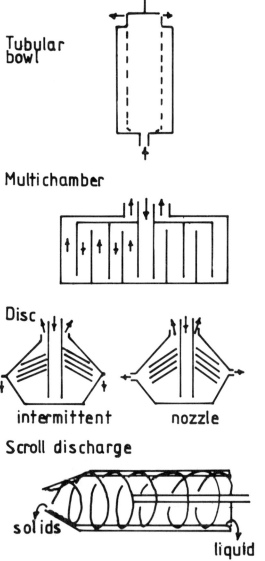

Fig. 7. Types of centrifuges available.

5.2.2. Separators

Other centrifuges (Fig. 7) depend on a density difference between solid and liquid and the ability of a centrifugal field to separate them.

Large-Scale Methods

The commercial advantages are low operating costs, lack of contamination (by filter aids), and the opportunities for continuous operation. The disadvantages are that most large-scale equipment was designed for the separation of small amounts of inorganic solids from dilute (Newtonian) solutions. This is exactly contrary to the requirements for protein separation, in which high solid contents of a colloidal nature are to be separated from saturated solutions. Not surprisingly these new requirements have required a reinvestigation of centrifuge efficiency (15). In practice, separators are used in series; usually a low-speed device followed by a high-speed "polisher."

A check-list of centrifuge capabilities is given in Table 5.

Their efficient performance, however, is not predictable except by consideration of mechanical properties of the precipitate. Protein sludges are gel-like and adhesive, and excessive build-up in an intermittant discharge disk centrifuge can prevent the chamber from opening, let alone discharging its contents. Air entrainment upstream can result in precipitate densities *less* than those of the liquid component, completely eliminating separation. It is the author's experience that such complications are extremely embarrassing on a plant operating on a ton/h scale.

Table 5
Types of Centrifuge Applicable

Advantages	Disadvantages
Tubular bowl	
High centrifugal force	Limited solids capacity
Good dewatering	Foaming
Easy to clean	Recovery of solids
	Noncontinuous
Multichamber	
No loss of efficiency as bowl fills with solid	No solids discharge
	Recovery of solids
Large solids holding volume	Noncontinuous
Good dewatering	
Disk	
Ability to discharge solids	Poor dewatering of solid
Liquid discharge under pressure	Difficult to clean
Scroll discharge	
Continuous solids discharge	Low centrifugal force
High input solids concentration	

Apart from such obvious difficulties, more subtle effects of the upstream conditions on separation efficiency are identified. These effects relate primarily to the particle size distribution entering the separator, which is a function of the particular protein type, the shear history, and the method of precipitation. Of these effects, the shear history is particularly important and the highest shear field encountered by precipitates occurs in the separator itself. This results in apparently anomalous separation efficiency. For example, in some cases, scroll discharge centrifuges appear more efficient than disk centrifuges, the reason being that the higher centrifugal fields developed in the latter are counteracted by the higher shear during injection of the feed and mechanical disruption by the rotating disk stacks. The principles of fluid mechanics in centrifuges and attempts to control liquid shear by efficient mechanical design are reviewed by Bell et al. (15).

6. Liquid/Liquid Separation

6.1. Ultrafiltration

Ultrafiltration is a process in which a porous membrane, or filter, separates components on the basis of size and shape. The development of controlled pore sizes in the 1–100 μm range allows for the separation of molecular colloids "dissolved" in an aqueous solution. In principle this enables one to avoid the problems associated with precipitation and solid/liquid separation, and as such represents a major step forward in biopolymer isolation. Despite its advantages, ultrafiltration on a large scale has its own problems. A comprehensive review of the technology has been provided by Beaton and Steadly (20).

Two types of equipment dominate large-scale processing; tubular and hollow fiber design. In the former, membranes are cast onto the inside of a mechanically strong, but porous, support (21,22). In the latter, hollow fibers with internal diameters of 0.5–1 mm are spun into a gelling bath. Precipitation is initiated in the center of the fiber, which produces an asymmetric membrane. Bundles of fibers are collected in parallel to produce a very high membrane area in a compact unit (23). Both of these designs, as well as simple flat plate systems, have been commercially successful. Tubular designs are robust, allowing for high liquid flow rates, and simple clean-in-place operation, whereas hollow fiber systems offer greater flexibility of pore sizes and operate at bench to full-factory scale.

Membrane characteristics are normally described in terms of their retention characteristics. This is quoted as the molecular weight of a globular protein that is retained (<90%) by the membrane. Solute retention (R) is formally defined as:

$$R = 1 - Cp/Cf$$

where Cp is the permeate concentration and Cf is the feed concentration.

It is important to recognize that these data are obtained in a low concentration filtration of single components where the membranes operate optimally. The figures should not be regarded as membrane specifications for practical plants, in which other factors influence transmembrane molecular flux and membrane selectivity (see below). For example, the high permeate flux rates obtained for dilute solutions are not maintained as the concentration of solute increases (Fig. 7). This is because of the phenomenon of "concentration polarization," in which solute accumulates at the surface, reducing the water flow through the membrane. For macrosolutes the effect is extreme since the concentration at the wall may exceed that at which a "gel" forms, i.e., the macromolecules form a structure that exhibits the rheology of a Bingham plastic. Since the shear at the wall is low, the yield stress may not be exceeded, and a stationary second membrane of macrosolute is produced.

Mathematical modeling of this phenomenon is described by Blatt et al. (24) and accounts for the observations that:

1. Increasing the pressure across the membrane results in only a transient flux increase—thereafter the gel layer thickens.
2. The reduction of flux is related to the macrosolvent hydration and gel-forming abilities—dissolved polysaccharides form gel layers at 1%; dissolved globular proteins, at ~10%; and precipitates, at 10 30%.

Ironically, therefore, the highest efficiency of ultrafiltration occurs for *precipitated* proteins, reducing some of the advantages of the method proposed above.

The prevalent formation of a second gel membrane causes the selectivity of separation to be governed by the macrosolutes, and not by the original membrane. This has been demonstrated by Butterworth and Wang for solutions of amylase and β-lactoglobulin (25) (*see* Fig. 8).

In practice, large-scale fractionation of proteins by ultrafiltration now appears unlikely (Dunnill, personal communication), though concentration and recovery is widely practiced using very

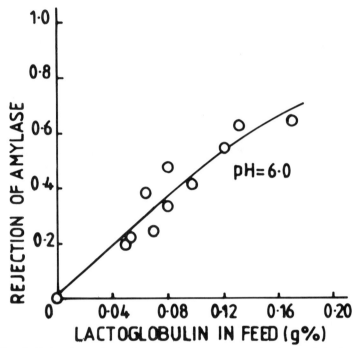

Fig. 8. Rejection of amylase in the presence of β-lactoglobulin (from ref. 25).

turbulent flow (in tube membranes) or back-flushing (hollow fibers) to minimize gel polarization.

6.2. Chromatography

The essential principles of chromatography have been discussed in previous chapters. The range and subtlety of separation is enormous and shows great promise for upscaling. The main problems in scaling up are:

1. Zone spreading, which can be minimized by the geometry of columns and packing.
2. Physical stability of the particle bed; the majority of solid phases are relatively soft. Decreasing the particle size improves resolution, but increases the pressure drop and the compaction of the bed. Recent developments of microporous glass—for molecular weight separation, granulation of alumina and hydroxylapatite, and cross-linking of ion exchange resins—all show promise in creating mechanically stronger systems.

3. Successful separation on the 1 g–1 kg scale are already possible with reasonable efficiency and high selectivity (Fig. 9). For high-value proteins, required in relatively small quantities, chromatography will probably be the preferred separation method of the future.

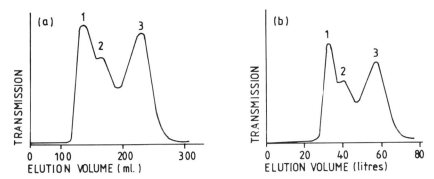

Fig. 9. Separation of proteins from human serum in (a) a typical laboratory scale separation, and (b) a separation using a five-section stack. Nearly 200 times more material was treated in (b), but the time was the same.

	a	b
Sample	8 mL Human placental serum	1.5 L Human placental serum
Column	K 25/100, Bed height 76 cm	Prototype KS 450, 5 sections, Bed height 75 cm
Gel	sephadex G-150	Sephadex G-150
Eluant	1% NaCl, 0.02% NaN_3	1% NaCl, 0.02% NaN_3
Flow rate	39 mL/h (7.7 cm/h)	13.2 L/h (8.3 cm/h)
Peak 1	γM-immunoglobulin, and macroglobulins	Peak 2 γG-globulin Peak 3 Albumin and transferin

7. Concentration and Drying

For laboratory and fine chemical use, proteins may be marketed as wet precipitates or frozen solutions, but on a large scale, further concentration and drying is required.

7.1. Concentration

Equipment is available for concentrating solutions or dispersions in bulk by evaporative methods. Shell, tube, and falling film types are used for food materials, but for isolated proteins solutions, the applied vacuum produces foaming and surface denaturation (1).

Enzyme solutions have been concentrated by heating to ~40°C in plate heat exchangers (1), but because of the low temperatures employed, the efficiency is low. In energy terms, ultrafiltration is emerging as the most efficient method of protein solution concentration.

7.2. Drying

Freeze drying represents the safest way of maintaining protein native states. Even with this process, freeze denaturation and irreversible aggregation can lead to a loss of solubility and activity. The principles of freeze drying are described in chapter 19.

At larger scales, roller drum, fluid bed, and spray drying are employed. The first two are energy efficient, but rely for this on high temperature and high evaporative rates. Denaturation frequently occurs, but for bulk proteins, when only the nutritional function of the protein is required, the methods are widely employed.

Spray drying still represents the preferred method for proteins for human food use and for some enzymes. The engineering principles of heat and mass transfer in spray drying have been extensively studied (26). Atomization of the feed stream by spray nozzles and rotating cups creates a large surface area and rapid evaporation. The higher the drying temperature and solid content of the feed, the more efficient the drying. For example, at an outlet temperature of 85°C, increasing the inlet temperature from 130 to 650°C causes the efficiency to increase from 44 to 90%. For proteinaceous materials, high efficiencies cannot be reached without considerable thermal degradation. The state of the dried protein is determined by its thermal history in the drier, and its history for droplet sprays can be calculated (27). Two phases are identified (Fig. 10).

1. Rapid evaporation, in which the droplet temperature is constant and approximates the wet bulb temperature of the pure solvent.
2. Slower evaporation, in which the droplet temperature rises to just below the outlet air temperature.

Thermal denaturation of proteins shows a marked dependence on water content, at levels below 0.7 g water/g protein. The

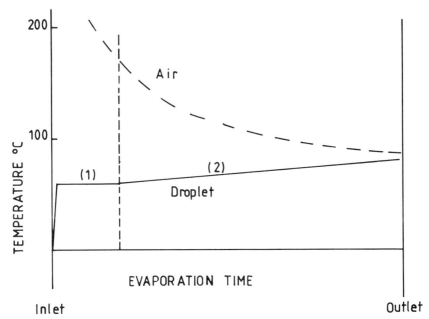

Fig. 10. Schematic of temperatures during spray drying (for explanation of stages, see text).

resulting curve of denaturation temperature versus water content shows an exponential increase in thermal stability as drying occurs (28).

It is arguable that the moisture content at which the denaturation temperature begins to increase is the same as that at which evaporative losses decrease. If this is so, then the protein is self-protected during drying, provided that:

1. the wet bulb temperature at the inlet does not exceed the denaturation temperature at high moisture, and
2. the outlet temperature is below the denaturation temperature at low moisture 5 g/100 g solids.

Detailed results on protein drying are not available, and most processes have been empirically established. Frequently a degree of damage (reduced solubility, activity loss, and so on) occurs, and at least two other factors must be considered:

1. If surface evaporation rates are too high, mass transfer of water through the surface film becomes rate limiting. The droplet temperature then *exceeds* the wet bulb temperature at *high* moisture contents.

2. In the concentrated state, intermolecular association and oxidation occur, which can be *accelerated* as the moisture content decreases.

As a result, for very high value proteins, milder freeze drying is preferred, but even then complete recovery of the native protein function is rarely achieved (chapter 19).

References

1. C. G. Heden, ed., *Fermentation and Enzyme Technology* John Wiley, New York (1979).
2. J. W. T. Wimpenny, *Process Biochem* **2** (7), 41–49 (1967).
3. L. Edebo, in *Fermentation Advances* (D. Perlman, ed.) Academic, New York (1969).
4. N. Papamichael, M. Hoare, and R. B. Leslie, in *Third International Drying Symposium* (J. C. Ashworth, ed.) Drying Research Ltd., Wolverhampton (1982).
5. P. R. Foster, P. Dunnill, and M. D. Lilley, *Biotechnol. Bioeng J.* **18**, 545–580 (1976).
6. D. J. Salt, R. B. Leslie, P. J. Lillford, and P. Dunnill, *Eur. J. Appl. Microbiol. Biotechnol* **14**, (3), 144–148 (1982).
7. P. J. Lillford, unpublished results.
8. E. J. Cohn, L. E. Strong, W. L. Hughes, D. J. Mulford, J. N. Ashworth, M. Melin, and H. L. Taylor, *J. Am. Chem. Soc.* **68**, 459–475 (1946).
9. E. Carling, ed., *Methods of Plasma Protein Fractionation* Academic, New York (1980).
10. D. H. Atha and K. C. Ingham, *J. Biol. Chem.* **256**, 12108–12117 (1981).
11. K. J. Ives, ed., *The Scientific Basis of Flocculation* (Sijthoff and Noordhoff, Alphen aan den Rijn, The Netherlands (1978).
12. M. Smoluchowski, *Z. Physik. Chem.* **92**, 155–168 (1917).
13. N. Tamb and H. Hozum, *Water Res.* **13**, 421–427 (1979).
14. D. J. Tomi and D. F. Bagster, *Trans. I. Chem. Eng.* **56**, 9–18 (1978).
15. D. J. Bell, M. Hoare, and P. Dunnill, *Adv. Biochem. Eng. Biotech.* **26**, 2–72 (1983).
16. Anon., *Process Biochem.* **17** (2), 23–28 (1982).
17. M. Hoare, *Trans. I. Chem. Eng.* **60** (2), 79–87 (1982).
18. W. R. Jewett, US Patent No. 3826740 (1974).
19. R. A. Boyer, J. Crupl, and W. T. Atkinson, US Patent No. 2377853 (1945).
20. N. C. Beaton and H. Steadly, in *Recent Development in Separation Science* vol. VII (N.N. Li, ed.) CRC Press Inc., New York (1982).
21. Abcor Ultrafiltration Bulletin, Abcor Inc., Washington, Massachusetts.

Large-Scale Methods 451

22. P. C. I. Membrane Systems, PCI Ltd., Laverstoke Mill, Whitchurch, Hants, UK.
23. B. R. Breslau, E. A. Agranat, A. J. Testa, S. Messinger, and R. A. Cross, *Chem. Eng. Prog.* **71**, 74–80 (1975).
24. W. F. Blatt, A. David, A. S. Michaels, and L. Nelson, in *Membrane Science and Technology* (J. E. Flinn, ed.) Plenum, New York (1970).
25. T. A. Butterworth and D. I. C. Wang, in *Fermentation Technology Today* (G. Terns, ed.) Society of Fermentation Technology, Japan (1972).
26. K. Masters, *Spray Drying* Leonard Hill Books, London (1972).
27. J. Dloughty and W. H. Gavvin, *Am. Inst. Chem. Eng. J.* **6** (1), 29–34 (1960).
28. P. J. Flory and R. R. Garrett, *J. Am. Chem. Soc.* **80**, 4836–4845 (1958).

Chapter 17
Large-Scale Processing of Plant Proteins

Chester Myers

1. Raw Materials

The choice of processing and the properties of the final protein isolate are partly determined by the source material, i.e., species, seasonal variations in the crop, internal structure of the particles, and any previous processing. Of cereals, legumes, and oilseeds, soybean proteins have received the most study. Soy is a major source of vegetable oil and the meal left after oil extraction is a convenient source of protein. The storage proteins of oilseeds and legumes are generally extractable into aqueous media, whereas those of the cereals are more resistant to such extraction: for the latter, one must resort to the use of surfactants or organic solvents.

1.1. Species Variation

Soya beans are considered a bulk commodity [$3.5 billion in the US in 1974 (1)]. However, it has been only for oil extraction, both for yield and oil content, and not for distribution of protein types, that the choice of species has been determined. Variable ratios of the 7S and 11S storage proteins exist between species (Table 1), and, after the same processing, the product properties are often quite different. The economic dominance of the oil fraction does

Table 1
Differences in Proportions of Protein Components Among Soya Varieties[a]

Variety	2S	7S	11S	15S	7S/11S
Hakuho	17.5	46.4	36.3	0	1.28
Akasaya	13..7	37.9	45.6	1.8	0.86
Aobata	16.6	33.9	49.5	0	0.69
Shirofsurunoko	12.9	29.9	53.2	3.9	0.56

[a]From ref. 13.

not apply to peas and beans. For these, there is variation in the distribution of vicilin and legumin (7S and 11S) fractions of the storage proteins, so that much greater potential flexibility in the final protein-enriched products exists.

Rapeseed is a major oilseed crop in several countries. It is Canada's largest oilseed crop and ranks second only to wheat in terms of acreage grown (3.4 million ha in 1979). Both *Brassica campestris* (turnip rape) and *Brassica napus* (Argentine type) are grown commercially and contain at least two major proteins (1.7S and 12S globulins). The variation in distribution is neither used nor considered in choosing species for growth. Regent, a "napus" variety, Altex, and Tower have been the main varieties grown. High oil yield, low levels of glucosinolates (which cause enlarged thyroid glands), phytic acid and erucic acid (which upset the lipoprotein balance), and early harvest have been of concern in plant breeding.

Although rapeseed contains good quantities of protein (24% of the whole seed, 39–45% oil), the commercial value of the oil has very much kept the protein at a lower interest. During processing for oil production, flaking is normally followed by cooking to deactivate the myrosinase enzymes, but this is detrimental to the storage protein. Moreover, solvent extraction of the press cake, after expelling, is followed by a desolventizing/roasting step, again involving heat. In spite of present processing methods, the technology does exist to overcome these problems, and rapeseed protein should become a major plant protein for food use both for nutritional and "functional" purposes. Its major asset may be in complementation with other plant proteins to achieve superior properties not possible with single plant proteins.

1.2. Seasonal Variation

If vegetable protein is to be recovered from seeds, production by nature of the mature seed provides a convenient control process for arresting biological change. Examination of the developing seed (1), or of its subsequent germination (2), shows significant changes in the protein distribution, but as yet neither of these processes has been used to control food material properties.

In the case of vegetable protein derived from leaves, or other "growing" parts of the plant, greater intrinsic variation between new and old growth is significant (Table 2). However, yield and nutritional value determine the cropping, rather than the potential "functional," properties of the combination of proteins present.

Table 2
Yields of Protein (in g/kg of Dried Material) in Leaf Protein Concentrate From Mixed Pasture[a]

Season	Regrowth, wk			
	3	4	5	6
Spring	112	77	59	34
Summer	77	46	32	24
Autumn	93	62	50	33

[a]From ref. 14.

1.3. Preprocessing

The bulk of oilseeds is still primarily processed for oil. For the larger seeds (soya, groundnut), hexane extraction involves relatively low temperatures (60–80°C). Neither the temperature nor the use of the nonpolar solvent threatens the denaturation of the protein, so that defatted meals of high protein extractability are readily available.

Rapeseed, because of its size, is one of the most difficult sources to oil extract. Most current processes maximize extraction by compressing the residual cake at temperatures of greater than 100°C. The denaturation temperatures can be elevated by as much as 40°C if the protein is dry rather than hydrated. One of the most economic methods of providing heat, however, is still via "live" steam, which also supplies the moisture to lower the denaturation temperature.

2. Refining Processes

An important constraint on refining is its cost. Only if the value, added to the refined product, is sufficiently high is the refinement process commercially feasible. Refining processes are capital- and energy-intensive: Benefits therefore derive from economy of scale. These further constraints favor simple and few process steps, low energy consumption, and a stable supply of raw material in large tonnages. The latter constraint is readily fulfilled by vegetable protein. There are still considerations, listed below:

> Mandatory—Removal of antinutritional factors, e.g., phytohemagglutinin, goitrogen saponins.
> Desirable—Removal of indigestible sugars, i.e., the flatulence factors raffinose and verbascose.
> Optional—Protein isolation, fractionation, specific modification.

These requirements are not mutually exclusive, so a route to an isolate may intrinsically contain steps that fulfill other requirements.

Refinement processes currently in operation can be further classified as nonaqueous. Although providing admirable solubility properties for the separation of components, a major disadvantage of water is that, since most refined products are transported and sold dry, every drop added must be removed, *at a cost*. Mechanical separation of dry materials, or the use of more volatile solvents, is obviously desirable.

2.1. Mechanical Separation—Air Classification and the Liquid Cyclone

Carbohydrate- and protein-enriched fractions have been produced by direct air classification of milled legume flours (Tables 3 and 4). Constraints on the process appear to be (1) low lipid content in flours—otherwise agglomeration results, (2) carbohydrate in the form of starch granules—otherwise classification fails. The

Table 3
Protein and Starch Fraction Compositions of Flour[a]

	% Protein			% Starch		
Source	Flour	Protein fraction	Starch fraction	Flour	Protein fraction	Starch fraction
Chick pea	19.5	28.9	15.3	50.0	30.3	57.3
Pea bean	24.7	52.4	15.2	38.4	1.4	51.6
Northern bean	24.0	53.5	15.6	40.3	1.4	51.5
Faba bean	29.8	66.6	14.4	42.4	1.4	57.5
Mung bean	26.5	60.4	12.3	50.0	6.1	67.7
Soybean	52.5	54.2	50.5			

Table 4
Air Classification of Soya Bean Meal.[a]
Size Was Deduced From the Described Process

Fraction	% Yield	% Protein	Size, μm
Flour		57.0	
1st coarse	25.6	50.4	40–60
2nd coarse	40.9	63.7	4–15
2nd fine	33.5	50.5	<4

[a]From ref. 16.

process does not modify the native state of either the carbohydrates or proteins. Provided the materials are used in products in which heating will destroy the labile inhibitors and phytohemagglutinins, however, the hazards to health or growth are no worse with these refined products than with the whole seed or meal. If fractions of higher nutritional values are required, combination of air classification with dry roasting processes, such as those described for navy bean (3), might prove interesting. Further work on alternative milling procedures might also produce classifiable fractions from defatted soya, groundnut, lupin, and so on.

The liquid cyclone is a process for "sorting" the gossypol-rich pigment glands of cottonseed from finely ground meal. Differential sedimentation of the pigment glands and protein bodies in hexane is exploited. (Fig. 1). The objective is to achieve detoxification of the glanded seed. A high-protein (66% protein) flour results. Although the cyclone process should not alter the protein, subsequent removal of solvents, drying, and sterilization can cause damage. Because of different morphology of the glands and the protein bodies, it is possible that a simpler air classification process would achieve the same results, with less chance of damaging the storage proteins.

2.2. Mixed Solvents

The use of alcohol/water mixtures is well known for concentrate production: low molecular weight sugars, flavor precursors, and saponins are removed. Thermal stability of the storage proteins may be reduced by such media (4), and the final product may therefore be less useful. Rapeseed protein is greatly underutilized as a food source material because of several toxic constituents (5).

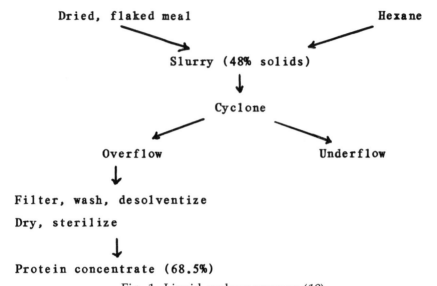

Fig. 1. Liquid cyclone process (19).

However, a proprietary process for oilseed processing, using aqueous alcohol extraction, carefully avoids conditions that denature the protein (6).

2.3. Refinement by Aqueous Extraction

Refinement by aqueous extraction provides an enormous variety of processes for producing concentrates and isolates from many sources. All depend on the differential solubility of the various source components (protein, carbohydrates, pigments, and so on) as a function of either temperature, pH, or ionic strength. Figure 2 presents a general scheme. Aqueous processes for the production of concentrates require insolubility of the protein to allow for removal of sugars—hence, protein denaturing temperatures and/or pH values corresponding to the protein p*I* are often used.

3. Protein Isolate Preparation

Most vegetable proteins have typical pH/solubility profiles with minimal dispersion at the isoelectric pH. Extremes of pH ("extremes" definable per protein) cause denaturation and functionally (and nutritionally) adverse reactions: Isolates obtained will vary

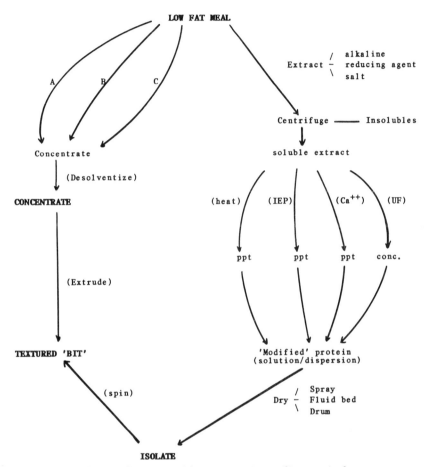

Fig. 2. Refining for vegetable protein ingredients. A, heat + water; B, acid; C, alkali; IEP, isoelectric precipitation; ppt, precipitate; UF, ultrafiltration.

in functional properties—heat gelation negatively and emulsification often positively. For most sources, pH values between 3 and 8.5 should not be directly detrimental to the protein. Binding to impurities, such as phytic acid, may cause problems. Ionic strength is another variable that controls solubility, particularly for the so-called globulins. A minimum solubility, at about $0.1M$ ionic strength, is usually observed, this being modulated by the particular pH. Rupture of disulfide bonds may also afford higher yields.

Extraction of protein from a source material is followed by clarification and then follows any intermediate steps, such as ultrafiltration (diafiltration to "flush" out small molecule contaminants, or concentration to raise protein level and lower the volume of solvent). Extracted proteins can be concentrated to levels as high as 25%, making direct drying of retentate commercially competitive. Control of pH is critical if hollow fiber ultrafiltration is used: Proper control of electroviscous effects, and hence shear thinning behavior, will preclude major fouling of membranes. Processes for production of isolates from soy, sunflower, rapeseed, cottonseed, and faba bean have all been described.

Large-scale "salting-out" methods, for plant seed protein recovery, are not employed commercially since byproduct recovery costs are too high. Precipitation by specific cations (e.g., Ca^{2+}) is used and forms the basis of Japanese tofu manufacture. Precipitates formed with soya are rubbery and not easily resolubilized, making this route undesirable for isolates. Heat coagulation is also possible, but the isolate is necessarily denatured and subsequent functional use is limited. When nutrition or digestibility are the only required properties, however, such materials are quite appropriate: heat coagulation is widely employed in leaf protein recovery.

Isoelectric precipitation is by far the most commonly used recovery method. The study of soya has been exhaustive, and data regarding yields are available for other protein isolates (Table 5).

Table 5
Yields of Precipitation, at pH 4.5, for Isolated Protein From Various Plant Sources[a]

Species	Isolate yield, g/100 g meal	Nitrogen content of whey, % total
Soybean	36–40	10–12
Rapeseed	25–27	30–32
Turnip rape	24–26	26–29
Flax	36–40	23–29
Sunflower (dehulled)	52–54	17–20

[a]From ref. 17.

Precipitation and subsequent centrifugal concentration of the proteins can lead to extensive aggregation and modification of solubility properties. This is mostly caused by protein/protein noncovalent interactions (Table 6).

Drying the isolate is potentially dangerous if heat is used. In general, the 7S globulins of legumes and oil seeds are denatured at temperatures as low as 70°C: Higher temperatures are allowable for cereal storage proteins because of a higher content of nonpolar amino acid residues (e.g., oats at 110–118°C, depending on the ionic environment). Spray drying is the most common economical method for producing a thermally undamaged isolate. The temperature of the drying droplet does not increase until the water content decreases to values at which the denaturation temperature is elevated. Even so, the high concentrations produced during drying generally lead to some irreversible aggregation and loss of subsequent dispersibility.

Table 6
Protein Aggregation Caused by Isoelectric Precipitation[a]

Sensitivity	Cure	Molecule	Percent of total
pH Induced by contact with reducing agent	Water extract or titrate to pH >7.6 low ionic strength	2S + Hemaglutinin	5–10
Concentration Not reversible by elevating ionic strength	Extract or dissolve in reducing media or elevate pH >7.6	7S + 11S globulins	50
Time Not reversible by elevating ionic strength or at pH >7.6	Extract at reducing conditions or dissolve at pH 7.6 with reducing agent		90

[a]From ref. 18.

Consideration of the differential preparation of soy proteins (2S, 7S, 11S, 15S) demonstrates the variety of processing procedures. Generally, the 11S fraction is preferentially precipitated either by divalent cation treatment (7), or by low temperature (8,9). Isoelectric methods favor isolates high in the 7S, as well as the 11S. Purer 7S fractions are collected by first precipitating 2S fractions with 32% ammonium sulfate and then increasing the salt concentration to 50%. The 2S protein is further purified by reprecipitation with ethanol. Ionic strength reduction favors the 11S precipitation, and a stable 18S aggregate has been detected in dispersions of isolate collected by this method.

Extraction of meal	Aqueous alkali
	Salting-in
↓	Aq. reducing agents
	Total aq. extract
Processing of extract	Ultrafiltration
↓	
Precipitation and fractionation	Ionic strength reduction
	Ca^{2+}, Mg^{2+}
	Heat, cold, or acidification
	$(NH_4)_2SO_4$, ethanol
↓	Recovery and redissolution
Chromatographic purification	Molecular sieving/ion exchange
	Hydroxylapatite/antibody affinity
	2S globulin or conglycinin or glycinin

3.1. Isoelectric Precipitation

Alkaline extraction at pH values as high as 11 achieves admirable protein extraction from many source materials. Following clarification, the pH is brought back to the isoelectric pH, usually 4.5 to 5. Loss of certain functional properties, production of antinutritional factors such as lysinoalanine, and retention (or even concentration!) of natural antinutritional components, such as phytic acid, are drawbacks of this isoelectric process. Solubility at high pH makes control of both temperature and time quite critical and slight variations in either can produce significant variation in final isolate quality: In fact, variation in quality between different lots of the one product has probably been the most serious liability in the use of protein isolates. In addition, precipitation in the pH 4.5 to 5 region tends to keep phytate levels in the isolate high unless other, potentially expensive, steps are included in the process. Residual phytase activity from the source material may be of some help.

3.2. Separation by Ionic Strength Manipulation

One process (10) produces a protein extract by salting-in of the protein: After clarification (Fig. 3), ionic strength reduction leads to the precipitation of the protein. This process is based on the effects of ionic strength and pH on protein solubility, as shown in Fig. 4. Control of pH is critical in order to obtain the isolate as a gelatinous, "gluten-like," mass via formation of initial fluid protein droplets (referred to as "protein micelles"): Coalescence of the droplets produces the wet isolate. Yields are comparable to isoelectric yields, and levels of phytic acid are much lower since the pI is avoided during precipitation. The protein is not denatured and LAL is not detected in the isolate. Albumins, the water-soluble low-molecular-weight proteins, tend to be excluded for most source materials: their content tends to be low in any case—however, possible loss of essential amino acids must be monitored.

An analogous process (11) is also described in which initial extraction is carried out without any added salt: The low levels of

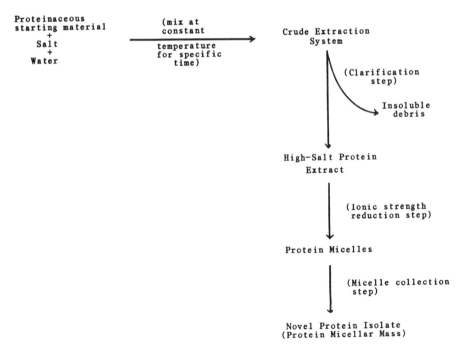

Fig. 3. PMM isolate production schematic (10).

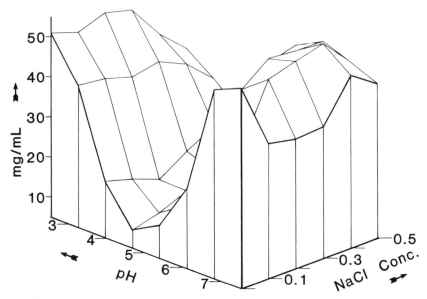

Fig. 4. Fababean storage protein solubility (mg/mL) as a function of pH and ionic strength. Source material was an air-classified concentrate containing about 60% protein. Dispersions of 20% wt/vol concentrate were stirred for 30 min at 25°C.

salt, indigenous to the source material, apparently supply sufficient ionic strength for solubilization. Subsequent dilution still produces an undenatured isolate. A complete process starting with the "just-harvested seed" is shown in Fig. 5.

Another process (12) describes the preparation of a "mesophase," a protein dispersion containing 15–50% "dissolved undenatured plant protein and sufficient water-soluble salts to keep the protein dissolved and having a pH in the range of about 4–6." An ionic strength of at least 0.2M is required.

3.3. Separation by Temperature Control

Temperature variation is not presently used commercially for protein isolation. However, the technology exists for the preparation of pure protein isolates by this method. Protein solubility often shows considerable sensitivity to temperature: Fractionation of different protein species is likely, this depending on the relative proportion of surface hydrophobic and hydrophilic amino acid residues.

Large-Scale Processing of Plant Proteins

Dehulling of seed (pneumatic dehulling, 3 passes)

Separation of Meats from hulls (air conveyance)

Expelled with Simon Rosedowns mini-press (fan cooling of cake)

Cold solvent extraction (Crown Iron Works)

Air dried

↓

Extraction in water

↓

Centrifugation (Westphalia, decanter)

Ultrafiltration (Romicon fibre unit, 50,000d cut-off)

(volume reduction, 10 to 15 fold)

↓

Dilution of concentrate into cold tap water

(dilution, 8 to 15 fold)

↓

Collection of isolate

Fig. 5. Flow diagram for the preparation of a protein isolate from rapeseed, using solvent extracted meal, after air drying, for the source material of a proprietary process (11).

Preparation of the 11S species from soy is possible with the aid of temperature manipulation (8). In Fig. 6, the whey fraction will contain the "albumin" fraction. Although the yield is low, oligosac-

Defatted soybean flakes

(Aqueous extract, pH 7, 50 °C)

↓

Cool to 3-4 °C (72 hr) and centrifuge

↓

Precipitate is washed, redissolved

↓

Dry extract after removal of insolubles

Fig. 6. Flow diagram for the preparation of protein fraction from soybean flakes (8).

charides were noted to be exceptionally low in comparison to isolates prepared either by isoelectric precipitation or by precipitation with calcium.

In each of the above processes, the emphasis is on the prevention of protein denaturation during isolation. Although emulsification applications may benefit from a denatured protein, it is still better to perform this denaturation *after* isolation, when a purer form of the protein should lead to a more controlled (hence more reproducible and better "tailored") process.

References

1. F. E. Horan, *J. Am. Oil Chem. Soc.* **51**, 67A–73A (1974).
2. N. Catsimpoolas, C. Ekenstam, D. A. Rodgers, and E. W. Meyer, *Biochem. Biophys. Acta* **168**, 122–131 (1968).
3. N. R. Yadar and I. E. Leiner, *Legume Res.* **9**, 17–25 (1977).
4. D. Fukushima, *Cereal Chem.* **46**, 156–163 (1969).
5. J. D. Jones and J. Holme, US patent no. 4,158,656.
6. J. D. Jones, *J. Am. Oil Chem. Soc.* **56**, 716–721 (1979).
7. A. G. A. Rao and M. S. N. Rao, *Prep. Biochem.* **7**, 89–101 (1977).
8. H. M. Bau, B. Poullain, M. J. Beaufrand, and G. Debry, *J. Food Sci.* **43**, 106–111 (1978).
9. W. J. Wolf and D. A. Sly, *Cereal Chem.* **44**, 653–668 (1967).
10. E. D. Murray, C. D. Myers, and L. D. Barker, Canadian patent no. 1028552 (1978).
11. J. J. Cameron and C. D. Myers, US patent no. 4,366,097 (1982).
12. M. P. Tombs, US patent no. 3,870,801 (1975).
13. R. Saio, M. Kamiya, and T. Watanabe, *Agr. Biol. Chem.* **33**, 1301–1308 (1969).
14. H. T. Ostrowski-Meissner, *J. Sci. Agric.* **31**, 177–187 (1980).
15. F. W. Sosulski and C. G. Youngs, *J. Am. Oil Chem. Soc.* **56**, 292–295 (1979).
16. R. B. Swain, US patent no. 3,895,003 (1975).
17. F. W. Sosulski and A. Bakal, *CIFST J.* **2**, 28–32 (1969).
18. P. J. Lillford and D. J. Wright, *J. Sci. Food Agric.* **32**, 315–327 (1981).
19. H. L. E. Vix, P. H. Eaves, H. K. Gardner, Jr., and M. B. Lambou, *J. Am. Oil Chem. Soc.* **48**, 611–615 (1971).

Chapter 18

Characterization and Functional Attributes of Protein Isolates

Biochemical Applications

Peter J. Lillford

1. Introduction

In this chapter we consider the properties of proteins isolated for their native function. The focus is on the native state of the protein and the subtle way in which evolution has tailored polypeptide chains to produce highly specific secondary, tertiary, and quarternary structure. For some proteins, this characterization is complete to atomic resolution (e.g., lysozyme, hemoglobin). This is because of the major efforts and technical advances in molecular biology in recent decades. It is important to recognize that for biochemical use, however, isolated material can be equally well characterized in terms of its "activity," i.e., the ability to perform its intended function. The latter approach does not necessarily refer to any molecular configuration or even to the presence of molecules at all. This is particularly true for pharmacology, in which functional characteristics are largely pragmatic. For example, the active macromolecule in antisera is identifiably proteinaceous, but the British Pharmacopeia characterizes scorpion venom antiserum as a serum obtained from healthy animals having not more than 17.0 wt/vol % protein, and sufficient potency to neutralize the maximum amount of venom from a single sting (1).

Hence for many biochemical uses, protein isolates are characterized by their preparative route and the *absence* of undesirable components. For example, Cohn fractionation of blood plasma allows for the isolation of factor VIII, but the screening of donors is required to eliminate hepatitis virus or the more recently discovered threat of acquired immunodeficiency syndrome (AIDS).

The advent of "genetic engineering" for proteins has brought molecular and functional characterization closer together. For example, expression of a particular gene is valuable only if the folding sequence produces an active molecular conformation, and the substitution of engineered for naturally produced proteins may require demonstration not only of safety and activity, but also molecular identity.

2. Hierarchy of Characterization Methods

Whereas proteins can be classified in terms of their biochemical activity, e.g., enzymatic, hormonal, structural, antigenic, and so on, their characterization normally entails further measurement of their structure. It is not the intention here to review the methodology used, but rather to indicate the normal sequence by which detailed structural characterization is achieved. For a practical guide to the theory and practice of these methods, readers are directed to standard texts, such as Haschemeyer and Haschemeyer (2).

2.1. Molecular Identity

Since proteins are always isolated from complex mixtures, characterization methods have been developed that test for molecular identity.

The least sophisticated, but widely used, method is that of ultracentrifugation. Because of their molecular uniqueness, each protein in a dilute mixture sediments at a fixed and measurable velocity in a centrifugal field gradient. Components of mixtures can be defined by their sedimentation coefficient in absolute terms or relative to a particular buffer system and temperature. For example, seed storage globulins, which have no simply testable biological function, are frequently named by their sedimentation coefficient (chapters 18 and 21). The method is not particularly sensitive; up to 5% impurity can be present in the sample without detection. Permeation chromatography suffers from the same inaccuracies (chapter

Protein Biochemicals

11). For detection of molecular uniqueness, electrophoresis has become the most potent and widely applied method (chapter 12), particularly in combination with immunological detection methods (chapter 15).

2.2. Subunit Composition

To determine subunit composition, polypeptide separation in dissociating conditions is employed. Electrophoresis in SDS solutions is the preeminent method; the speed of operation makes it preferable to ultracentrifugation or chromatography. The latter is gaining considerable ground, however, because of the advent of gas–liquid chromatography (GLC) columns with resolution capabilities for protein molecular weights (chapter 11).

The methods that allow for the identification of proteins and/or their subunit composition can usually be upscaled to produce sufficient material for the chemist to apply methods of amino acid analysis and primary sequence determination. This becomes key information in the more detailed examination of structure–function relationships.

2.3. Size and Shape Measurement

Both hydrodynamic and equilibrium methods are available. Although both give reasonable assessments of molecular weights, the assessments of size and shape are always questionable, particularly by hydrodynamic methods. This is because the transport of the macromolecule through solvent usually occurs with some accompanying solvent (hydration layer). Since this associated solvent is not independently measurable, the measured molecular radii, axial ratios, and so on, are open to question (3).

More direct information is obtained by scattering techniques, in particular, small angle X-ray scattering (SAXRS). For small angles, the intensity of scattering at angle h is dependent on the radius of gyration (R_G) of the scatterer:

$$\ln I(h) = \ln I(0) \frac{-h^2 R_G^2}{3} \tag{1}$$

where $I(0)$ is the zero angle intensity.

The same equation applies for neutron scattering (SANS) and, indeed, for light scattering, when the particle has a dimension comparable to 1/20 or greater of the wavelength of light. Thus, size information can be obtained directly, but shape factors must be obtained by fitting the angular dependence of the scattering intensity to the observed curve. For particles of the dimensions appropriate to light scattering (>20 nm), the electron microscope is capable of indicating shapes by direct visualization (chapter 16).

2.4. Secondary Structure by Spectroscopy

2.4.1. Ultraviolet Absorption

The ordered structures in native proteins produce perturbations in the absorption bands of amino acid chromophores. The perturbation may result in an increase (hyperchromism) or decrease (hypochromism) in absorptivity at a particular wavelength. Such effects are to be expected from ordered arrays of peptide groups such as those in the α-helix or β-pleated sheet structures. This phenomenon has been widely used to follow changes in protein conformation accompanying, for example, denaturation (4,5). The method is not diagnostic of the presence of a particular conformation, however.

2.4.2. Optical Rotary Dispersion (ORD) and Circular Dichroism (CD)

The phenomenon of optical rotation of amino acids and the specificity of the L-amino acids in protein structures has been well studied, and circular dichroism is the absorption event causing the rotatory dispersion effect. When its application to proteins was first considered, established practitioners believed that the result would be a simple sum of the constituent amino acids and yield no further structural information. Undeterred by theoretical considerations, Yang carried out ORD measurements on native protein and polypeptides and showed a significant structural dependence of the dispersion curve (6). This effect has been shown to result from the fixed and specific orientation of peptides in α-helical and β-sheet structures, and has allowed for extensive studies of native protein conformation and their susceptibility to environmental change. Calculation of the helix content is obtained from the b_0 parameter of the empirical equation derived by Moffitt (7).

$$[M'] = \frac{a_0 \lambda_0^2}{(\lambda^2 - \lambda_0^2)} + \frac{b_0 \lambda_0^4}{(\lambda^2 - \lambda_0^2)^2} \qquad (2)$$

where $[M']$ is the reduced mean residue rotation at wavelength λ and λ_0 is a fitted parameter, usually 212 nm.

For a right handed α-helix, $b_0 = -630$, whereas for a random coil, $b_0 \approx 0$.

The method appears to relate well to crystallographic helix content in proteins, provided that little other ordered structure is present, but β-sheet structures also produce dispersive effects that limit its use as a predictive tool.

Circular dichroism is the absorptive counterpart of ORD, but has the advantage that the optically active transitions are more easily and completely characterized. The α-helix exhibits characteristic absorption bands at 207 and 221 nm; the β-sheet, at 217 nm; whereas the random coil is weakly absorbant at normally accessible wavelengths.

Because of its greater selectivity for ordered conformation, CD has proved more useful as a general tool for structure analysis of a variety of proteins (8).

2.4.3. Infrared Spectroscopy

Some information on conformational order can be obtained from group absorption spectra. In particular the carboxyl stretching frequency (amide I) band is shifted to 1660/cm for the α-helix and 1640/cm for β-sheet structures. This has allowed some assessments to be made, but the most valuable contribution has been with high-concentration samples that are oriented and allow for analysis of the infrared dichroism (9). It may be that developments of Fourier transform methods in infrared, which allow for examination of surface reflectance and accurate difference spectra at high dilutions, will lead to a revival in the techniques' application.

2.5. Detailed Structure by X-Ray Diffraction

Without doubt, the dominant method of structural analysis and the reference point for all other methods is still X-ray diffraction and the determination of crystal structure resolved to atomic dimensions. The efforts have been heroic and interested readers are directed to the research reports of Phillips (10) and Perutz (11). The results are always dependent on the production of high-quality single crystals and of course give essentially static rather than dynamic information on protein structure. The true value of the results are that they provide a vital reference point from which deductions of the dynamic and functional attributes of native proteins can be elucidated (chapter 2).

2.6. Application of Nuclear Magnetic Resonance to Protein Structure and Dynamics

2.6.1. Principles of the Method

One of the problems and limitations of X-ray crystallography is that the structure obtained is a static "snapshot" of the protein structure in a highly condensed (crystalline) state. The phenomenon of nuclear magnetic resonance (NMR) allows for at least a partial structural study of short-range order in the molecule, and further information on the dynamics of amino acids in a dissolved protein can be obtained.

The NMR method derives from the fact that common nuclei such as the proton ^1H, carbon ^{13}C, and phosphorus ^{31}P have more than one spin energy level when in a high magnetic field. Second, because electronic configurations in a molecule induce slight differences in the magnetic field experienced by each nucleus (the chemical shift), the frequencies at which resonance occurs are "labeled" by the bonding arrangement in the molecule. This allows for the resolution of the spectrum and assignment of the observed peaks to particular amino acid residues. In addition, neighboring nuclear spins can be "coupled" to each other via the bonding electrons. For highly mobile molecules, this results in a splitting of resonance signals into multiplets by neighboring nuclei. The magnitude of the splitting is independent of the applied field, but dependent on the number of neighboring nuclei, chemical bond type, and bond angles. Spin–spin coupling constants (J) are known for several conditions of the above and, when clear assignments can be made, allow for the detailed geometry of molecules in solution to be determined (12). Studies of single amino acids allow for the construction of a synthetic spectrum for a protein of known amino acid composition (13). The spectrum of native proteins never directly matches these idealized spectra (14) (Fig. 1).

The reasons for this are twofold. First, the synthesized "random coil" spectrum assumes that amino acids in dilute solution have the same translational and rotational mobility as in the protein native state. This is rarely the case, and the width of the resonance line is inversely proportional to the mobility. The lines can also be broadened by exchange (of protons) between macromolecule and solvent. Thus, if carefully analyzed, molecular dynamics can be derived from the NMR spectrum. Second, the complex folding of the peptide chains brings groups into proximity

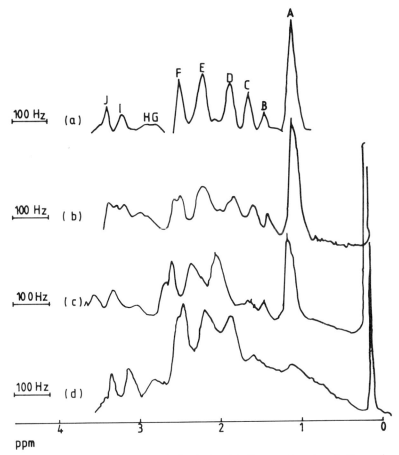

Fig. 1. Proton spectrum of glycinin: (a) Computer simulation of random coil spectrum, (b) denatured in 8M urea, (c) denature by heat, and (d) native (from ref. 14).

and results in additional chemical shifts that could not be predicted from the random coil spectrum. Hence a degree of structural information can be deduced, provided resonances can be assigned to particular amino acids in the sequence.

2.6.2. Protein Dynamics by NMR

Results using NMR show that the crystallographer's view of proteins as ordered structures is, to a first approximation, maintained in solution. Unfortunately, this implies that the mobility of many groups are close to that of the whole molecule and this results in linewidths so broad that the individual resonances are not resolvable and appear to be "missing" from the spectrum.

The fact that some resolved lines are visible implies that a proportion of the residues even in very large globular proteins are conspicuously mobile. These are normally assigned to surface located residues and is the first necessary modification of static structures.

Detailed studies have tended to focus on proteins or fragments with molecular weights of less than 25,000 daltons. In these samples, the whole spectrum is in the high-resolution range, and differences in mobility of groups exhibiting assignable resonances can be equated with restricted side chain motion caused by internal structure.

Further dynamic information can be derived from studies of "exchange" phenomena, which may be of three distinct types (15):

1. Isotope exchange, in which, for example, a ^1H signal slowly disappears on replacement by a deuteron.
2. Rapid proton transfer, between solvent and an ionized group causing line broadening.
3. Positional exchange, a residue occupying a number of positions (and therefore magnetic environments) during the NMR experiment.

An example of the latter is given in Fig. 2, in which in cytochrome c, a tyrosine exhibits a flip about its $C-C$ bond (16).

2.6.3. Protein Structures by NMR

For a globular protein, the preferred conformation in solution gives rise to secondary shifts that can be used in structural assignment. In general, detailed studies are necessarily limited to small molecular weight species or to certain special cases in which the macromolecule has essentially a low degree of order in its native state. Two types of secondary shifts have been used to study conformation.

2.6.3.1. Aromatic Ring Current Shifts

Protons on aromatic rings resonate at anomalously low fields. This is because of the reinforcement of the external magnetic fields by the field induced by flow of the delocalized electrons, shown schematically in Fig. 3.

As well as allowing for easy identification of the aromatic residues, any group brought into proximity of an aromatic ring by folding of the peptide chain will also experience a reinforcement or reduction of the external field. Approximate calculations can be

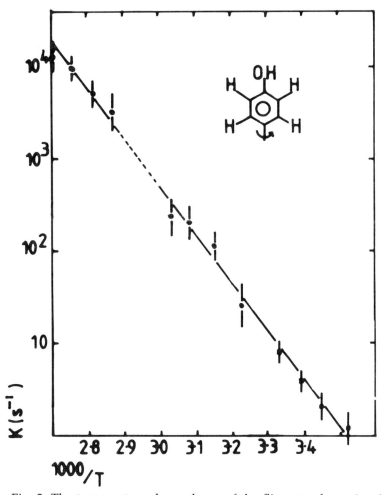

Fig. 2. The temperature dependence of the flip rate of tyrosine 46 in horse ferrocytochrome c; by line shape analysis (○) and cross saturation experiments (●) (data from ref. 16).

performed that allow for the relative distances of resonating nuclei from aromatic residue to be determined, allowing for accurate mapping of binding sites of enzymes and antibodies (see below).

2.6.3.2. Use of Paramagnetic Ions

When bound to proteins, paramagnetic ions and particularly lanthanides produce large local fields or a dramatic increase in linewidths. These phenomena, which shift the resonance position of

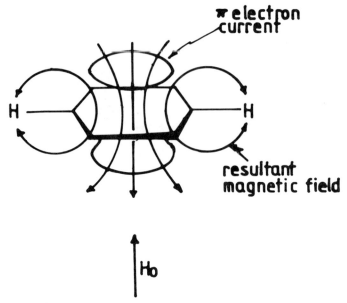

Fig. 3. Schematic representation of the induced π electron current responsible for "ring current shifts."

affected nuclei or broaden their linewidths, can be used to obtain highly specific interatomic distance estimates within a protein molecule (17). The limitation of the method is that the binding must be specific or the effects of multiple binding must be separable. The example of lysozyme, for which the method has proved extremely powerful, will be mentioned later.

Ideally, protein structure should be examined by direct assignment of chemical shifts and through space spin coupling of adjacent nuclei. This would be the equivalent of the solution structural studies of small molecules (12). As mentioned above, there are intrinsic limitations to the approach since the molecular motion and linewidths prohibit detailed resolution, even at extremely high field strengths. In proteins with molecular weight of up to ~20,000 daltons, the resolution of a globular protein spectrum allows the attempt to be made. Furthermore, the development of automated and selective pulse sequences in modern Fourier Transform equipment removes much of the tedium from selective radiation experiments.

Although no significant attempt has yet been made to perform a structural and dynamic study of a protein of unknown structure in solution, the advent of automated NMR machines and macro-computers allows a strategy to be developed for such an approach (18).

In the following sections we will examine the application of these techniques to understanding the relation of structure and function of particular systems.

3. Serum Albumin

Serum albumin comprises ~50% of the total serum protein of blood and its primary function appears to be osmotic regulation, i.e., it allows fluid to be driven into tissues on the arterial side and extracts fluid on the venous side. Saline solutions can also osmoregulate, but ions of small hydrated radius distribute throughout the entire extracellular space. The size of the protein molecule restricts it to half this volume. Its value, therefore, in transfusion makes it the dominant commercial component in blood plasma. Characterization of human serum albumin in terms of its molecular weight, size, and shape was carried out over 30 yr ago. Interestingly enough, however, final agreement on molecular weight was not reached until a crystal structure determination was attempted, when Low (19) calculated a value of 65,200 and 65,600 daltons for two types of crystals. This was confirmed by ultracentrifugation by Creeth (20). Previous hydrodynamic and osmotic pressure measurements were all shown to be in error, either experimentally or in interpretation. Since that time, serum albumins have been used as a calibration standard for faster, but more approximate, techniques such as chromatography and electrophoresis.

4. Insulin

Insulin is a low molecular weight protein hormone used clinically in the treatment of diabetis. It is biosynthesized and stored as a single chain (proinsulin), which is subsequently cleaved to the hormonally active insulin (Fig. 4).

Insufficient human insulin is available for clinical treatment, so that characterization has focused on the molecular similarity with insulin from other sources. The sequence appears to be highly conserved in nature and only a few differences in amino acid sequence (and therefore virtually none in structure) appear (*see* chapter 1 and 2).

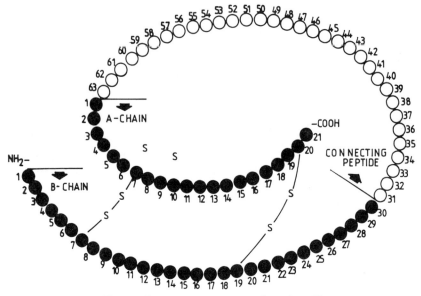

Fig. 4. Primary structure of proinsulin.

Animal insulin is successfully used in the clinical treatment of diabetis. Even so, allergic and resistant responses tend to be induced. Genetic engineers are highly active in this area. Their characterization necessarily involves proof of the amino acid composition and sequence *and* removal of trace materials likely to cause allergic responses.

5. Oxytocin

5.1. Function and Primary Structure

Oxytocin is a peptide hormone inducing uterine contraction in mammals. It has the amino acid sequence:

H — Cys — Tyr — Ile — Gln — Asn — Cys — Pro — Leu — Gly — NH$_2$
　　　1　　　2　　　3　　　4　　　5　　　6

5.2. NMR Studies

Its size, and the geometrical constraints of the 20-atom ring, has allowed a detailed study of its solution conformation to be made by high-resolution NMR (21). Early studies of proton resonance used dimethyl sulfoxide (DMSO) as a solvent, so that amide NH peaks were not obscured by water protons (22). A single conformation was proposed involving 2β-turns, stabilized by intramolecular hydrogen bond in both the cyclic and tail regions of the molecule. Subsequently, detailed aqueous solution spectra have been obtained, by both ^{1}H and ^{13}C resonance, that show a much more flexible conformation and little or no evidence for intramolecular hydrogen bonding.

In particular, resonances from Cys-1 and Tyr-2 amide protons are shifted from their positions in DMSO. Furthermore side chain resonances of Gln-4 and Pro-7 are less superposed in water than DMSO (23).

The flexibility of oxytocin in water can be assessed by comparing the magnitude of its coupling constants with those of a competitive inhibitor [1-L-penicillamine] oxytocin. When motion occurs in the peptide backbone, coupling constants between amide (NH) and α-protons on adjacent carbons are reduced. Oxytocin shows smaller coupling constants for all of the cyclized residues than does the derivatized inhibitor (24). This leads to the proposal that flexibility in oxytocin permits the formation of the activated hormone–receptor complex; whereas for the more rigid inhibitor, binding occurs, but activation is prevented.

6. Antibody Structure and Function

6.1. Chemical Studies

Antibody molecules have been shown to be multichain in nature, and the folded structures show domains that are responsible for antigen binding. Figure 5 shows the sequence of studies that have identified the domain structure shown (25,26).

6.2. Crystal Structure

To obtain more detailed information, crystal structures are required and these have been obtained by raising homogeneous antibodies from cancerous cells (27,28). Structures of the binding fragments (Fab) show that each chain forms two layers of antiparallel β-sheet folded at a connecting disulfide bridge (the immunoglobulin fold). One domain from each chain associates to form the globular

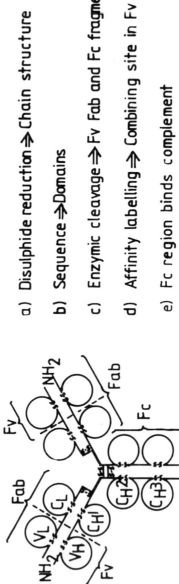

Fig. 5. The structural features of immunoglobulins as deduced from chemical studies (from ref. 30).

a) Disulphide reduction ⇒ Chain structure
b) Sequence ⇒ Domains
c) Enzymic cleavage ⇒ Fv Fab and Fc fragments
d) Affinity labelling ⇒ Combining site in Fv
e) Fc region binds complement

Protein Biochemicals 481

Fab and Fv regions. The selectivity and specificity of antibodies appears to be arrived at by alterations in amino acids that occur in the loops of peptide chain between the structured parts of the immunoglobulin fold. This implies that there should be conservation of sequence in other parts of the globular heads. This has been confirmed and allows direct comparison of structures with totally different molecular binding specificity to be made (29). For example, protein 603 (28) has a negatively charged combining site for the trimethylammonium part of phosphoryl choline. No such requirement is necessary for the binding of Vitamin K_1 by protein New (27), so binding site residues are changed without the disruption of the immunoglobulin fold.

6.3. NMR Studies

The binding site fragments of antibodies are sufficiently small to exhibit rapid rotational motion and therefore a high-resolution proton resonance spectrum. By examination of the spectral shifts accompanying hapten binding, an extremely detailed architecture of the binding site "in action" has been obtained (30). This work shows very elegantly how the cavity for binding can be modified in size, shape, and distribution of charged and stacking amino acids to accommodate the specific tight binding of a particular hapten without totally disrupting the antibody structure.

7. Lysozyme

Lysozyme is probably the most intensively studied and best characterized protein. An extensive review (31) in 1974 records 2500 references and work has continued in the ensuing decade.

7.1. Biological Function

The ability of secretions containing lysozyme to lyse the bacterial cell wall was described by Fleming (32). Subsequently, primary identification and activity of the enzyme was agreed to be by its lytic action on *Micrococcus lysodiekticus*. Formally, the enzyme is a β-1,4-glycan hydrolase, which relates to the specific cleavage reaction that is catalyzed.

7.2. Molecular Identity

Single lysozymes have been prepared from hen, goose egg white, and normal human tissue and have the same chromatographic and electrophoretic profiles and amino acid compositions (33). However, two lysozymes from duck egg white are electrophoretically distinguishable, as are their amino acid sequence and composition (31).

7.3. X-Ray Crystallography

The relatively easy access to sources of hen egg lysozyme and its crystals led to the first electron density map of a protein by Phillips (34), which, given the relatively small size of the molecule and its published sequence, permitted the direct assignment of the secondary and tertiary structure to be made (Fig. 6).

Subsequent studies of lysozyme–inhibitor complexes showed the location of the substrate binding site in a groove in the enzyme surface that runs approximately from leucine 56 to aspartate 101 (35).

7.4. NMR Mapping of the Active Site

Binding of inhibitors to lysozyme produced a downfield shift in the aromatic region of the spectrum (36). This is entirely consistent with the crystal structure, which shows tryptophan 62, 63, and 108 in the active site cleft. Campbell et al. (30) used lanthanide ions to map the active site. Fortunately, crystallography shows that Gd(III) binds in the active site between the carboxyl groups of Asp 52 and Glu 35. Because of the severe distance effect of Gd III broadening, which decays as r^{-6}, the methyl groups of Val 109, Ala 110, and the C_2 proton of Trp 108 are preferentially broadened, which exactly fits with their predicted position from the X-ray structure. Ring current shifts are also observed in lysozymes, methyl groups resonating upfield of their normal (random coil) position. Assignments of many of these residues have been made, but again knowledge of the crystal structure was necessary (37).

7.5 Catalytic Mechanism

The crystal structure of protein–inhibitor complexes suggests that glutamate 35 and aspartate 52 are specifically involved in the reaction intermediates (*see* Fig. 6); the former residue by proton

Protein Biochemicals

transfer to the substrate and the latter by charge stabilization of the carbonium ion intermediate. Subsequent solution titration showed that these groups had highly anomalous pK values, both in human and hen-egg white lysozyme, which otherwise differ in 40% of the 130 amino acid positions (38).

Clearly the folded structures of various lysozymes are all designed to maintain the stereochemistry of the binding site and the particular energetics of the catalytic residues.

8. Interferons
8.1. General Description

Finally, we examine the isolation and characterization of a class of proteins discovered as recently as 1957 (39). Interferons are pro-

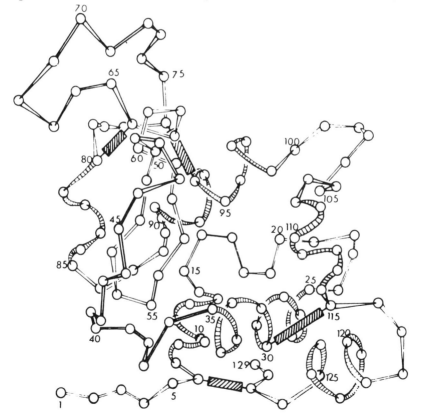

Fig. 6. Early representation of lysozyme structure (from ref. 31).

teins (or glycoproteins) produced by the body in defense against viral attack. Their classification is complex, however, since unlike antibodies, which are molecularly specific and act by a definable mechanism, interferons are part of a broad spectrum defense process of organisms. They derived their name initially from a fluid capable of "interference" with multiplication of viruses in healthy cells from which they were subsequently isolated. Recently, however, proteins with some structural homology with interferons were discovered that exhibited no specific antiviral activity, but still limited their proliferation. These proteins are now also regarded as interferons.

The study of these proteins has taxed the methodology of protein purification and separation techniques to the limit, largely because of the incredible activity of molecules. Biological activity appears at levels as low as one molecule per cell, equivalent to an in vivo concentration between 10^{-11} and $10^{-14}M$.

A comprehensive review of the development of interferon research was compiled by Pestka (40,41); and much of the information referred to here is drawn from these sources.

8.2. Isolation and Separation of Natural Interferons

Interferons are produced by activation of cells in cultures. Three basic types of interferon are identified that relate to the cell type, namely:

α-interferon, from white blood cells (leukocytes)
β-interferon, from connective tissue fibroblasts
γ-interferon, from T-cells (a type of white blood cell)

More recently, large-scale methods for production of interferons from lymphoblast cells has also been described (40, p. 69).

The type of interferon induced depends upon the activator. Viruses generally induce α-interferon, even in fibroblast and lymphoblast cells; β-interferon is induced in fibroblasts by double-stranded RNA; and γ-interferon in T-cells by phytohemagglutenin, concanavalin, or staphylococcal enterotoxin A.

The proteins are concentrated by precipitation with acid, salts of alkali earths, or by salting out.

Purification is achieved by sequential chromatography, usually by an adsorption of interferon onto glass beads followed by elution at acid pH. Subsequently, further adsorbtion/desorption or immunoadsorbent procedures are used. Alternatively, high-performance liquid chromatography (HPLC) can be employed (e.g., ref. 40, p. 472).

8.3. Characterization of Interferons

Initially, the α, β, and γ interferons were identified by their different antigenic properties. However, this can lead to confusion of types without concurrent structural and functional information. Fortunately, two further developments in methodology allowed some structural characterization to take place. First, electrophoresis of peaks separated by HPLC in SDS buffers allowed for demonstration of the heterogeneity of interferons and an estimate of their molecular weights (e.g., 40, p. 472). Typical results are shown in Fig. 7.

Second, development of amino acid analysis methods at picomolar levels has allowed for comparison of the various peptides (Table 1).

The next stage of characterization was to determine the primary structure (sequence) of amino acids. This again required improvement of the techniques so that subnanomolar quantities could be analyzed (41, p. 31). Comparison of the sequences for human leucocyte interferons derived from gene mapping and peptide sequence analysis shows agreement, except for ten amino acids missing at the carboxyl terminal end (42).

To proceed further, determination of the secondary and tertiary structure of interferons will be necessary. This requires the application of physical, rather than chemical techniques, which are less readily increased in sensitivity. Fortunately, the possible application of interferons as antiviral drugs requires their production in bulk. Also, cloning of genes in *Escherichia coli* has permitted production of recombinant forms in large quantities (41, p. 599). Functional studies show that the various natural interferons exhibit differences in their activity, whereas the corresponding natural and recombinant forms are identical. It appears, therefore, that functional differences derive from different amino acid sequences, presumably resulting in different tertiary structures and hence modification of the interaction with a variety of cells. Large-scale purification has reached the stage of crystallization of single interferons (41, p. 3), which sets the scene for X-ray structure and other molecular conformation studies.

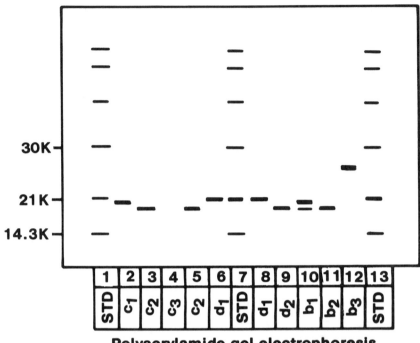

Fig. 7. Polyacrylamide gel electrophoresis of interferons (from ref. 40).

Table 1
Amino Acid Compositions of KG-1 Interferons[a]

Amino acid	Species					
	b_2[b]	b_3	c_1	c_2	d_1	d_2
Asx	9.9	8.7	7.7	9.4	8.7	9.9
Thr	4.9	3.1	5.8	4.7	3.2	4.7
Ser	6.0	6.8	7.7	6.4	6.8	5.9
Glx	16.0	17.3	17.3	15.3	17.3	16.7
Pro	3.9	3.3	3.8	4.6	3.3	4.3
Gly	2.5	1.4	3.6	3.0	1.6	2.8
Ala	5.8	5.5	5.3	6.6	5.5	5.8
Val	4.3	4.2	4.4	4.1	4.3	4.6
Met	3.9	4.6	3.2	4.0	4.5	3.6
Ile	4.6	6.1	4.8	4.4	6.1	4.7
Leu	14.7	14.1	13.7	14.3	14.0	14.3
Tyr	2.6	3.2	3.2	2.7	3.2	2.4
Phe	5.4	6.5	6.2	5.3	6.4	5.1
His	1.9	1.9	1.8	1.8	1.9	1.8
Lys	5.5	6.5	6.1	5.3	6.5	6.0
Arg	8.0	6.8	5.2	8.0	6.7	7.2
Cys	ND[c]	ND	ND	ND	ND	ND
Trp	ND	ND	ND	ND	ND	ND

[a]Values are given in mol %.
[b]Analysis was performed on material having two biologically active bands on polyacrylamide gel electrophoresis in sodium dodecyl sulfate.
[c]ND, not determined.

8.4. Action of Interferons

Much remains to be clarified with respect to the molecular biology and mechanisms of action of interferons. This will require the application of structural studies not only of the proteins, but protein–cell complexes. No doubt, the potential stock market value of drugs derived from such understanding will provide a sufficient driving force.

There is certainly agreement that interferons initiate their action at cell surfaces. Confidence is sufficient that mechanisms appear displayed in journals not normally noted for their scientific content. The scheme in Fig. 8 appeared recently in the *Economist* (43).

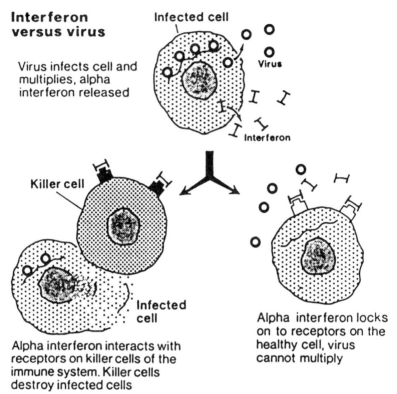

Fig. 8. Mechanisms of interference with virus multiplication (from ref. 43).

References

1. *Br. Pharmacopoeia* **vi**, 863 (1980).
2. R. H. Haschemeyer and A. E. V. Haschemeyer, in *Proteins—A Guide to Study by Physical and Chemical Methods* Wiley, New York (1973).
3. C. Tanford, in *Physical Chemistry of Macromolecules* Wiley, New York (1961).
4. D. B. Wetlaufer, in *Advances in Protein Chemistry* (C. B. Anfinsen, C. B. Anfinson, M. L. Anson, K. Bailey, J. T. Edsall, eds.) Academic, New York and London (1962).
5. W. B. Gratzer, in *Poly-α-Amino Acids* (G. D. Fasman, ed.) Marcel Dekker, New York (1967).
6. J. T. Yang, personal communication.
7. W. Moffitt, *J. Chem. Phys.* **25** (3), 467–478 (1956).

Protein Biochemicals 489

8. S. Beychok, in *Poly-α-Amino Acids* (G. D. Fasman, ed.) Marcel Dekker, New York (1967).
9. E. J. Ambrose and A. Elliott, *Proc. Roy. Soc.* **A205**, 47–60 (1951).
10. D. C. Phillips, *Proc. Natl. Acad. Sci. USA* **57**, 484–495 (1967).
11. M. F. Pentz, *Eur. J. Biochem.* **8**, 455–466 (1969).
12. T. L. James, in *Nuclear Magnetic Resonance Biochemistry* Academic, New York (1975).
13. C. C. McDonald and W. D. Phillips, *J. Am. Chem. Soc.* **91**, 1513–1521 (1969).
14. P. J. Lillford, in *Plant Proteins* (G. Norton, ed.) Butterworths (1978).
15. I. D. Campbell, in *N.M.R. in Biology* (R. A. Dwek, ed.) Academic, London (1977).
16. I. D. Campbell, S. Lindskog, and A. I. White, *J. Mol. Biol.* **98**, 597–614 (1975).
17. B. A. Levine and J. R. P. Williams, *Proc. Roy. Soc. Lond.* **A345**, 5 22 (1975).
18. K. Wuthrich, Abstracts I.U.P.A.B. 8th International Biophysics Symposium, Bristol, UK (1984).
19. B. W. Low, *J. Am. Chem. Soc.* **74**, 4830–4834 (1952).
20. J. M. Creeth, *Biochem. J.* **51**, 10–17 (1952).
21. O. Jardetsky and G. C. K. Roberts, ed., *NMR in Molecular Biology* Academic, New York (1981).
22. D. W. Urry and R. Walter, *Proc. Natl. Acad. Sci. USA* **68** (5), 956–958 (1971).
23. A. I. Richard Brewster and V. J. Hruby, *Proc. Natl. Acad. Sci. USA* **70** (12), 1306–3809 (1973).
24. J. P. Meraldi, V. J. Hruby, and A. I. Richard Brewster, *Proc. Natal. Acad. Sci. USA* **74** (4), 1373–1377 (1977).
25. J. B. Fleischman, R. H. Pain, and R. R. Porter, *Arch. Biochem. Biophys. (suppl.)* **1**, 114–180 (1962).
26. G. M. Edelman, B. A. Cunningham, W. E. Gall, P. D. Gottglieb, U.R.S. Rutishauser, and M. J. Waxdal, *Proc. Natl. Acad. Sci. USA* **63**, 78–85 (1969).
27. R. J. Poljak, L. M. Amzel, H. P. Avey, L. N. Becka, and A. Nisonoff, *Nature New Bio.* **235**, 137–140 (1972).
28. S. Rudikoff, M. Potter, D. M. Segal, E. A. Padlan, and D. R. Davies, *Proc. Natl. Acad. Sci. USA* **69**, 3689–3692 (1972).
29. E. A. Padlan, D. R. Davies, I. Pecht, D. Givol, and C. E. Wright, Cold Spring Harbor Symposium (1976).
30. R. A. Dwek, I. D. Campbell, R. E. Richards, and R. J. P. Williams, ed., *NMR in Biology* Academic, London (1977).
31. E. F. Osserman, R. E. Canfield and S. Beychok, eds., *Lysozyme* Academic, New York (1974).
32. A. Fleming, *Proc. Roy. Soc. Lond.* **B93**, 306–317 (1922).
33. P. Jolles, *Angew Chem.* **8** (4), 227–294 (1960).

34. C. C. F. Blake, D. F. Koenig, G. A. Mair, A. C. T. North, D. C. Phillips, and V. R. Sarma, *Nature* **206**, 757–761 (1965).
35. C. C. F. Blake, L. N. Johnson, G. A. Mair, A. C. T. North, D. C. Phillips, and V. R. Sarma, *Proc. Roy. Soc. Lond.* **B167**, 378–388 (1967).
36. J. S. Cohen, *Nature* (Lond.) **223**, 43–46 (1969).
37. H. Sternlicht and D. Wilson, *Biochemistry* **6**, 2881–2892 (1967).
38. G. P. Hess and J. A. Rupley, *Ann. Rev. Biochem.* **40**, 1013–1044 (1971).
39. A. Isaacs and J. Lindenmann, *Proc. Roy. Soc. B Lond.* **B147**, 258–263 (1957).
40. S. P. Colowick and N. O. Kaplan, eds., *Methods in Enzymology* vol. 78, Academic, New York (1981).
41. S. P. Colowick and N. O. Kaplan, eds., *Methods in Enzymology* vol. 79, Academic, New York (1981).
42. W. P. Levy, M. Rubenstein, J. Shively, V. del Valle, C-Y. Lai, J. Moschiva, L. Brink, L. Cresber, S. Stein, and S. Prestka, *Proc. Natl. Acad. Sci. USA* **78** (10), 6186–6190 (1981).
43. *Economist* **293**, 86 (1984).

Chapter 19

Functional Attributes of Protein Isolates

Foods

Chester Myers

1. General Considerations for Evaluation of Protein Isolates

The highly specific structural requirements of proteins used for medical and pharmaceutical applications (1,2) are usually replaced by nutritional and functional considerations in food systems. The latter have often not taken into consideration that changes in the native biological structure will have significant impact on properties in a food matrix.

General classes of functional properties for proteins, important in food applications, include organoleptic, kinesthetic, hydration, structural, and textural. A cursory analysis shows that the descriptors are far from scientific. Most describe perceived properties that cannot be related to objective physical and chemical measures.

Examination of protein performance by food type, however, can be used as a starting point, as in Table 1.

1.1. Solubility

Solubility has already been considered with respect to extraction, where the extraction medium includes nonprotein solutes

Table 1
Types of Foods and the Related Functional Properties
for Proteinaceous Ingredients

Food type	Functional property
Beverages	Solubility, grittiness, color
Baked goods	Emulsification, complex formation, foaming, viscoelastic behavior, matrix/film formation, gelation, hardness, absorption
Dairy substitutes	Gelation, coagulation, foaming, fat-holding capacity
Egg substitutes	Foaming, gelation
Meat emulsions	Emulsification, gelation, liquid-holding capacity, adhesion, cohesion, absorption
Meat extenders	Liquid-holding capacity, hardness, chewiness, cohesion, adhesion
Soups/gravies	Viscosity, emulsification, water-holding capacity
Toppings	Foaming, emulsification
Whipped desserts	Foaming, emulsification, gelation

dissolved from the source. The *isolate* solubility must therefore be determined independently since the extraction medium may now contain only *known* components. The latter information is relevant for those applications, such as emulsification, in which the end use of the protein relies on solubility.

"Solubility" of proteins is used somewhat "tongue in cheek." Proteins are really colloidal species and, as such, produce "hydrophilic sols" or colloidal dispersions. Even allowing the term "solubility," however, other problems still exist. Thermodynamic (equilibrium) solubility is independent of the route by which it is achieved:

$$\underset{\text{crystal}}{A\ (25\,°C,\ 1\ \text{atm})} \rightleftharpoons \underset{\text{solution}}{A\ (25\,°C,\ 1\ \text{atm},\ \text{pH}\ 7,\ I = 1)}$$

This is rarely observed practically, except for simple low concentration solutions of single, stable proteins. It is always interesting, for example, to take an apparently "saturated" protein "solution," and to further concentrate it by ultrafiltration and observe no precipitation. As a result, "useful" definitions of Protein Dispersibility Index (PDI) and Nitrogen Solubility Index (NSI) have arisen: These measure dispersed protein under closely defined conditions.

Protein Isolates

Whether an isolate needs to be highly soluble or not depends on the end use. In most cases, for ease of use it is convenient to be able to readily disperse the protein. Solubility is normally better for proteins that have not been denatured during isolation. Thus not only is it easy to work with such an isolate, but one also has the choice of selective denaturation at a later stage in order to elicit a desired functional property. In a few cases, however, it is observed that, for example, alkaline denaturation, during isolation, aids later dispersibility of the isolate. Emulsification properties tend to be enhanced. Thermal gelation, though, usually results in gels of lower mechanical strength.

Examination of the protein "solution" by techniques already discussed—spectral, chromatographic, and so on—is important. Even for food use, for example, the relative proportion of the 7S and 11S proteins in a soy isolate controls textural characteristics of Tofu gels. The 11S species, when gelled, gives higher tensile strength, greater thermal expansion, and higher water-holding capacity. Low levels of contaminating phytic acid cause softening (3).

1.2. Electrostatic Charge/pH Titration

Since interaction between molecules is important in the use of proteins, both for gelation and emulsification, electrostatic charge effects are important. Being zwitterionic, isoelectric pH determination is necessary. Table 2 gives pI values for several proteins.

A range of pI values provides potential for "engineering" electrostatic charge behavior. However, the charge on a protein at any particular pH value, removed from the pI, will depend also on the content of titratable groups in the pH range involved (pI − pH). Figure 1 shows typical pH titration curves for some proteins and, again, striking differences are noted. For example, between pH 5.5 and 8.5, ovalbumin has a relatively flat titration curve, indicating it should be more insensitive to change of pH in this range than are the other proteins shown. This is reflected, for example, in thermal gelation behavior, in which only slight change in gel integrity occurs within the same range of pH values.

During thermal gelation, for example, as the temperature increases, the entropically controlled hydrophobic interactions are strengthened. In the absence of electrostatic repulsive forces, aggregation may occur prior to thermal denaturation—thus "islands" of aggregates are formed and, upon thermal denaturation, the re-

Table 2
Isoelectric pH Values (pI) for Proteins,
Measured by Column Isoelectric Focusing[a]

Protein	pI
A commercial gelatin	9.8
Ribonuclease	9.5
Gelatin I (sigma)	8.7
Gelatin II (sigma)	8.5
Rapeseed 12S	7.0
Bovine serum albumin	5.4, 5.1, 4.7
Lactoglobulin	5.3
Lactalbumin	5.1
Gelatin III (sigma)	5.0
Gelatin IV (sigma)	5.0
Soy 18S	4.6
15S	4.6–5.1
11S	4.9–5.3
7S	5.0–7.5
Casein	4.6
Ovalbumin	4.5–4.9

[a]Rapeseed and soy proteins were collected from protein isolates by gel filtration.

sulting "gel" will be granular and opaque. If the protein is highly charged, then aggregation, at temperatures lower than the denaturation temperature, will be precluded. Increase in exposure of nonpolar domains, during denaturation, will then result in a clearer, more isotropic gel than in the former case. Van der Waals forces will be weakened as the temperature increases since they are under enthalpic control.

$$N \text{ (native)} \rightleftharpoons D \text{ (denatured)} \rightarrow \text{gel}$$
$$\downarrow$$
$$\text{Aggregates} \rightleftharpoons D' \text{ (denatured)} \rightarrow \text{gel}'$$

If used as an aid to emulsion stability, the protein can impart a significant contribution to the zeta potential of the emulsion disperse phase. Although the adsorbed protein may extend beyond the effect of the electrical double layer, so that charge stabilization may not directly help to stabilize the emulsion, the charge still helps determine the rheological properties of the emulsion: Thus stability may be either increased or decreased depending on the complex interaction of the electroviscous effect with the effect of particle size,

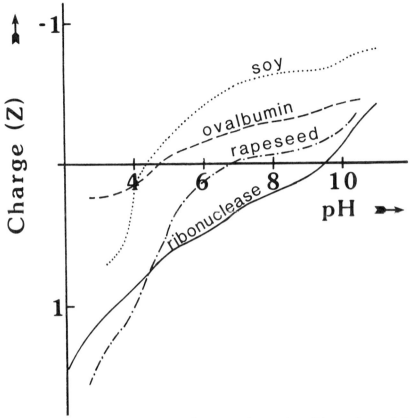

Fig. 1. pH Tritration curves for several proteins. Dispersions of the respective proteins were prepared at 0.8% (wt/vol) and titrated with appropriate NaOH or HCl solutions using Radiometer titration equipment (ABU 12 auto burette, TTT 80 titrator, TTA 80 titration assembly, REC 61 servograph, REA 160 titrigraph module, and a pHM 64 pH meter).

particle size distribution, and disperse phase concentration on the emulsion viscosity (4,5). Moreover, it is important to know if the adsorbed protein layer is likely to provide predominantly electrostatic or steric stability, since these two influences result in quite different considerations for emulsion stability behavior (6).

1.3. Criteria of Purity and Denaturation

In order to best evaluate the efficiency of isolation (yield of isolation, quality of isolate), a variety of laboratory techniques is essential to characterize the proteins of the source material and the isolate. Optimal protein yield is desirable for commercialization of any process. Lack of denaturation is demanded for biological function, and will normally also be an asset when the isolate is incorporated into a food product. For the former, the particular biological activity is the best criterion of lack of denaturation during processing.

It is possible, during instrumental monitoring of changes in protein structure, that the detected change is not significant for biological activity. Conversely, it is possible that the technique is insensitive to another structural change that is important for activity. For example, studies on antithrombin III (7) have shown that protection against thermal denaturation by sugars and their derivatives, as measured by a fluorescent probe technique, does not necessarily correlate with preservation of activity.

A number of facile methods is available, however, that when used together, do give useful information with respect to the viability of the isolate for certain uses. Particularly in the food industry, calorimetric, chromatographic, and spectroscopic examinations can be very informative for isolates that are intended for emulsification and/or thermal gelation.

1.3.1. Column Chromatography and Electrophoresis

Column chromatography (see chapter 11) is recommended as an initial technique to examine any isolate, and to compare it with the source material. The particular mode of chromatographic examination will be partly determined by the quantity of protein isolated. Gel filtration methods give an estimate of protein molecular weight, and separate individual species for spectral and other analyses. Figure 2 is a profile of a plant material extract and the corresponding isolate. It is readily apparent from this profile that certain species, present in the source, are essentially absent in the isolate. Subsequent UV examination identified these as nonproteinaceous. Ion exchange columns require less protein: Affinity methods are even more sensitive and recovery is virtually quantitative. The intermediate-pressure Pharmacia Fast Protein Liquid Chromatography equipment is very useful. Techniques using HPLC are also fast and efficient. Both FPLC and HPLC offer a full spectrum of modes of separation—various hydrophobic columns, strong and weak ion exchangers, a range of affinity supports and so on.

More than one method of separation is advisable in order to determine homogeneity of a particular column fraction. For example, the vicilin and legumin storage proteins of fababean have

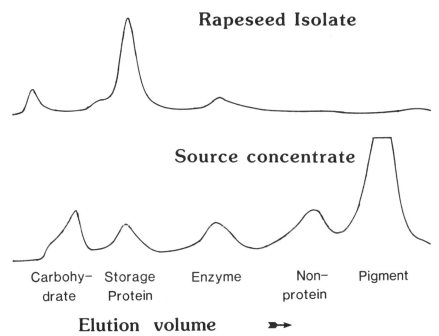

Fig. 2. Gel filtration profile for an extract of both the source concentrate and an isolate prepared therefrom by a patented procedure (81). A Pharmacia K26/100 column, packed with Sephacryl S-300, was eluted with 0.5M NaCl. The profile was recorded at 280 nm by an ISCO UV monitor equipped with a flow cell of 2 mm path length. Fractions were collected 100 drops per tube.

molecular weights of 140,000 and 340,000 daltons, respectively. However, they are not readily separated by gel filtration since they both have the same effective molecular diameter. As determined by transmission electron microscopy, the vicilin molecule is horseshoe shaped, whereas the legumin is globular. Moreover, by regular electrophoresis at pH 7.4, these two species are also not readily separated. Hydrophobic chromatography, on the other hand, does separate the two proteins quite readily.

Once separated and collected, each protein may then be subjected to further examination. Determination of the UV spectrum, and its first derivative, should be the first method of subsequent examination. Figure 3 shows the UV spectra of typical legume storage proteins: The absence of tryptophan and high levels of phenylalanine are apparent for the vicilin. Fluorescence properties are also important to examine. Changes in tyrosine-to-tryptophan energy

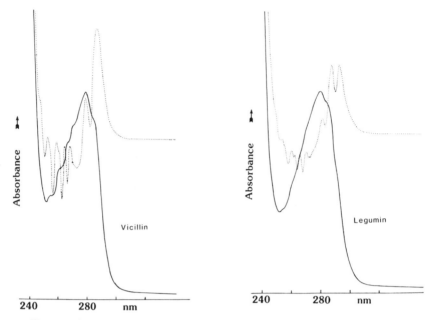

Fig. 3. Ultraviolet spectrum of chromatographically purified fababean vicilin and legumin storage proteins, dissolved in 0.5M NaCl. The solid line is the absorbance spectrum and the dotted line is the (mechanically derived) first derivative of the absorbance. Spectra were obtained with a Cary 118C spectrometer operated in Auto Gain mode.

transfer monitor conformational changes: Shifts in the position of tryptophan fluorescence monitor its environment and can indicate protein unfolding, particularly exposure of hydrophobic portions of the molecule.

Once UV/fluorescence methods have established the chromatographically separated species as proteins, then sodium dodecyl sulphate polyacrylamide electrophoresis (SDS-PAGE) is useful to establish "finger-printing" subunit patterns. These provide a geography that enables electrophoretic analysis of either the source materials or products, using the isolates. Subunit patterns may be useful for following changes resulting from denaturation, either during isolation or use of the isolate: Denaturing conditions are often used during electrophoresis, however, and these preclude determination of prior denaturation. It is customary to use staining techniques to identify protein bands: UV scans of the unstained gels at 280 nm are preferred, and a full wavelength scan of the protein "280" peak will better ensure that the separated bands are proteinaceous. Unfortunately, there are few gel scanners that permit the determination of the protein spectrum from the gel: Most provide only a densitometric scan of the stained gel. Thus, if electrophoresis is to be used, preparative scale may be worthwhile in order to extract sufficient quantity of the separated bands into solution for UV examination. Plant source materials other than proteins tend to be stained by so-called protein stains. Staining, however, is probably adequate to identify subunits of a purified protein.

Immunoelectrophoresis techniques are more selective and are particularly useful for isolates that are available in only small quantities (8,9).

Analytical ultracentrifugation, electron microscopy, light scattering, and so on afford more definitive characterization of chromatographically purified fractions. The protein literature already addresses these quite adequately (10).

1.3.2. UV and Fluorescence Spectroscopies

The study of proteins by UV spectroscopy has been well substantiated and several papers have covered the general usefulness of the technique (11,12). Tyrosine, tryptophan, and phenylalanine are the major contributors to the 280 nm peak, whereas the backbone peptide linkages produce a much stonger absorbance at about 205 nm. The latter is usually about 30 times more intense than the 280 nm peak, the actual ratio depending on the proportion of the protein that is made up of the 280 nm contributors. Any protein source, and isolate derived therefrom, should be characterized by its UV absorbance spectrum. Any modern instrument also affords at least a first derivative and this provides more definitive "fingerprinting," as shown in Fig. 3. Table 3 gives the positions of the derivative peaks for phenylalanine, tyrosine, and tryptophan. If the instrument determines the derivative mechanically, exact positions may vary, depending on instrumental parameters. Such uncertainty is avoided with newer instruments that give a derivative calculated from the absorbance spectrum by means of a microprocessor.

Both tyrosine and tryptophan have derivative peaks at about 284 nm, whereas only tryptophan has a peak at about 290 nm. The derivative intensity, I, may be measured at these two wavelengths, and the I^{284}/I^{290} ratio calculated. Use of a calibration curve (Fig. 4), in which known ratios of tyrosine and tryptophan have been measured, provides values for the ratio of these two amino acids in a protein. Since the tyrosine content is easily determined by

Table 3
Maxima for First Derivatives of UV Absorbance Spectra for
Phenylalanine (1.70 mM), Tyrosine (0.223 mM),
and Tryptophan (2.23 mM)[a]

Amino acid	Major maxima, nm	Minor maxima, nm
Phenylalanine	267	242
	263.5	247
	257.5	
	252	
Tyrosine	284.5	
	277	
Tryptophan	290.5	272.5
	283	

[a]Spectra were recorded with a Cary 118C spectrometer. A slit of 0.1 nm, a pen period of 1 s, a scan speed of 2.0 nm/s, and a sensitivity of one setting more sensitive than that used for the absorbance were used.

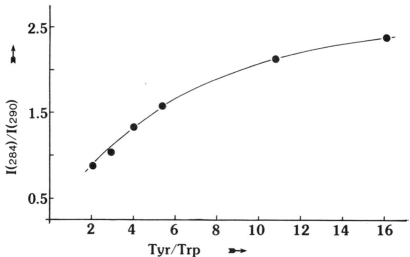

Fig. 4. Ratio of the first derivative peak intensities for solutions containing various molar ratios of tyrosine and tryptophan in 0.5M NaCl. Values along the ordinate axis are the measured ratios of the "284" derivative peak to the "290" derivative peak.

amino acid analysis, then the tryptophan content is conveniently determined from the UV first derivative spectrum.

Difference spectroscopy has also become well accepted (13–15). Subtle changes in the environment of tyrosine may result in a difference in the degree of ionization: the phenolate anion gives a differently shaped band than does the corresponding protonated species. Not only pH changes, but also changes in solvent polarity and temperature, have been monitored by difference spectroscopic methods. Such methods yield more information at the molecular level and help identify changes in conformation that involve specific amino acid residues. Figure 5 shows thermal difference spectra for ribonuclease and rapeseed storage proteins. The substantial increase between 54 and 62°C, for ribonuclease, detects the thermal denaturation of that protein: For the rapeseed protein, in the same temperature region, the UV spectrum is obliterated as turbidity develops. Thus thermal difference spectroscopy is precluded by the aggregation of the protein that occurs with increasing temperature. The aggregation presumably results from hydrophobic interactions between protein molecules. A corollary of this observation is that turbidimetric studies are useful for monitoring protein aggregation: Turbidity resulting from aggregation does not require the sensitivity of a light scattering instrument, and a UV instrument may be a sufficiently sensitive monitor.

Determination of protein content in solution is easy with a 280 nm absorbance measurement. An extinction coefficient, for the particular protein, must have been previously determined if absolute protein contents are to be determined. The usual method is to lyophilize the chromatographically purified protein, and then to make a solution from which the absorbance will yield the extinction coefficient. Usually, the protein will have been collected in a buffer and the solution must then be dialyzed, prior to lyophilization, to remove the buffer salts. Not only is this procedure often tedious, but the dried protein may be difficult to solubilize quantitatively: In particular, lyophilization, at temperatures above the collapse temperature, is likely to cause some denaturation, which may lead to the formation of insoluble aggregates.

An alternative method for determining extinction coefficients is to use differential refractometry to directly determine the protein content of the solution collected by chromatography. Although protein concentration for light scattering experiments (16) is traditionally determined by this method, it does not seem to be widely practiced otherwise. The UV absorbance of the same solution will then readily give the desired extinction coefficient. Since refractive index results from polarizability of the molecule, the temperature must be controlled, and the ionic strength of the medium must be known. Figure 6 shows the dependence of refractive index on

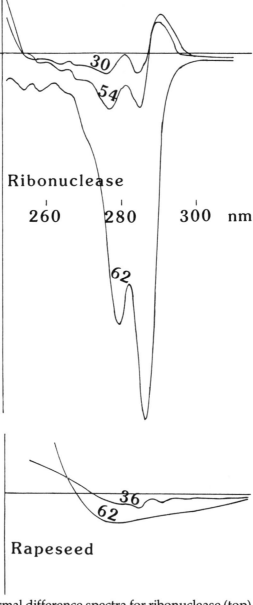

Fig. 5. Thermal difference spectra for ribonuclease (top) and rapeseed 12S (bottom) proteins. The reference cell contained a protein solution identical to that in the "sample" beam and was maintained at 20°C. Proteins were in 0.3M NaCl: Difference spectra were recorded at temperatures (°C) as indicated.

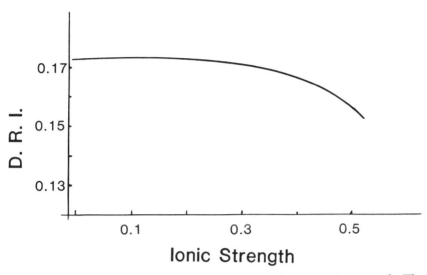

Fig. 6. Dependence of protein refractive index on ionic strength. The ordinate axis is differential refractive index ($\Delta n/\Delta c \times 10^{-4}$), where n is the refractive index and c is the protein concentration in mg/mL.

ionic strength. Such a calibration need be done only once, and as shown, it is at levels above $0.3M$ that the greatest sensitivity is exhibited. Table 4 gives extinction coefficients for a series of plant seed storage proteins that were collected by gel filtration.

When a pure protein is characterized by UV, it is also useful to calculate the ratio of the maximum, at about 280 nm, to the minimum at about 260 nm. For most proteins, a value of greater than 2 is found. Table 5 gives A^{280}/A^{260} values for several proteins. A low value of 1.97 for fababean vicilin reflects the absence of tryptophan. These values are then useful in later examination of isolates, in order to assess possible contamination.

Ultraviolet absorbance measurements may be performed directly on isolate solutions, without any chromatographic purification. The A^{280}/A^{260} ratio may be used to empirically monitor impurities that absorb at 260 nm. If the isolate source material contained nucleic acids, then these will be the most likely contributors to absorbance at 260 nm. For isolates prepared from plant seed sources, however, carbohydrates, phytic acid, and other materials are more likely contaminants. These also give a greater contribution at 260 than at 280 nm, although there is no 260 nm chromophore *per se*. Figure 7 gives UV absorbance spectra for two soy protein isolates. Although the isolate prepared by ionic strength manipulation gives a spectrum

Table 4
One Percent Extinction Coefficients For Proteins[a]

Protein	mg/Absorbance unit	$E^{1\%}$, 280 nm
Faba 340,000d (legumin)	1.313	7.62
140,000d (vicilin)	1.903	5.25
Isolate	1.62	6.17
Pea isolate	1.67	5.98
Soy 7S	2.6	3.8
11S	1.75	5.71
18S	1.31	7.66
Rapeseed 12S	1.25	7.97
17S	1.38	7.22

[a]Protein contents of solutions were determined by differential refractometry (C. N. Wood Model RF-500). All proteins were in 0.5M NaCl solutions. Each was microfiltered through 0.45 µm filters into the refractometer cell. Ultraviolet absorbance was determined with a Cary 118C spectrometer, at 0.1 nm slit. Faba and soy proteins were collected chromatographically from isolates prepared according to ref. (81). Rapeseed proteins were collected chromatographically from isolates prepared according to ref. (18). Isolates were prepared according to ref. (81).

Table 5
A^{280}/A^{260} Values for Chromatographically Purified Proteins Collected From Isolates[a]

Protein	A^{280}/A^{260}
Faba 340,000d (legumin)	2.34
140,000d (vicilin]	1.97
Soy 7S	2.30
11S	2.36
18S	2.36
Rapeseed 12S	2.46
17S	2.30

[a]Prepared according to either ref. (18) or (47). Recorded absorbance values were the maximum and minimum of the respective "280" peak and the "260" valley for each protein spectrum.

that resembles that of a chromatographically purified protein, the isoelectrically prepared isolate spectrum does not resemble that of a protein. Table 6 gives A^{280}/A^{260} values for a series of soy isolates. It is possible, of course, that the contaminating material may not

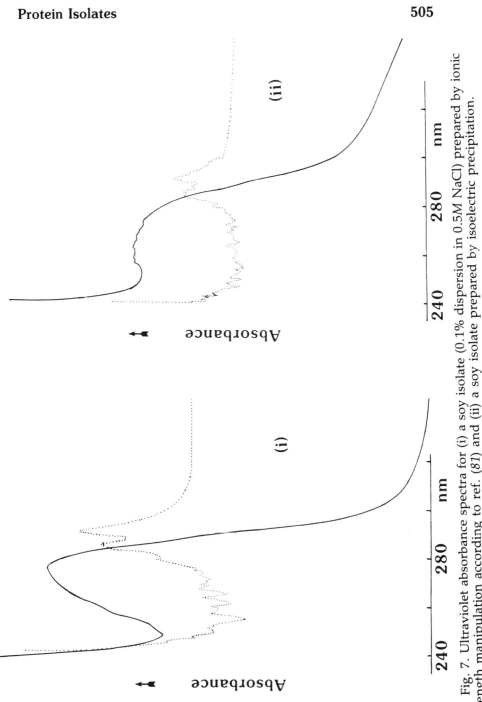

Fig. 7. Ultraviolet absorbance spectra for (i) a soy isolate (0.1% dispersion in 0.5M NaCl) prepared by ionic strength manipulation according to ref. (81) and (ii) a soy isolate prepared by isoelectric precipitation.

interfere with the functional use of the protein. It was observed, however, that when the isolates of Table 6 were dispersed into aqueous media and heated to form "thermal" gels, the gel integrity was inversely related to the "260" absorbing contaminating material. Gel integrity was quantitated as a "hardness" value, with a Texturometer (17).

Table 6
A^{280}/A^{260} Values for a Series of 0.1% (wt/wt) Dispersions of Soy Isolates in 0.5M NaCl, pH 6.0

Isolate	A^{280}/A^{260}
Supro 620	0.54
Promine D	0.67
U4-102	1.00
PMM	1.45

Many protein isolates collected from plant sources are likely to absorb at about 320 nm. For example, isolates prepared from oilseeds, such as sunflower and rapeseed, will possibly be contaminated by phenolic compounds such as ferulic acid, chlorogenic acid, sinapic acid, sinapine, and so on. Many of these tend to form insoluble complexes with the protein after dispersion, so that functional use is thwarted in applications such as emulsification and thermal gelation. Again, UV examination of a 1% (wt/wt) dispersion in 0.5M NaCl provides a facile monitor for such contamination, and a A^{320}/A^{280} ratio is another convenient value to measure. Table 7 gives both A^{280}/A^{260} and A^{320}/A^{280} values for a series of rapeseed isolates. Figure 8 is the UV spectrum for a rapeseed isolate prepared by ionic strength manipulation (18). A very pure protein is indicated and Kjeldahl analysis with a nitrogen conversion factor of 6.25 indicated 97% protein content.

Standard fluorescence spectroscopy has also been well established for protein studies (19–23). Phenylalanine excites in the region of 200 to 285 nm, with the most useful excitation between 240 and 275 nm. Tyrosine excites between 200 and 295 nm with the useful region being 250 to 292 nm. Tryptophan excites from 200 to 302 nm, the useful region being 250 to 300 nm. In aqueous media, phenylalanine emission occurs at 282 nnm, tyrosine at 303 nm, and tryptophan at 348 nm: Tryptophan emission is sensitive to its environment and in nonpolar media occurs at 325 nm (22). The quantum yield for phenylalanine is sufficiently small that it is not a significant contributor to the fluorescence spectra of most proteins. In

Table 7
A^{320}/A^{280} and A^{280}/A^{260} Values for a Series of Rapeseed 12S Proteins Collected Chromatographically From Various Isolates

Protein	A^{320}/A^{280}	A^{280}/A^{260}
12S	2.2007	0.1373
12S	2.1269	0.0984
12S	2.1031	0.0942
12S	2.0570	0.0641
12S	1.9261	0.0907
12S	1.9032	0.1194

undenatured proteins, tyrosine emission is not usually seen, as a result of either quenching by nearby groups such as carboxylate anions in the protein or by resonance energy transfer to tryptophan. Fluorescence polarization can also yield useful information (24). Lifetimes (25) monitor processes on the nanosecond timescale. For studies that are intended to monitor protein conformational changes, without determining exact details of the molecular environment of the chromophore, both fluorescent quantum yield and polarization values are followed quite easily with a number of commercially available instruments.

It is useful to excite at 270, 280, 290, and 297 nm when recording intrinsic emission spectra of proteins. Assuming no significant phenylalanine contribution, at 270 nm only tyrosine excites, and if its emission is efficiently relayed to tryptophan, then only the tryptophan emission is apparent: The half band width (area divided by maximum height) monitors the efficiency of the energy transfer. At 297 nm excitation, only tryptophan is excited. If the tryptophan is in a polar environment, i.e., exposed to the aqueous medium, its emission will be at about 350 nm. If, however, the tryptophan is buried in a nonpolar region of the protein, its emission will be at shorter wavelengths, as low as 325 nm. Conformational changes of the protein that result in tryptophan shifting to regions of different polarity may thus be monitored by the intrinsic fluorescence. In addition, if tyrosine moves away from tryptophan, then the energy relay may be broken, and at 280 or 290 nm excitation, both emission bands may be evident. Figures 9 and 10 give the emission spectra for fababean storage proteins vicilin and legumin, respectively. The absence of tryptophan in vicilin is evident. There is no suggestion of individual emission bands. Figures 11 and 12 show the emission spectra for rapeseed 12S storage protein and one of its subunits. Although the fluorescence spectra were re-

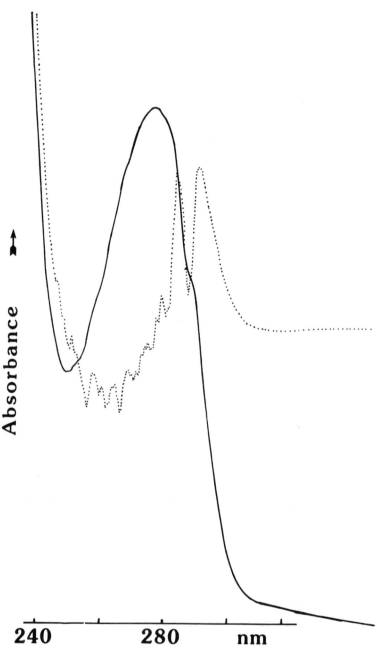

Fig. 8. Ultraviolet absorbance spectrum for a rapeseed isolate 0.1% dispersion in 0.5M NaCl. The isolate was prepared according to ref. (18).

Protein Isolates

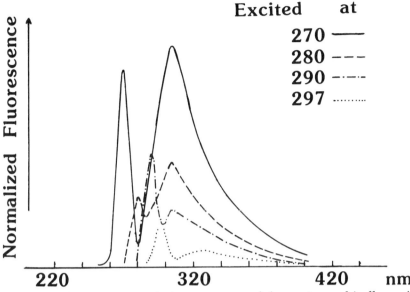

Fig. 9. Fluorescence emission spectrum of chromatographically purified fababean vicilin storage protein. The spectrum was recorded at 25°C with a Perkin Elmer MPF-2A spectrofluorimeter, using an excitation slit of 4 nm and an emission slit of 10 nm. The protein was dissolved in 0.5M NaCl in a 1 × 1 cm Hellma Suprasil cuvet.

corded under identical conditions, it is apparent that the emission bands for both tyrosine and tryptophan are present for the subunit. Either the subunit has unfolded from its conformation in the "mother" 12S molecule, so that the tyrosine and tryptophan are farther apart, or the energy transfer in the 12S moiety was between adjacent subunits. When recording fluorescence spectra, the absorbance of the sample should be kept below 0.1 in order to prevent significant self-absorbance. Cells of path lengths <1 cm may be useful if dilution is undesirable.

The initial sharp bands, present in all the emission spectra, are the respective Rayleigh scattering peaks resulting from scattering of the incident beam by the protein molecule. This is a very useful monitor of aggregation. Hydrophobic association with increasing temperature, as noted above, may be monitored by this technique since it is conveniently sensitive to such changes in aggregation. For changes in conformation, however, where changes in hydrodynamic dimensions of the protein molecule are less dramatic than the changes in size with aggregation, greater sensitivity is desirable, and low angle light scattering provides better monitoring.

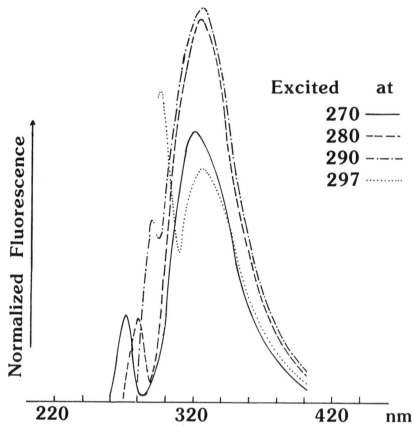

Fig. 10. Fluorescence emission spectrum of chromatographically purified fababean legumin storage protein. The spectrum was recorded as in Fig. 9.

Fluorescent probes are useful for obtaining information about protein characteristics (26). Such studies yield information about the polarity of parts of the protein, distances between specific sites of the protein, protein flexibility, and the rate of conformational changes (27). Figure 13 is a binding plot for the fluorophore 2-p-toluidinyl-naphthalene-6-sulfonate (TNS) with fababean legumin storage protein. This molecule undergoes a well-known change in its fluorescent properties as a result of noncovalent interactions with proteins (26). The method of Wang and Edelman (28) was used to calculate the dissociation constant, K_d, of the protein–fluorophore complex and the value n, the number of fluorophores bound per unit weight of protein. Table 8 gives data for several proteins

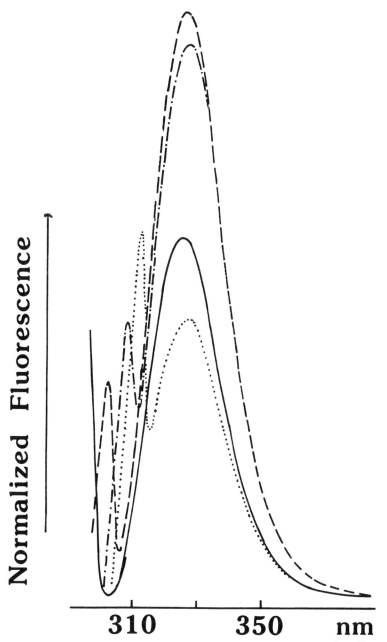

Fig. 11. Fluorescence emission spectrum of chromatographically purified rapeseed 12S storage protein. The spectrum was recorded as in Fig. 9.

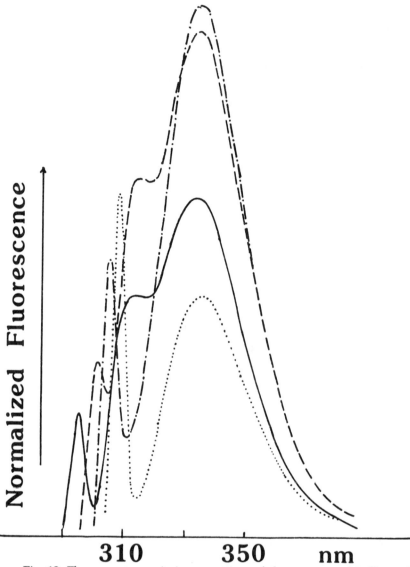

Fig. 12. Fluorescence emission spectrum of chromatographically purified rapeseed 12S storage protein subunit. The 12S protein had been collected by gel filtration on sepharose 6B, 0.5M NaCl, pH 6. The subunit was collected by dissociating the protein at pH 3 and collecting the subunit on Sepharose 6B, 0.5M NaCl, pH 3. The spectrum was recorded as in Fig. 9.

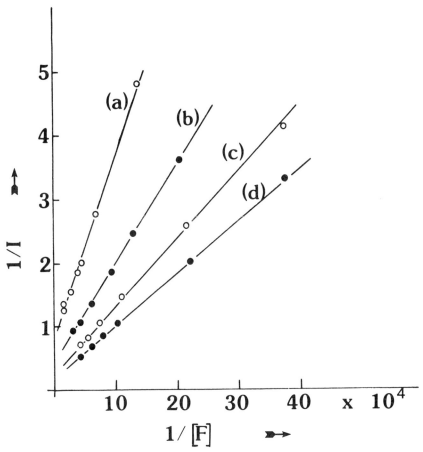

Fig. 13. Fluorescence titration of fababean legumin storage protein with TNS at pH 6. [F] represents the concentration of TNS in mol/L. I is the normalized fluorescence emission peak height at λ = 445 nm. The dissociation constants were (a) 1.04×14^{-6} at 0.117 mg/mL, (b) 1.24×10^{-6} at 0.225 mg/mL, (c) 1.11×10^{-4} at 0.373 mg/mL, and (d) 1.12×10^{-4} at 0.704 mg/mL protein concentration.

with this probe. Temperature dependence of K_d would make van't Hoff evaluation possible. Trends in K_d with temperature, ionic strength, or pressure would provide information about the relative degree of polar and nonpolar interactions. Any temperature dependence studies would be best at temperatures below the thermal denaturation temperature, in order to avoid complex trends

Table 8
Data for TNS Binding to Various Proteins[a]

Protein	Concentration	K_d
Faba 340,000d	0.117 mg/mL	1.04×10^{-4}
	0.225 mg/mL	1.24×10^{-4}
	0.489 mg/mL	1.11×10^{-4}
	0.704 mg/mL	1.12×10^{-4}
140,000d	0.069 mg/mL	2.47×10^{-6}
	0.138 mg/mL	6.46×10^{-7}
	0.274 mg/mL	4.75×10^{-7}
	0.413 mg/mL	3.17×10^{-7}
BSA	$4.85 \times 10^{-6} M$	6.13×10^{-5}
	$9.61 \times 10^{-6} M$	8.46×10^{-5}
Ovalbumin	$6.93 \times 10^{-6} M$	7.33×10^{-5}
	$2.08 \times 10^{-5} M$	5.01×10^{-5}

[a]Calculations were according to Wang and Edelman (34). All spectra were recorded with a Perkin-Elmer MPF-2A Fluorescence Spectrophotometer, in the ratio mode. Excitation slit width was 4 μm and emission slit was 10 μm.

resulting from additional binding sites becoming exposed upon cooperative thermal unfolding.

Nakai and coworkers (29–34) have developed a fluorescent probe technique that appears to be useful in predicting the relative value of various proteins for applications such as foam stability, emulsification, and thermal gelation. The technique resembles that described for TNS, except that the probe is cis-parinaric acid. The results are discussed in terms of "surface," or "effective," hydrophobicity. In combination with other characteristics, such as solubility and sulfhydryl content, good correlations are obtained with technological attributes.

1.3.3. Differential Scanning Calorimetry (DSC)

The stability of proteins has been the subject of several excellent reviews (35,35a) (see also, chapter 4). Most proteins undergo a cooperative thermal transition at a particular temperature that is measurable by several commercially available differential scanning calorimeter instruments: The transition is usually endothermic, as shown schematically in Fig. 14. The area under the peak is ΔH, the heat associated with the transition; the width at half-height ($\Delta T_{1/2}$) is a measure of the cooperativity ("sharpness") of the transition, and the area divided by the height ("half band width," or

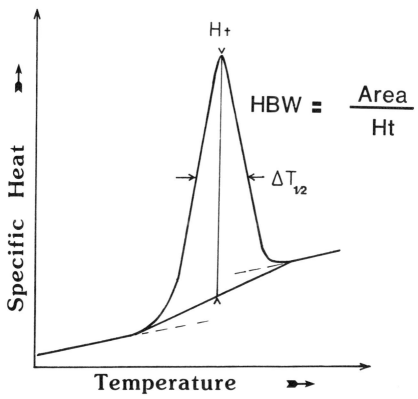

Fig. 14. Schematic of differential scanning calorimetric thermogram. The $\Delta T_{1/2}$ value is the width of the peak at half height. HBW, the half band width, is the peak area divided by the peak height. Both $\Delta T_{1/2}$ and HBW are in the units of the ordinate axis (°C). The peak area represents the enthalpy of the transition, i.e., the specific heat integrated over the temperature range of the transition.

HBW) represents another useful value when either more than one protein is present or when the protein interacts with another material in the matrix to give an additional thermal transition. In both cases, a side band may broaden the main transition. If this side band is minor, the $\Delta T_{1/2}$ value may not change, whereas the HBW value will increase. Both $\Delta T_{1/2}$ and HBW often increase if the protein has been partly denatured by other than heat. For example, at extremes of pH, at which the protein is highly charged, an increase in transition width is observed if the interactions between the domains, involved in the thermally effected transition, are weakened. Although the "band width at half height" has been rigidly defined in terms of the broadening of the profile from an homogeneous protein, the "half band width" allows for a more empirical, but useful, approach for those proteinaceous materials that must be examined as rigorously as possible, but that are not pure homogeneous preparations.

At the high concentrations required for most commercial calorimeters, thermal denaturation is usually accompanied by aggregation. Thus irreversibility of the thermal denaturation prevents true thermodynamic treatment of the thermogram. This may result from several sources:

(1) High concentrations encourage aggregation
(2) Complexity of the quaternary and tertiary structure
(3) Plant seed proteins, being synthesized in a low water environment, will possibly be in a metastable state at higher water levels and should not be expected to return to their native conformation even *if* there were a "reversal" from the thermally denatured state.

The DSC enthalpy value is a composite. In general, for the transition

$$\text{Native (N)} \rightleftharpoons \text{Denatured (D)} \rightarrow \text{Aggregate (A)}$$

the N-to-D transition will entail disruption of both hydrophobic interactions (exothermic) and polar interactions (endothermic)—overall, a net endotherm is usually observed. For the D-to-A transition, the formation of hydrophobic interactions will be mainly endothermic and the formation of polar interactions, exothermic: overall, a net exotherm normally results. For the complete N-to-A transition, then, one may expect net exo-, and endo-, or a-thermal transitions: usually, net endotherms are observed.

As a result of hydrophobic interactions being under entropic control (36) they behave oppositely to polar interactions in response to temperature, ionic strength, and pressure.

	Polar	Nonpolar
Increasing ionic strength	Weaken	Strengthen
Organic nonpolar additives	Strengthen	Weaken
Increasing temperature[a]	Weaken	Strengthen
Increasing pressure[b]	Compress	Expand

[a]Up to 60–80 °C.
[b]Approximate behavior, actually quite complex.

Although temperature and ionic strength effects on hydrophobic interactions are well substantiated (see chapter 5), the effects of pressure are less well understood. However, both melittin (37) and elastin (38) have been shown to demonstrate expected trends—i.e., increasing pressure causes expansion of the proteinaceous moieties.

Modulation of behavior with organic additives is also in accordance with the effect of concentration of the additive on the hydrophobic interactions (39). Below about 10–20°C and at low concentrations ($\leqslant 10\%$) of organic reagents such as methanol, ethanol, and propylene glycol, hydrophobic interactions are stabilized (40).

Figure 15 shows the DSC profiles of fababean vicilin and legumen for two different salt levels, but at the same pH. The increased

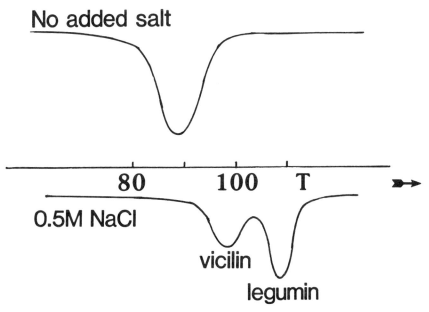

Fig. 15. DSC thermograms for faba legumin and vicilin storage proteins, recorded at NaCl levels of 0 and 0.5M, respectively (DuPont 990 Model 990 thermal analyzer with a 910 DSC cell base).

thermal stability, at higher levels of salt, is a manifestation of the role of hydrophobic interactions. Table 9 lists more DSC data that substantiate the above comments. Section 2.1 also indicates the value of DSC analysis.

Table 9
DSC Data for Several Proteins at Various Conditions[a]

	T_d, °C	ΔH	HBW, °C
Apo-ferritin	106	9.9	16.0
Ribonuclease A	62	7.5	8.5
(5% ethanol added)	60	8.9	6.9
(10% ethanol added)	57	10.3	6.9
Bovine serum albumin	62	5.2	12.3
Faba 340,000d (0.5M NaCl)	107	7.1	9.0
140,000d (0.5M NaCl)	107	3.4	9.0
Rapeseed 12S (0.5M NaCl)	85	5.0	11.0
Soy 11S (0.5M NaCl)	98	5.0	6.5
Oat storage protein	113	7.2	9.0
(0.5M NaCl)	118	7.5	8.6
(10% dioxane)	103	6.1	9.4
(10% ethylene glycol)	110	7.2	9.0
(20% ethylene glycol)	107	7.2	8.6
(20% propylene glycol)	99	7.1	8.6

[a]Analyses were performed with a Dupont model 990 thermal analyzer with a 910 cell base, using 4–15% (wt/vol) of protein at pH 6 in water.

1.4. Amino Acid Profile

Amino acid content is important whether for assessing the identity of a protein or for indicating possible changes in functional behavior. Single residue substitutions can have quite dramatic results. For example, replacement of a partially exposed arginine by histidine in a mutant lysozyme from bacteriophage T4 results in no significant structural change, but a drop of 14°C in thermal stability (thermal denaturation temperature, at pH 3, drops from 57 to 43°C), apparently from destabilization of the hydrophobic core (41). Such variation would be very significant for proteins used in foods for thermal gelation properties: emulsion and foam stabilization properties would likely also be altered.

Once the amino acid composition is known, then certain calculations are possible, giving quantities that describe bulk properties of the protein. A theoretical hydrophobicity value can be arrived at via several procedures (42–45). Unfortunately, these rely only on total amino acid content and do not take into consideration any level of protein structure. Another, probably more accurately descriptive, evaluation (46) takes into consideration the primary sequence. A charge frequency factor (42) indicates the electrostatic charge possible on the protein: this may relate to solubility in aqueous media. Correct nitrogen conversion factors, for subsequent use in Kjeldahl analysis, and UV extinction coefficients, may also be derived from the amino acid profile. For sources in which nonprotein nitrogen is difficult to remove from the protein, the amino acid content provides the most accurate method for calculating protein content. Removal of about 90% of the potential peptide bond water, and use of 85% of the ammonia generated during hydrolysis, will give protein values that are more accurate than any of the standard colorimetric procedures. Table 10 gives results of calculations from the amino acid analysis of a protein solution.

Amino acid analysis is also important in assessments of the nutritional value of the protein. Plant proteins are often deficient in lysine or methionine, both essential amino acids. Loss of nutritional value through degradation of the protein during processing will be detected: Failure to extract a particular protein from the source material may result in the nutritional value of the isolate being lower than the starting source material.

1.5. Nutritional Evaluation

Nutritional evaluation is related to the amino acid profile for pure protein isolates: Particularly for those isolates prepared from plant sources, the presence of antinutritional factors (proteinaceous and others) modulates this relationship.

The digestive process usually produces hydrolysates down to the levels of single amino acids that are subsequently transported and resynthesized to the required protein components. The criteria for nutritional quality relate to amino acid availability, which is either calculated from the chemical composition or assayed biologically in terms of growth or nitrogen retention.

Certain amino acids are considered essential—they are not synthesized by the animal from nonprotein nitrogen sources. For man, the essential amino acids are isoleucine, leucine, methionine (modulated by cystine content), phenylalanine (modulated by tyrosine content), threonine, tryptophan, valine, and lysine. A variety of methods has been used for assessing protein nutritional value (47).

Table 10
Example of Values Derived From Amino Acid Data Collected by
Ion Exchange Separation After 6N HCl Hydrolysis,
at 205°C for 24 h, of Soy 11S Storage Protein

Amino acid	Molecular weight	Sample, µmol/mL	Sample, µg/mL	Protein in sample, %
Lys	146.19	4.9312	72.089E 01	5.194
His	155.16	2.2899	35.530E 01	2.578
Ammonia	17.04	3.1210E 01	45.206E 01	0.036
OH-Arg	190.20	0.0000	0.000	0.000
Arg	174.20	6.7301	11.723E 02	8.617
Asp	132.61	1.1412E 01	15.133E 02	10.766
Thr	119.12	3.5285	42.031E 01	2.943
Ser	105.09	5.4800	57.589E 01	3.947
Glu	146.64	1.8121E 01	26.572E 02	19.156
Pro	115.13	7.8541	90.424E 01	6.296
Gly	75.07	7.5323	56.544E 01	3.593
Ala	89.09	5.5550	49.489E 01	3.281
Cys/2	121.16	1.1100	13.448E 01	0.944
Val	117.15	5.3210	62.335E 01	4.353
Met	149.21	0.8424	12.569E 01	0.908
Ile	131.17	5.1667	67.771E 01	4.813
Leu	131.17	7.7413	10.154E 02	7.212
Tyr	181.19	2.5142	45.554E 01	3.361
Phe	165.19	4.3942	72.587E 01	5.305
Trp	204.22	2.0100	41.048E 01	3.062
Am Adipic	161.16	0.0000	0.000	0.000
Allo-Ile	131.17	0.0000	0.000	0.000
OH-Pro	131.13	0.0000	0.000	0.000
Glu and Asp	(calculated)	3.0044		3.062
Percent ammonia used in calculations				85.000
Percent water removed for calculations				90.000
Total number µmol of residues in sample				10.253E 01
Total weight of protein in sample (µg)				12.337E 03
Percent protein in sample				1.233
Average amino acid residue molecular weight				13.213E 01
Hydrophobicity calculated as per Bigelow and Channon, avg. hydrophobicity (cal)				97.060E 01
Charge frequency				0.165
Absorbance at 280 nm				6.810
Wt. extinction coefficient (1%)				5.526

Protein Isolates

1. Chemical assessment (the "chemical score") is a comparison of the amino acid composition with a standard (ovalbumin or casein). The amino acid showing the lowest ratio, relative to the control, is referred to as the "limiting amino acid" and its ratio is the "chemical score" of the test protein. Chemical assessment takes no account of antinutritional factors nor the efficiency of in vivo hydrolysis, which may be different from that of the chemical hydrolysis that precedes amino acid analysis (Table 11).

Table 11
Essential Amino Acid Patterns for Soy and Hen's Egg Proteins

Amino acid	Flour	Concentrate	Isolate	Egg proteins
Isoleucine	119	115	121	129
Leucine	181	188	194	173
Lysine	161	151	152	125
Total "Aromatic AA"	209	220	227	195
Phenylalanine	117	125	134	114
Tyrosine	91	95	93	81
Total sulfur AA	74	73	60	107
Cystine	37	40	34	46
Methionine	37	33	27	61
Threonine	101	100	93	99
Tryptophan	30	36	34	31
Valine	126	118	120	141
Protein score	68	68	56	100

2. Biological methods are nutritionally valid but time consuming and rarely practiced on human subjects. Rats, rabbits, and chickens are usual "participants"!

    Protein efficiency ratio (PER) is a growth assay in which the test protein is incorporated as the sole nitrogen source into the otherwise protein-deficient diet. The PER is then "weight gain/protein intake": it is usually compared with milk casein or egg as a standard.

    Nitrogen retention is the basis of the Biological Value (BV) and Net Protein Utilization (NPU) methods.

    $$BV = \text{adsorbed nitrogen/retained nitrogen}$$
    $$= [I - (F - F_K) - (U - U_K)]/[I - (F - F_K)]$$

where I = nitrogen intake, F = fecal nitrogen, F_K = endogenous fecal nitrogen, U = urinary nitrogen, U_K = endogenous urinary nitrogen, NPU = retain nitrogen/ingested nitrogen, and digestibility = NPU/BV.

3. There is considerable interest in establishing in vitro methods for digestibility, employing enzyme combinations that simulate in vivo conditions. The results are frequently quoted, but no standards are universally accepted.
4. Table 12 demonstrates discrepancies between calculated and actual nutritional values. Inhibitors of the digestive enzymes are a major cause. For example, soy bean trypsin inhibitors are well known: Heat denaturation is used to destroy these, or they may be removed through differential solubility in comparison to the desired proteins.

Since the digestive tract is a continuous flow enzyme reactor with permeable walls, the kinetic passage of nutri-

Table 12
Nutritional Values of Protein Sources

	BV	NPU	PER
Meat	82	78	3.2
Egg	97	96	3.8
Milk	77	71	2.1
Fish	79	77	—
Wheat	49	48	1.8
Corn	50	45	1.2
Oats	66	—	2.2
Rice	67	63	1.7 (cooked)
Field beans	56	48	1.2
Peas	64	47	1.6
Lentils	45	30	0.9
Potatoes	68	55	—
Soya bean	73	61	0.5 (raw)
			2.3 (cooked)
Groundnut	55	43	1.7
Cottonseed	67	58	2.3
Sesame	62	53	1.8
Rapeseed	—	69	2.6
Copra	69	55	2.1
Algae (*Spirulina maxima*)	73	61	—
Baker's yeast (*Saccharomyces cerevisiae*)	67	56	2.2
n-Alkane yeast (*Candida utilus*)	73	62	1.0
Microfungi (*Fusarium*)	—	60	2.7
Bacteria (*B. pseudomonas sp*)	79	76	—

ents influences digestibility. In vivo proteolysis shows that substrate conformation influences hydrolysis rate, e.g., casein/extended structure—rapid hydrolysis, soy globulin/compact structure—slow hydrolysis. Therefore, chemical score is modified by true bioavailability of amino acids. Heat denaturation of native soy globulins also accelerates their hydrolysis and increases bioavailability.

For ruminant feeding, protection of protein against bacterial hydrolysis at neutral pH in the reticulum/omasum may be used. The "protected" protein is crosslinked with formaldehyde at pH 7 and these crosslinks are labile to the stomach (abomasum) acid (pH 2). Without crosslinking, ruminants convert plant protein to bacterial and protozoal protein—the latter protein is then digested in the abomasum. Protein nutritional concerns for ruminants, however, are obviated provided a nitrogen source is available—urea, for example.

5. Effects of processing on isolate nutritional value, therefore, are important both from a negative (toxicology) and positive (nutrition) aspect. Figures 16 and 17 demonstrate some practical examples for soybean meal.

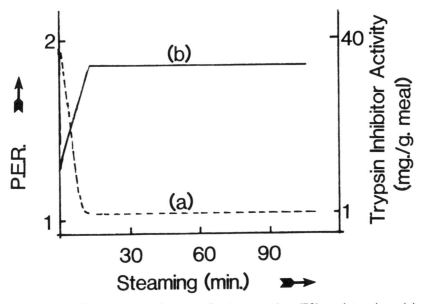

Fig. 16. Effect of time of atmospheric steaming (5% moisture) on (a) trypsin inhibitor activity, and (b) protein efficiency ratio (PER) of raw soybean meal. For reference, casein has a PER of 2.15 (from ref. 88, with permission).

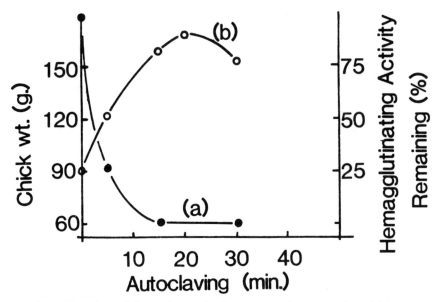

Fig. 17. Effect of time of autoclaving raw soybean meal on (a) hemagglutinin content, and (b) growth response for chicks (weight at 16 d) (from ref. 89, with permission).

Lysinoalanine formation, as a result of alkali/high temperature combinations, has received considerable attention. Plant proteins, particularly the legumes, are already deficient in lysine, and destruction during processing detracts from the nutritional value. In addition, LAL has been associated with nephrotoxic reactions in rats ("renal cytomegalia"): hence, the spectre of an "added" negative factor.

Racemization of amino acids is another problem associated with heat/alkali treatment. Four protein sources have been reported to give the following order of lability to racemization (48):

Promine D > casein > wheat gluten > lactalbumin

This study also noted almost completely racemized methionine in fish protein concentrate after 95°C, 0.2M NaOH treatment, for 20 min. Phenylalanine was reported to be significantly racemized, whereas leucine and valine were racemized only in trace amounts.

Antinutritional factors that are indigenous to the source material should be excluded from the isolate. (1) Phytic acid is believed to chelate with metal ions required by the digestive system, resulting in decreased digestibility of food. Enzyme degradation, ultrafiltra-

Protein Isolates

tion, and solubilization are methods for its removal. Milling is also used to remove phytate rich portions of the seed. Both phytates and phenolic pigments (ferulic acid, chlorogenic acid, and so on) are believed to lower digestibility of proteins by forming "insoluble" complexes with them. (2) Protein enzyme inhibitors have already been mentioned. (3) Lectins (carbohydrate binding proteins) occur in plants (phytolectins), animals (zoolectins), and fungi (mycolectins), and from at least some sources these seem to impair intestinal absorption, giving large losses of nitrogen in the feces. (4) Gossypol is a pigment in cottonseed that reacts with lysine: Its levels appear to be inversely related to protein content (49). (5) Cyclopropenoid fatty acids such as sterculic and malvalic acids (2-octyl-1-cyclopropene-1-octanoic and 2-octyl-1-cyclopropene-1-heptanoic, respectively) are found in cottonseed and other plants of the order *Malvales*. Pink coloration in chicken egg yolk, mutation and reduced fertility in rats, high saturated fatty acid levels, and carcinogenicity in trout have all been blamed on these materials (50).

2. Proteins as Food Ingredients

2.1. Protein Gelation/Aggregation

Protein gelation provides mechanical integrity to many food matrices by forming a molecular network that confers solid-like properties. This gelation impacts on kinesthetic, structural, textural, and rheological properties. At high concentrations (20%), most protein dispersions exhibit this phenomenon (raw meat is an anisotropic gel): either lower concentrations, or protein molecules that are of small molecular dimensions with little tendency for intermolecular interaction, will decrease this solid-like property.

Wheat gluten represents a special case of protein gelation: at low levels of water addition, hydrated, but "insoluble" glutenins, aggregate to form an extensible network. Correlation has been established between genetic variants of glutenins and gluten properties, but the molecular structure of the network is essentially unknown. Gelatin, a very common "functional" agent, has a highly specialized gelling mechanism. The gel structure is maintained by hydrogen bonded triple helices, which "melt-out," conferring thermoreversible gel-sol behavior. Animal muscle proteins (actin and myosin) have gelling properties that are exploited in emulsions such as sausages. Globular proteins produce gelled systems of variable textural/rheological properties, depending on pH, ionic strength, and temperature/time of gelation. Elevated temperatures are required, the particular temperature depending on the protein used. Figure 18 gives DSC thermograms for several protein isolates.

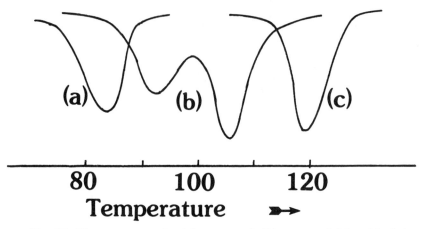

Fig. 18. Thermograms for (a) rapeseed, (b) pea, and (c) oat isolates. The two peaks for pea are vicilin and legumin, respectively.

Normally heat is used to "set" a proteinaceous formulation: Gelatin, however, "cold sets." A variety of techniques detects such protein thermal transitions: Difference spectroscopy (13–15), ORD-CD (51,41), viscometry (52), fluorescence (14,26), pH difference (53), enzyme kinetics (54), dilatometry (55,56), and so on may be applicable, usually only for ideal cases in which intermolecular interactions are prevented by dilution.

Differential scanning calorimetry is a valuable technique for following protein thermal transitions, provided there is a measurable amount of heat (absorbed or evolved) associated with the transition. A good practical technique for examining gelation/aggregation phenomena combines calorimetry with some method of directly assessing the aggregation. A simple procedure is to increase the temperature of a protein dispersion and to take aliquots at several temperatures up to the denaturation temperature (up to 100°C for aqueous systems): Cooling to room temperature followed by centrifugation to remove aggregates, and measurement of the remaining dispersed protein by OD_{280} measurement will monitor aggregation. An even more direct method is to follow turbidity development as a function of temperature. If an isotropic gel is desired, then aggregation before the highly cooperative unfolding, that accompanies thermal denaturation, should be avoided: "Islands" of aggregates at temperatures below the T_d would thus be precluded (section 1.2).

Thermal gel analysis (57) involves electrophoretic examination of dispersed nonaggregated protein as a function of temperature: disulfide bond formation is monitored through the use of dithiothreitol. Again, DSC data are used along with electrophoresis information in order to follow irreversible protein aggregation.

Response to environment is an important consideration for the use of isolates (section 1.3.3): pH and ionic strength are the most important. Other additives, particularly those with "surfactant" properties (often added as processing aids or textural modifiers), need to receive individual study since generalization is not presently possible. Figure 19 gives a typical response of thermal denaturation to ionic strength variation, Fig. 20 for pH variation.

Measurements on gelled systems concentrate on texture and stability. In the absence of better-defined rheological terms, "hardness," "brittleness," 'springiness," and so on are used to describe elastic/viscous properties. Figure 21 demonstrates such response to pH and ionic strength and indicates the value of measurements such as those in the two previous figures. A further interesting observation is that surfactants provide substantial potential for manipulation of thermal denaturation properties: Complex behavior results from the opposing effects of temperature on those interactions under enthalpic control (polar) and those under entropic control (nonpolar). Figure 22 gives the effects of two nonionic detergents, respectively, on the thermal denaturation temperature of rapeseed 12S storage protein.

How, then, does a solution of globular polymers form a continuous network? Multiple bond formation is involved and the balance is highly specific to the protein type—e.g., ribonuclease gels only after prolonged heating above its T_d, with disulfide bond formation. Ovalbumin, BSA, and soya 15S gels are dominated by weak forces that have marked pH/ionic strength dependence—the kinetics of strand formation and final gel properties are therefore affected. Table 13 gives possible contributions of various interactions.

General theories for network formation are now emerging (58). Although it seems that a combination of polymeric network formation and colloidal aggregation (59) could be responsible, for proteins it seems that the former will dominate. Since gelation represents a type of demixing in which polymer–polymer (i.e., protein–protein) interactions are favored over polymer–solvent interactions, then application of modifications of the Flory-Huggins treatment (60) is valid. Such study of proteins is not presently apparent in the published food protein research literature.

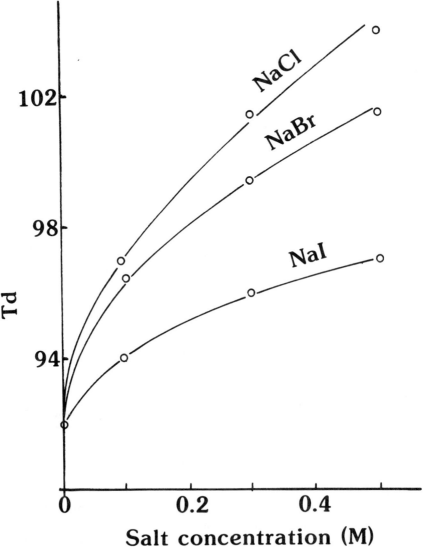

Fig. 19. Response of temperature of denaturation (°C) to NaX (X = Cl, Br, I) concentration (fababean isolate). There was no significant change in the heat associated with the denaturation.

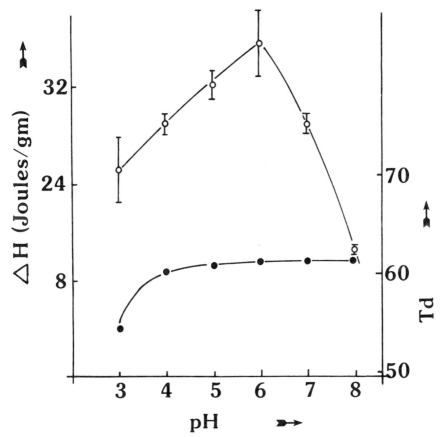

Fig. 20. Response of enthalpy (open circles) and temperature of denaturation (°C, solid circles) to pH for ribonuclease.

2.2. Surface Activity

Sols and emulsions are heterogeneous systems characterized by an *interfacial boundary* that separates two or more *phases*. They are not thermodynamically stable, with the interfacial region being "a large 'sink' of surface free energy, mainly electrostatic in origin" (59). Most studies of disperse systems consist of trying to find ways to either stabilize the interfacial region (i.e., remove the thermodynamic driving force toward decreased interfacial area) or slow down the rate at which the interfacial area decreases (i.e., slow the kinetics of destabilization).

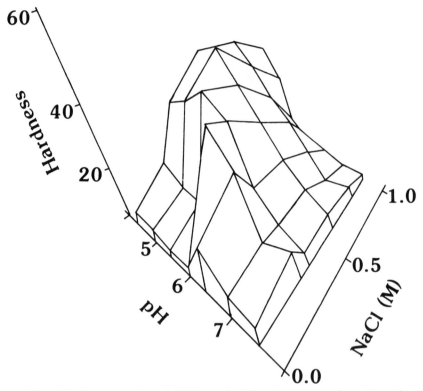

Fig. 21. Soy protein gel (20%, wt/wt) hardness as a function of pH and ionic strength. Measurements were with a Texturometer (17).

The distinction between sols and emulsions is in the size of the disperse phase particle. For sols, the disperse phase particle is from 5 nm (50 A) to about 0.1 µm in diameter. For an emulsion, it is >0.1 µm in diameter. Emulsion and sol properties are described thermodynamically only approximately in terms of the "interfacial" free energy. In comparison, solutions are homogeneous, i.e., one phase: They are considered to be chemically and physically uniform throughout, although the boundaries dividing solute from solvent may visibly scatter light. Solution properties are defined thermodynamically by the "chemical potential," in which the chemical potential for the ith component is expressed in terms of temperature, pressure, and the mole fraction of molecules of the ith kind.

Emulsions of relatively low disperse phase volume have properties similar to lyophobic sols (61): "Addition of electrolytes affects the zeta-potential and the electrophoretic velocity in the same man-

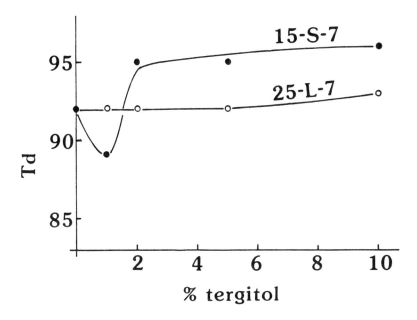
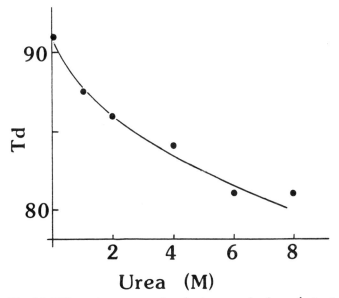

Fig. 22. Effect of concentration for two nonionic surfactants (Tergitols) and for urea, on the thermal denaturation (T_d, °C) temperature of rapeseed 12S storage protein.

Table 13
Proposed Crosslink Bonding of Protein Gel Structures
and Their Properties

Type	Energy kcal/mol	Interaction distance, Å	Groups involved	Role in gel matrix
Covalent bonding	80–90	1–2	$-SS-$	Bridging/ordering
Hydrogen bonding	2–10	2–3	$-NH...O=C-$	Bridging/stability
Hydrophobic (nonpolar) interactions	1–3	3–5	Nonspecific	Strand thickening, strengthens, stability
Ionic	10–20	2–3	$-NH_3^+$, $-COO-$, and so on	Solvent interactions, salt links

ner as for a lyophobic sol with negatively charged particles...." Emulsions of higher disperse phase volume have properties similar to the lyophilic sols, "e.g., high viscosity, relatively high concentrations, and stability to electrolytes."

Information relevant to the study of disperse systems relates either to their formation, stability, or end use.

Surface activity of proteins refers to emulsification and aeration properties via an induced "asymmetry of hydrophobicity" (62): i.e., whereas in the native protein structure most proteins have hydrophobic and hydrophilic domains, the presence of an oil–water or air–water interface can induce conformational change that maximizes exposure of hydrophobic domains to the oil (or air) and of hydrophilic domains to the aqueous medium—a degree of asymmetry results. In foams, the ability to hold water in a "gel" surrounding the air "particle" is important. Stability, for both emulsions and foams, partly results from the rheological properties of the interfacial film (section 2.2.3). These are difficult to investigate experimentally, particularly in compositions relevant to food (i.e., appropriate phase volume ratio, protein concentration, aqueous phase rheology).

2.2.1. Sources of Emulsion Stability

Although emulsions are thermodynamically unstable, there are thermodynamic quantities that may be optimized in order to lower the driving force toward instabilization.

Protein Isolates

1. Interfacial tension should be lowered: extremely low tensions, however, give a weak interface that destabilizes. Lipid or soap-like moieties provide such surfactancy. These small molecule surfactants (SMS) are divided into nonionic and ionic species. The latter are subdivided according to whether they are anionic, cationic, or zwitterionic. Proteins and carbohydrates also aid in lowering interfacial tension, although they are not as efficient as the SMS. Proteins are of the zwitterionic type, although carbohydrates may be either nonionic or anionic.
2. Electrostatic charge is another stabilizing force. This is expressed in terms of the zeta (ζ) potential, and it is usually referred to as the double layer. This double layer (Fig. 23) has a tightly held layer (the Stern layer) adjacent to the disperse phase droplet, and an outer, loosely held layer (the Guoy-Chapman layer). High electrostatic charge provides repulsion between particles and discourages coalescence. Depending on the pH relative to the isoelectric point, a protein may aid in providing significant electrostatic charge to the emulsion disperse phase. This may not, however, be directly reflected in increased emulsion stability if the protein layer has "tails" that extend beyond the effect of this double layer [depending on the relative strength of tail–solvent and tail–tail interactions, "bridging flocculation" may result (59)]. If the value of ζ results in increased emulsion viscosity (5), then, indirectly, an increase in *kinetic* stability results.

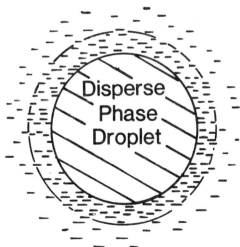

Fig. 23. Schematic representation of electrical double layer surrounding an emulsion disperse phase droplet.

3. Steric stabilization has conventionally been qualitatively addressed in terms of a physical barrier that forms about the disperse phase droplet, thus preventing coalescence of droplets by means of the mechanical, viscoelastic properties of the resulting "skin." Figure 24 shows the commonly accepted domains of a macromolecule at an oil–water interface. Only the "train" sections are in intimate contact with the oil. Studies have concentrated on determining the amount of adsorption at the interface, and, for macromolecules such as proteins, the likelihood of conformational changes in the three-dimensional structure of the adsorbate (63–69). Various reviews (e.g., 70) adequately cover the nature of the current status in this area.

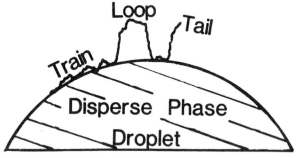

Fig. 24. Schematic representation of an adsorbed polymer at an oil–water interface.

Thermodynamically, steric stabilization is one of the most complex forces that determine emulsion stability. Only recently has an extensive treatment been attempted (6,71,72). There are two considerations here. The first has to do with the adsorption of the "surfactant" at the interface, this being a function of the Flory-Huggins X parameter. Adsorption therefore depends on the relative tendency for polymer–interface and polymer–solvent interactions. If a range of sizes of "surfactants" is available, the smaller moieties will diffuse more rapidly and adsorb more quickly: with time, however, these will be displaced by the larger species, which have greater affinity for the interface. The second consideration is the relative tendency of the loops and tails of the adsorbed layer to be involved in adsorbate–solvent interactions or adsorbate–adsorbate interactions. The latter will encourage flocculation, whereas the former will be detrimental to the initial adsorption. Obviously, a balance must be achieved.

Protein Isolates

Empirical approaches have been useful, although the proposed mechanisms are not based on thermodynamics. Emulsifiers were assigned numbers that approximately correlate with their relative hydrophilic-lipophilic balance (HLB) values. Another expression of the HLB was the temperature at which the emulsifier "demixes" from solution: This temperature became known as the cloud point. Similarly, emulsions were defined by the temperature at which phase inversion occurs, i.e., the phase inversion temperature (PIT). More recent investigations consider adsorption in terms of osmotic pressure and entropic effects (71,73), i.e., thermodynamics.

Steric stabilization has been noted as having certain advantages over charge stabilization (6). These are low sensitivity to electrolytes, low viscosity at high levels of dispersed solids, and equal effectiveness in both aqueous and nonaqueous media. Whereas electrostatic (repulsive) and van der Waals (attractive) forces have been used extensively in predicting stability of disperse systems, it is noted that this treatment is inadequate (71) when particles are close enough for overlap of steric layers to occur, i.e., when adsorbate-adsorbate interactions are possible. Thermodynamically stable emulsions are predicted as being possible for sterically stabilized disperse systems (6). Hydrophobic interactions are also overlooked: These have been clearly shown to have a different dependence on distance than do either electrostatic or Van der Waals forces (74): A comprehensive theory must also consider the effects of the latter.

Consideration of the above readily shows that protein aids in all mechanisms of stability. Studies of emulsions need to consider the processes involved in breakdown—flocculation, coalescence, and creaming.

2.2.2. Mechanisms of Emulsion Instability

The three principal mechanisms of emulsion instability are creaming, flocculation, and coalescence. Qualitatively, they are easy to rationalize in terms of differential density between the disperse and continuous phases, gravity, viscosity, temperature, concentration (number of droplets), and size.

1. The driving force for creaming is the buoyancy of the disperse phase droplet (i.e., differential density). Small particle sizes, low gravity, and high viscosity all help retard rates of creaming.
2. High temperature and high numbers of droplets are the driving forces for flocculation. High viscosity, again, is a

retarding influence. Brownian motion and the ζ potential are important for particles <1 μm.
3. Coalescence requires the breakdown in integrity of the "steric skin." In the extreme, phase separation of oil and water results. It is obvious that the rheological properties of the interfacial skin are important here. Coalescence has been less well described quantitatively than have creaming and flocculation.
4. A fourth, but less well recognized, form of emulsion destabilization mechanism is gelation. It would appear that gelation could be an asset, since creaming, flocculation, and hence coalescence rates, will all be retarded. It is for those emulsions in which fluid properties are required that gelation would be a liability. In studying this phenomenon, those considerations necessary for protein thermal gelation (section 2.1) will again be of value. How good a solvent the continuous phase is for the tails and loops of the adsorbate layer is very important for "polymer network" formation: The value of the ζ potential and the particle size/distribution are very important for colloidal aggregation.

2.2.3. The Various Roles of Viscosity

In the formation of the emulsion, viscosity of both the oil and aqueous phases is important. Generally, lower viscosity enables more efficient utilization of the energy that is put into the system to form as large an interfacial area as possible (i.e., small droplets).

Oil viscosity is normally in the region of 40 to 60 mPa×s at 25°C, and Newtonian behavior is observed. Only an increase in temperature will help to lower the oil viscosity: An Arrhenius type relationship should be expected (61).

The viscosity of the continuous phase is much more varied. It may be that, for other reasons, a level of ingredients is necessary that results in an undesirably high viscosity of the aqueous phase. The emulsion should then be formed first, in the absence of those additives that substantially increase the viscosity, adding them later. Alternatively, oil-insoluble, but oil-dispersible, ingredients may be added via the oil, allowing diffusion, after emulsion formation, into the continuous phase. Both proteins and carbohydrates may conveniently be added in this manner. In fact some of the carbohydrates that are more difficult to "wet" by water, are "wetted" very readily by oil (the surface of the carbohydrate is hydrophobic): Migration into the aqueous continuous phase occurs later.

At any time after the start, the homogenization process is the break-up of larger droplets to form smaller droplets. The viscoelastic properties of the interfacial region are now of great significance. Particularly for macromolecular surfactants, such as carbohydrates and proteins, a skin forms at the interface and this is normally more rigid than the interface is in their absence. Break-up to smaller droplets now requires greater energy input. It seems that most homogenization processes have sufficient energy to overcome such interfacial rigidity: However, with small laboratory instruments such as a Polytron, significantly larger droplet sizes result in the presence of macromolecules.

After the initial homogenization, another aspect of viscosity becomes significant. For emulsions of disperse phase volumes above approximately 35%, the viscosity of the emulsion is often much higher than that of the pure continuous phase. This can also result in increased energy input being necessary for sufficiently small droplet sizes to be attained. The increase in viscosity results from particle–particle interactions: These can be of an electrostatic nature, i.e., the electroviscous effect, or of a steric nature.

Mechanical stabilization of emulsions refers to the slowing of the destabilization processes by increased viscosity of the emulsion and increased viscoelasticity of the interfacial "skin." The viscosity relevant to both creaming and flocculation rates is the *low shear* viscosity. It must be remembered that most emulsions are pseudoplastic (viscoelastic). Thus, shear thinning produces a lower viscosity, usually essential for pourability, whipability, mastication, and swallowing. At very low shear rates, for example $10^{-5}/s$ (75), however, high viscosity aids stability.

2.2.4. Important Measurements

Certain measurements are useful in studies of the role of proteins in emulsification.

1. A practical evaluation of the kinetic stability—so-called creaming measurements—is useful. For those emulsions of practical interest, the stability against creaming will normally be such that an accelerated aging method must be used. Increased gravitational force in a centrifuge is the most realistic method. A force of $2500g$ for 20–30 min is conveniently sensitive for many food emulsions.
2. Particle-size analysis is the key to any emulsion characterization. This measurement is important for the formation, stability, and use. For formation, size analysis provides an estimate of the interfacial area generated, and this indicates the efficiency of the homogenization process. For stability,

the smaller the particle size, the more stable will be the emulsion: moreover, the narrower the distribution of particle sizes, the greater the stability. A few large particles provide a strong destabilizing force. In the practical use of emulsions, rheology is one of the most important characteristics. This depends very substantially on particle size, for the same reasons as noted with respect to emulsion formation. There are various methods for measuring particle sizes. Direct microscopic observation with appropriate image analysis (76) is the best method since it is not subject to major artifacts. A variety of other methods is available, however, and may be used with discretion. For *relative* estimates of particle sizes, turbidity measurements, via apparent absorbance in the visible spectrum, may have some usefulness (77). Light scattering techniques have also approached a level that, with appropriate calibration, provides greater ease in obtaining this important information.
3. Electrical potential is an important attribute. Even though macromolecular emulsifiers may extend beyond the direct stabilizing effect of the double layer, the magnitude of the charge gives an indication of the extent of coverage of the oil by the protein. Rheological properties are also dependent on the charge via the electroviscous effect: pseudoplastic behavior results partly from proper utilization of the electroviscous effects.
4. Interfacial rheology has been more difficult to assess experimentally than has bulk viscosity (78). Too "elastic" an interfacial skin will prevent droplet breakup during formation, but some elasticity should aid in the maintenance of an emulsion, particularly with concentrated disperse phases. Gerson (79) notes, however, that a fairly simple apparatus may be assembled that will give useful information. Figure 25 demonstrates the change of the extensibility of an oil–water interface in the presence of protein.
5. Viscosity is extremely important. Both creaming and flocculation processes critically depend on viscosity at low shear rates (certainly, <1/s). Usually, higher viscosity is associated with a more stable emulsion for concentrated disperse phase volumes: however, non-Newtonian behavior is a characteristic of such systems and Figs. 26 and 27 demonstrate the need for careful measurement here—shear dependence *must* be monitored! Flow properties, important in the use of an emulsion, depend on viscosity at higher shear rates (likely >30/s). In foods, texture is particularly dependent on rheological properties.

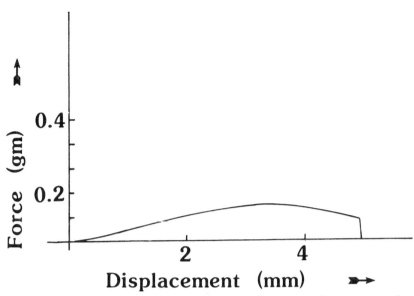

Fig. 25. Force versus displacement cure for corn oil–water interface with 1% (wt/vol) soy isolate protein dissolved in the aqueous phase. Measurement was with a Du Nouy ring suspended from a Mettler balance: Displacement and force were monitored with a computer.

6. Low interfacial tension is important for formation and sometimes for stability. For food systems, viscosity is usually the more important stabilizing factor. Proteins and carbohydrates lower interfacial tension values from 14 to 20 mJ/m^2, for the oil–water interface, to about from 4 to 8 mJ/m^2 in the presence of the macromolecule. Values below 1 mJ/m^2 are achieved with the addition of small molecule surfactants: Whereas whipping properties may be enhanced, the emulsion stability tends to be lower because of a weaker, i.e., nonrigid, interface.
7. Measurement of coalescence is usually achieved with particle size analysis as a function of time. In order to eliminate the effects of creaming, weighting techniques are often used in order to match the densities of the disperse and continuous phases. A useful procedure is to use 3,3'-dimethylbiphenyl as the disperse phase. This oil has a density of 0.999. A parallel study with paraffin oil will then indicate the relative importance of creaming. For both oils, natural surfactants are absent, so that the study of the role of individual ingredients is simplified. This feature is particularly important in the presence of proteins. Figure 22 indicated

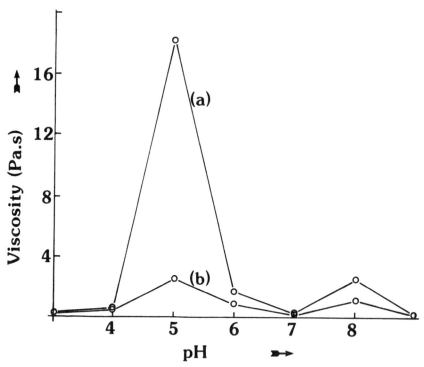

Fig. 26. Viscosity as a function of pH for a 70% corn oil-in-water emulsion stabilized with 1% lactalbumin. Shear rates are (a) 0.2/s and (b) 1.0/s.

substantially different effects on protein thermal gelation, by two subtly different nonionic small-molecule surfactants. Modulation of the protein surface activity must also, therefore, be expected by the natural surfactants in food grade oils.

Meat emulsions represent a challenge for any food scientist. These are used for frankfurters, pate, and so on, and rely not only on emulsification, but also on gelation. The aqueous phase contains a multiplicity of dispersed proteins and exhibits yield stress properties (this helps prevent oil migration in the undisturbed mixture): The nonaqueous phase contains oil and fat crystals. Suspended particles help increase oil droplet breakup during formation, and this aids stability—solid particles at the interface may also directly help increase shelf life. Reasonably high levels of protein are used in these systems and greater than 10%, by weight, ensures saturation of the oil–water interface. For meat emulsions, heat stability tests measure whether proper emulsification occurred, rather than droplet stabilizing effects of the protein.

2.3. Texturization and Rheology

Texturization of protein is the use of processing to impart certain kinesthetic and rheological properties that may be subjectively evaluated. A variety of methods and concepts is used. Traditionally, cooker/extrusion and spinning methods have dominated. The former is the extrusion under pressure of a liquid proteinaceous matrix through an orifice (dye) at a temperature sufficiently high to gel the matrix. Most raw materials containing >10% protein can be restructured by this technique: critical levels of strong-gelling protein are not normally required. "Spinning" of protein dispersions involves extrusion through a spinneret of an alkaline "dope" into an acid bath causing precipitation/coagulation and forming into a fiber: Towing is necessary for appropriate fiber formation. Both the Tombs mesophase isolate (*80*) and the PMM isolate (*81*) are also amenable to spinning: However, the "dope" is not alkaline and

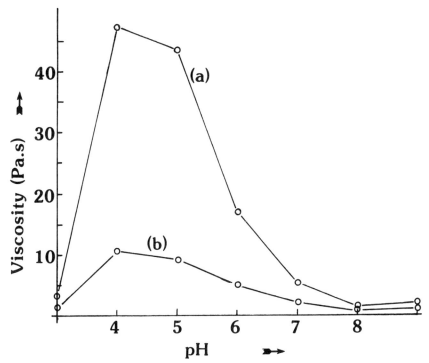

Fig. 27. Viscosity as a function of pH for a 70% corn oil-in-water emulsion stabilized with 1% lactoglobulin. Shear rates are (a) 0.2/s and (b) 1.0/s.

aggregation is effected by salt reduction/elevated temperature combinations in the former case, and by strictly thermal gelation in the latter. Spinning via the "thermal" method makes greater demand on the protein isolate, and prior denaturation is normally detrimental. Melt spinning (of casein) is another technique. In each case, the fibers formed are amorphous gels—crystallinity does not develop without extensive stretching.

Fiber size and variation of the "setting" process, particularly for the thermal method, give the ability to control the texture of foods in which the fibers may be incorporated. Stability of the fibers in the food matrix is necessary to maintain this texture: Contact angle measurements might be a suitable method of study, since destruction of the fiber by the surrounding matrix must depend on the "interfacial tension" between the two phases. Also of importance to the texture of foods is the swelling and water-holding capacity of the protein components. Tests to measure this are very empirical.

Emulsification is another aspect of texturization. Rheology, normally involving viscous and elastic components, is particularly sensitive to changes in ζ-potential, particle size distribution, thickness of adsorbed layers about a disperse phase droplet, and continuous phase properties, all of which may be modulated by the choice of protein, and the response of the particular protein to its pH/ionic environment.

Electron microscopy is applied to textural evaluations. Scanning methods predominate: however, transmission EM will become more widely practiced (82). Proper sample preparation is critical in order to minimize artifacts.

2.3.1. Viscosity of Dilute Solutions

The rheology of proteinaceous systems depends on concentration and whether a solution or a disperse system is being studied. For dilute polymer solutions, apparent viscosity is a function of size, shape, and rigidity of the protein molecule (chapter 3). Characterization of very pure proteins—those for "biochemical use"—will include this measurement. For a distribution of molecular weights, a viscosity average results, and characterization of an individual molecular entity is precluded.

Proteins are polyelectrolytes so that, in solution, apparent viscosities are influenced by pH and ionic strength. This is complicated by changes in conformation. The net effect is usually seen as in Fig. 28.

Protein Isolates

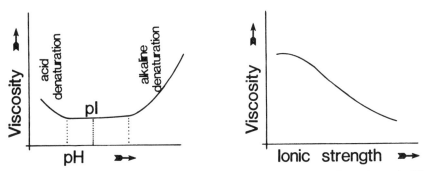

Fig. 28. Typical viscosity, of a protein solution, as a function of pH and ionic strength.

Temperature dependence (Fig. 29) is also much more complex than for simple polyelectrolytes, since specific conformations are stable over narrow temperature ranges.

Even for a simple protein in dilute solution, documented curves do not usually represent equilibrium viscosities since aggregation rates are comparable to measurement rates. Figure 30 shows viscosity as a function of temperature at different pH values.

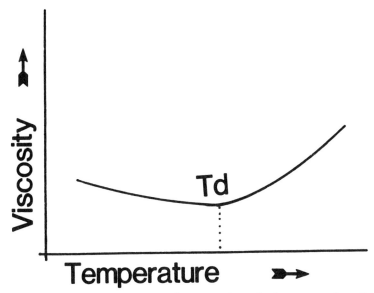

Fig. 29. Typical viscosity, of a protein solution, as a function of temperature.

For soluble isolates (e.g., soya), complex temperature dependence is observed with stages in viscosity increase corresponding to stepwise denaturation of the 7S and 11S globulins.

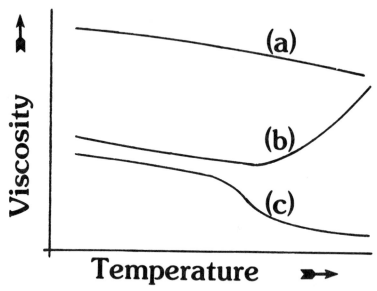

Fig. 30. Typical viscosity of a protein solution as a function of temperature, at different pH values: (a) pH > pH_d, (b) pI < pH < pH_d, (c) pH = pI.

2.3.2. Viscosity of Disperse Systems

Description of the viscosity of disperse systems is usually given by the Einstein equation:

$$\eta_{sp} = (\eta/\eta_0) - 1 = 2.5\phi$$

The viscosity coefficient of the system is η, η_0 is the viscosity coefficient of the solvent, and ϕ is the fraction of disperse phase volume.

Except at very low dilution, interactions between particles take place, and the Einstein relationship is not obeyed. Several models have been developed in order to mathematically describe observed data (4,5). In some cases, rheological behavior can be qualitatively rationalized in terms of a "ball-bearing" action of hard spheres rolling over each other. In other cases, sterically stabilized dispersions have potential for entanglement of the adsorbed layers of adjacent particles, and higher viscosity results at low shear rates. Under sufficiently high shear, the adsorbed layer may be forced to a more

Protein Isolates

compact conformation, resulting in a smaller effective diameter of the particle. Less interaction between particles results in lower viscosity at these higher shear rates—i.e., shear thinning behavior. In a similar fashion, electrostatically stabilized dispersions will resist motion as a result of overlap of adjacent electrical double layers. At sufficiently high shear rates, however, the more loosely held Guoy-Chapman layer may be either sloughed off or distorted to an elongated shape in the direction of flow and of smaller dimension perpendicular to it. The effectively smaller radius again gives a shear thinning behavior.

The measured viscosity is a function of the shear rate (protein dispersions tend to be shear thinning, i.e., viscoelastic, also called pseudoplastic). Non-Newtonian viscosity may also be either time dependent or time independent. Figure 31 demonstrates various viscous behaviors.

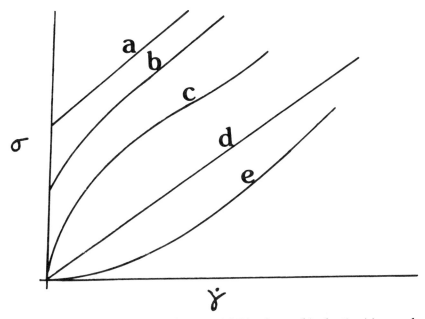

Fig. 31. Types of flow behavior: (a) Bingham, (b) plastic, (c) pseudoplastic, (d) Newtonian, and (e) dilatant. The ordinate axis is shear stress (σ) and the abscissa is shear rate ($\dot{\gamma}$).

Thixotropy occurs when there is a reversible decrease in shear stress and apparent viscosity at both constant shear rate and temperature (thixotropy is essentially pseudoplasticity with a time dependence). Thixotropy tends to be present when the dispersion contains large aggregates that interact via weak secondary bonding forces to form a network when at rest.

Thus, viscosity of Newtonian systems is expressed as $\sigma = \eta\dot{\gamma}$, where σ is shear stress (Pa), η is viscosity (Pa×s), and $\dot{\gamma}$ is shear rate (s^{-1}). Of great industrial usefulness has been a power-law model (83–87) for non-Newtonian behavior, expressing shear stress (σ) as

$$\sigma = m\dot{\gamma}^n$$

where m and n are called "consistency coefficient" and "flow behavior index," respectively. For Newtonian behavior, $n = 1$: $n < 1$ for pseudoplastic behavior, and $n > 1$ for dilatant or shear thickening behavior. For Newtonian behavior, $n = 1$ and m is put in place of η. Using both expressions then gives

$$\eta = m\dot{\gamma}^{n-1}$$

The consistency coefficient value is a measure of the viscosity at a shear rate of 1/s, as an extrapolation from a plot of viscosity against shear rate on logarithmic coordinates. This is

$$\log(\eta) = \log(m) + (n - 1)\log(\dot{\gamma})$$

Although the power law model adequately describes the flow behavior of many food dispersions, other models must be used for those systems that are strongly elastic.

References

1. F. Franks, ed., *Characterisation of Protein Conformation and Function* Symposium Press, London.
2. R. Lumry and R. Biltonen, in *Structure and Stability of Biological Macromolecules* (S. N. Timasheff and G. D. Fasman, ed.) Marcel Dekker, New York (1969).
3. K. Saio and T. Watanabe, *J. Texture Studies* **9**, 135–157 (1978).
4. P. Sherman, *Emulsion Science* Academic, New York (1968).
5. P. Sherman, in *Industrial Rheology* Academic, New York (1970).
6. D. H. Napper, in *Chemistry and Technology of Water-Soluble Polymers* (C. A. Finch, ed.) Plenum, New York (1983).
7. T. F. Busby and K. C. Ingham, *Biochim. Biochim. Acta* **799**, 80–89 (1984).
8. J. J. Scheidegger, International Archives of Allergy and Applied Immunology, **7**, 103–110 (1955).
9. P. Graber and C. A. Williams, *Biochim. Biophys. Acta* **10**, 193–194 (1953).

Protein Isolates

10. W. B. Gratzer and R. Mendelsohn, in *Techniques in Protein and Enzyme Biochemistry* Elsevier/North-Holland Biomedical, Amsterdam (1978).
11. J. W. Donovan, in *Physical Principles and Techniques of Protein Chemistry* Pt. A (S. J. Leach, ed) Academic, New York (1969).
12. D. B. Wetlaufer, *Adv. Prot. Chem.* **17**, 303–390 (1962).
13. V. S. Ananthanarayanan and C. C. Bigelow, *Biochemistry* **8**, 3717–3723 (1969).
13a. V. S. Ananthanarayanan and C. C. Bigelow, *Biochemistry* **8**, 3723–3728 (1969).
14. C. C. Bigelow, *Comp. Rend. Lab. Carlsberg* **31**, 305–324 (1960).
15. C. C. Bigelow and I. I. Geschwind, *Comp. Rend. Lab. Carlsberg* **31**, 283–304 (1960).
16. E. P. Pittz, J. C. Lee, B. Bablouzian, R. Townend, and S. N. Timasheff, *Meth. Enzymol.* **XXVII**, 209–256 (1973).
17. H. H. Friedman, J. E. Whitney, and A. S. Szczesniak, *J. Food Sci.* **28**, 390–396 (1963).
18. J. J. Cameron and C. D. Myers, US Patent #4,366,097 (1982).
19. R. F. Chen, H. Edelhoch, and R. F. Steiner, in *Physical Principles and Techniques of Protein Chemistry* Pt. A (S. J. Leach, ed.) Academic, New York (1969).
20. I. Feldman, D. Young, and R. McGuire, *Biopolymers* **14**, 335–351 (1975).
21. F. W. J. Teale, *Biochem. J.* **76**, 381–388 (1960).
22. F. W. J. Teale and G. Weber, *Biochem. J.* **65**, 476–482 (1957).
23. I. Weinryb and R. F. Steiner, *Biochemistry* **9**, 135–146 (1970).
24. G. Weber, *Biochem. J.* **51**, 145–167 (1952).
24a. G. Weber, *Adv. Prot. Chem.* **8**, 415–459 (1953).
25. J. R. Lakowicz, *J. Biochem. Biophys. Meth.* **2**, 91–119 (1980).
26. G. M. Edelman and W. O. McClure, *Acct's Chem. Res.* **1**, 65–70 (1968).
27. L. Stryer, *Science* **162**, 526–533 (1968).
28. J. L. Wang and G. M. Edelman, *J. Biol. Chem.* **246**, 1185–1191 (1971).
29. A. Kato and S. Nakai, *Biochim. Biophys. Acta* **624**, 13–20 (1980).
30. A. Kato, N. Tsutsui, N. Matsudomi, K. Kobayashi, and S. Nakai, *Agric. Biol. Chem.* **45**, 2755–2760 (1981).
31. E. Li-Chan, S. Nakai, and D. F. Wood, *J. Food Sci.* **49**, 345–350 (1984).
32. A.-A. Townsend and S. Nakai, *J. Food Sci.* **48**, 588–594 (1983).
33. L. P. Voutsinas, E. Cheung, and S. Nakai, *J. Food Sci.* **48**, 26–32 (1983).
34. L. P. Voutsinas, S. Nakai, and V. R. Harwalker, *Can. Inst. Food Sci. Technol. J.* **16**, 185–190 (1983).
35. P. L. Privalov, *Adv. Prot. Chem.* **33**, 167–241 (1979).
35a. P. L. Privalov, *Adv. Prot. Chem.* **35a**, 1–104 (1982).
36. F. Franks, in *Water: A Comprehensive Treatise* vol. 4, *Aqueous Solutions of Amphiphiles and Macromolecules* (F. Franks, ed.) Plenum, New York (1975).
37. R. B. Thompson and J. R. Lakowicz, *Biochemistry* **23**, 3411–3417 (1984).
38. C. J. French and J. M. Gosline, *Biochim. Biophys. Acta* **537**, 386–394 (1978).

39. J. F. Brandts, in *Thermobiology* (A. H. Rose, ed.) Academic, New York (1967).
40. D. G. Oakenfull and D. E. Fenwick, *J. Phys. Chem.* **78**, 1759–1763 (1974).
41. M. J. Grutter, R. B. Hawkes, and B. W. Matthews, *Nature* **277**, 667–669 (1979).
42. C. C. Bigelow, *J. Theoret. Biol.* **16**, 187–211 (1967).
43. J. Janin, *Nature* **227**, 491–492 (1979).
44. R. Wolfenden, L. Anderson, P. M. Cullis, and C. C. B. Southgate, *Biochemistry* **20**, 849–855 (1981).
45. B. Yu. Zaslavsky, N. M. Mestichkina, L. M. Miheeva, and S. V. Rogozhin, *J. Chromatogr.* **240**, 21–28 (1982).
46. E. Eisenberg, R. M. Weiss, and T. C. Terwilliger, *Nature* **299**, 371–374 (1982).
46a. E. Eisenberg, R. M. Weiss, and T. C. Terwilliger, *Proc. Nat. Acad. Sci. USA* **81**, 140–144 (1984).
47. A. B. Morrison, in *Symposium on Foods: Proteins and Their Reactions* (H. W. Shultz and A. F. Anglemier, eds.) AVI, Westport, Connecticut (1964).
48. P. M. Masters and M. Friedman, *J. Agric. Food Chem.* **27**, 507–511 (1979).
49. S. N. Pandey and N. Thejappa, *JAOCS* **52**, 312–315 (1975).
50. D. Selivonchick, Univ. Oregon, in seminar, Toronto (1979).
51. A. J. Adler, N. J. Geenfield, and G. D. Fasman, *Meth. Enzymol.* **XXVII**(D), 675–735 (1973).
52. C. Tanford, *Physical Chemistry of Macromolecules*, John Wiley, New York (1961).
53. H. B. Bull and K. Breese, *Arch. Biochem. Biophys.* **158**, 681–686 (1973).
54. M. Dixon and E. C. Webb, *Enzymes* 2nd ed., Longmans, London (1964).
55. J. C. Lee, K. Gekko, and S. N. Timasheff, *Meth. Enzymol.* **61**(H), 26–49 (1979).
56. J. C. Lee and S. N. Timasheff, *J. Biol. Chem.* **256**, 7193–7201 (1981).
57. K. A. Lysko, R. Carlson, R. Taverna, J. Snow, and J. F. Brandts, *Biochemistry* **20**, 5570–5576 (1981).
58. M. G. Bezrukov, *Angew. Chem. Ind. Ed. Engl.* **18**, 599–610 (1979).
59. E. Dickinson and G. Stainsby, *Colloids in Foods* Applied Science Publishers, New York (1982).
60. W. Burchard, in *Chemistry and Technology of Water-Soluble Polymers* (C. A. Finch, ed.) Plenum, New York (1983).
61. S. Glasstone, in *Textbook of Physical Chemistry* D. Van Nostrand, New York (1946).
62. E. Eisenberg, *Ann. Rev. Biochem.* **53**, 595–623 (1984).
63. D. R. Absolom, A. W. Neumann, and C. J. van Oss, in *Polymer Adsorption and Dispersion Stability* (E. D. Goddard and B. Vincent, eds.) ACS Symposium Series 240, ACS, Washington, DC (1984).

64. D. R. Absolom and Z. Policova, *J. Disp. Sci. Tech.*, in press.
65. D. E. Graham and M. C. Phillips, *J. Colloid Interface Sci.* **70**, 403–414 (1979).
66. D. E. Graham and M. C. Phillips, *J. Colloid Interface Sci.* **70**, 415–426 (1979).
67. D. E. Graham and M. C. Phillips, *J. Colloid Interface Sci.* **70**, 427–439 (1979).
68. D. E. Graham and M. C. Phillips, *J. Colloid Interface Sci.* **76**, 227–239 (1980).
69. D. E. Graham and M. C. Phillips, *J. Colloid Interface Sci.* **76**, 240–249 (1980).
70. F. MacRitchie, *Adv. Prot. Chem.* **32**, 283–326 (1978).
71. B. Vincent, in *Polymer Adsorption and Dispersion Stability* (E. D. Goddard and B. Vincent, eds.) ACS Symposium Series 240, ACS, Washington, DC (1984).
72. B. Vincent and S. G. Whittington, *Surfaces and Colloids* **12**, 1–117 (1982).
73. K. Furusawa, Y. Kimura, and T. Tagawa, in *Polymer Adsorption and Dispersion Stability* (E. D. Goddard and B. Vincent, ed.) ACS Symposium Series 240, ACS, Washington, DC (1984).
74. J. N. Israelachvili and R. M. Pashley, *J. Colloid Interface Sci.* **98**, 500–514 (1984).
75. V. V. Chavan, *J. Dispersion Sci. Technol.* **4**, 47–104 (1983).
76. D. R. Absolom, D. G. Wicks, C. D. Myers, and A. W. Neumann, submitted to *JAOCS*.
77. V. R. Kaufman and N. Garti, *J. Dispersion Sci. Technol.* **2**, 475–490 (1981).
78. M. Joly, in *Surface and Colloid Science* (E. Matijevic, ed.) Wiley-Interscience, New York (1972).
79. D. Gerson, *Immunol. Meth.* **II**, 105–138 (1981).
80. M. P. Tombs, US Patent #3,870,801 (1975).
81. E. D. Murray, C. D. Myers, and L. D. Barker, Canadian Patent #1028552 (1978).
82. A. H. Clark, F. J. Judge, J. B. Richards, J. M. Stubbs, and A. Suggett, *Intl. J. Peptide and Protein Res.* **17**, 380–392 (1981).
83. A. deWaele, *J. Am. Chem. Soc.* **48**, 2760–2776 (1926).
84. P. G. Nutting, *J. Franklin Inst.* **191**, 679 (1921).
85. W. Ostwald, *Kolloid Z.* **36**, 99–117 (1925).
86. W. Ostwald, *Kolloid Z.* **36**, 157–167 (1925).
87. W. Ostwald, *Kolloid Z.* **36**, 248–250 (1925).
88. J. J. Rachis, *Fed. Proc.* **24**, 1488–1493 (1965).
89. I. E. Liener, *Arch. Biochem. Biophys.* **54**, 223–231 (1955).

Protein Index

Actin, 19, 45, 47
Adenylate cyclase, 174, 209, 222, 227
Adrenocorticotrophic hormone (ACTH), 20, 180, 236, 237
Alcohol dehydrogenase, 13, 16
Aldolase, 42, 61
Aminocyclase, 6
Amylase, 5, 6, 445
Amyloglucosidase, 8
Antifreeze glygoprotein, 13
Antithrombin, 496
Apotransferrin, 143
Arachin, 57
Asparaginase, 3
Aspartase, 6
ATPase, 123
Autocyanase, 6
Avidin, 311

Bromelain, 7

Calcitonin, 3
Calmodulin, 92
Carbamyl phosphate synthetase, 123
Carbonic anhydrase, 16, 139
Carboxyhemoglobin, 29
Carboxypeptidase, 22, 110, 139
Casein, 13, 16, 19, 57, 494
Catalase, 5, 6, 108, 110, 169
Cellulase, 5, 6
Ceruloplasmin, 19
Cholesterol esterase, 18

Chymotrypsin, 18, 81, 105, 115, 142
Chymotrypsinogen, 16, 57, 69, 81, 108, 151
Collagen, 13, 16, 20, 34, 45, 57, 69, 104, 141, 148, 151
Conalbumin, 57, 83
Concanavalin, 484
Crystallin, 57, 353
Cytochrom c, 16, 19, 30, 57, 69 ⟩

Elastin, 16, 20, 47

Ferricytochrome c, 114
Ferritin, 13, 19
Ferrocytochrome c, 475
Fetoprotein, 198, 200
Fibrinogen, 19, 47, 50, 69, 110
Fibroin, 16, 20, 47
Ficin, 22
Flagellin, 46
Fumarase, 434

Gelatin, 57, 494, 525
Gliadin, 19
Globulin, 13, 18
Glucagon, 167
Glucoamylase, 5, 6
Glucose isomerase, 6–8
Glucose oxidase, 5–7
Glucose-6-phosphate dehydrogenase, 123, 151
Glutamate dehydrogenase, 18

Glutathione, 13
Glyceraldehyde-3-phosphatase, 42
Glycinin, 435, 473
Glycogen phosphorylase, 18
Guanylate cyclase, 174, 183, 184, 226

Hemerythrin, 39, 40, 57
Hemocyanin, 19, 72
Hemoglobin, 13, 18, 19, 151, 157, 169, 395
Hexokinase, 18
Histone H2A, 16

Iminopeptidase, 22
Immunoglobulin, 48
Insulin, 3, 18, 19, 28, 29, 43, 57, 224, 287, 477
Interferon, 3, 19, 484, 485
Invertase, 6

Keratin, 16, 20, 47, 57, 148

Lactalbumin, 108, 114, 119, 120, 494
Lactate dehydrogenase, 81, 108
Lactoglobulin, 16, 57, 58, 69, 108, 151, 157, 453, 494
Legumin, 517
Lysozyme, 5, 16, 18, 57, 69, 81, 93, 103, 105, 108, 111, 113, 121, 122, 143, 148, 149, 151, 281, 476, 481–483, 518

Melibiase, 6
Metmyoglobin, 108
Myogen, 57
Myoglobin, 16, 18, 19, 47, 57, 80, 81, 151
Myosin, 16, 18, 19, 45, 69

Nitrate reductase, 13
Nucleohistone, 49

Opsin, 19
Ovalbumin, 19, 57, 69, 114, 493, 494, 527
Oxytocin, 239, 478

Pancreatin, 7
Papain, 7, 22, 81, 140
Parvalbumin, 112
Pectinase, 8
Penicillinase, 115
Pepsin, 7, 8, 16, 22, 57
Phosphodiesterase inhibitor, 227
Phosphofructokinase, 123
Phosphoglycertate kinase, 98, 100, 106, 114
Plasmin, 50
Proinsulin, 240
Prolactin, 57
Prolidase, 22
Pyruvate carboxylase, 13, 123
Pyruvate dehydrogenase, 49

Rennet, 110
Resilin, 47
Rhodopsin, 57
Ribonuclease, 18, 19, 28, 55, 69, 81, 92, 108, 114–116, 118, 494
Ricin, 19
Rubredoxin, 39

Salmine, 57
Sclerotin, 20
Serum albumin, 3, 18, 19, 57, 69, 108, 202, 477, 494
Staphylococcal nuclease, 39
Succinate dehydrogenase, 13

Protein Index

Thrombin, 19, 50
Thymohistone, 57
Thyroglobulin, 57
Triose phosphate isomerase, 39
Tropomyosin, 45, 57, 69
Troponin, 45
Trypsin, 8, 19, 22
Trypsin inhibitor, 16, 41, 152
Tryptophan synthetase, 18
Tubulin, 44, 104
Tyrosine oxidase, 13

Ubiquitin, 20
Urease, 69, 110
Urokinase, 3

Vasopressin, 238
Vicilin, 517

Zein, 19

Subject Index

Acquired immunodeficiency syndrome (AIDS), 468
Adrenocorticotropic hormone (ACTH), 180, 236, 240
Aggregation, 501
 hydrophobic, 501
 intermediate states, 123
 orthokinetic, 440
 perikinetic, 439
Agonist, 218
Alpha-helix, 32
Amino acid
 acidic/basic, 10, 54
 analysis, 257
 aromatic, 77, 508
 classification, 10
 dansyl chloride, 264
 derivatization, 255, 258
 enantiomer separation, 252, 269, 271, 273
 essential, 519
 free energy of transfer, 100, 132
 hydrophobic, 10
 non-helix-forming, 34
 non-protein, 10
 pK_a values, 54
 polar, 10
 profile, 15, 257, 487
 racemization, 524
 rare, 10, 30
 sequence, 26
 titration, 54, 495
 transesterification, 255
Antibody
 binding of vitamin K, 481
 function, 415

 monoclonal, 419, 421
 purification, 424
Antigen
 binding, 417, 419
Antinutritional factor, 456
 phytohemagglutinin, 457, 484
 removal, 524
 trypsin inhibitor, 41, 522
Antiserum, 416
 polyclonal, 419
 reactivity, 416
Atherosclerosis, 203
Axial ratio, 68, 431

Beer-Lambert law, 75
 extinction coefficient, 75
Beta-sheet, 34
Binding
 affinity, 190, 193, 194
 capacity, 190
 conformational change, 190, 192
 membrane protein, 220
 serum albumin, 202
 steroid, 202
Blood–brain barrier, 180, 223, 241
Blood serum, 436
 fractionation (Cohn), 436, 468

Carrier protein, 157, 169, 203
Chaotropism, 98, 113, 116
Chemical messenger, 165, 178, 226
Chemical potential, 61
Chromatofocusing, 285, 311

555

Chromatography
  adsorption, 286
  affinity, 311
  band broadening, 372, 446
  bonded phase, 288
  column, 446, 496
  droplet counter-current, 287
  enantiomer, 252, 271
  exchange kinetics, 303
  field flow, 324
  gel filtration, see size exclusion
  gel permeation, see size exclusion
  high-performance TLC, 251
  hydrophobic interaction, 292, 296
  ion exchange, 294, 296, 301
  ligand exchange, 269
  metal chelate interaction, 307
  open bed, 247
  paper, 246
  partition, 287
  recycle mode, 373
  reversed-phase, 289, 293
  sample loading, 367, 375, 378, 447
  separation ratio, 370
  size exclusion, 294, 316, 468, 496
  temperature effects, 366
  thin layer (TLC) 249, 381
Cilia, 44
Circular dichroism (CD), 79, 98, 105, 470
  ellipticity, 79
  secondary structure determination, 80
Coacervation, 439
Cold inactivation, 104
Collagen, 45
  hydration structure, 141
  hydrogen bond, 47
  myofibril, 45
  polyproline helix, 46
  propeptide, 46
Column, 210, 446, 496
  cellulose, 273
  efficiency, 369, 377
  mixed-bed, 304
Computer simulation, 151
  potential energy minimum, 151
Configurational entropy, 138
Conformational transition, 190, 514
  cooperativity, 514
  thermal, 526
Cyclase activity, 175, 183, 209, 210, 222, 226, 227
  messenger, 226

Denaturation, 60, 97
  cold, 104, 105
  freeze, 107
  freeze drying, 501
  ionic strength effect, 111, 119, 517
  lyotropic effects, 60, 111, 119
  multi-state, 121
  pH, 111
  pressure, 107
  rate, 434
  shear, 108, 366
  solubility as indicator, 58
  solvent accessibility, 100, 134, 139
  sorption, 120, 289, 295, 349
  spectroscopic determination, 90
  sugar, 496
  temperature, 111, 449, 461
  thermal, 104, 119, 435, 448, 514, 516
  thermodynamics, 97
  two-state model, 97, 99
Denatured state, 119
Densitometry, 252, 360, 499
Desorption, 150
Detector, 358
  electron capture (ECD), 255
  flame ionization (FID), 255
  fluorescence, 265
  stain (Coomassie Blue), 360

Dialysis, 66
  ion binding, 66
Dielectric permittivity, 56, 60
Differential scanning calorimetry (DSC), 97, 102, 502, 514, 528
Diffraction, 38, 129, 471
  comparison with optical activity, 81
Diffusion coefficient, 68, 69
  by NMR, 143
Distribution coefficient, 301
Disulfide link, 28, 238
Donnan equilibrium, 67
Double labeling, 209

Edman process, 259
Electroendosmosis, 342, 392, 406
Electron paramagnetic resonance, 82
  g-factor, 82
  hyperfine splitting, 82
Electrophoresis, 469
  anticonvective medium, 342
  continuous, 395
  free-flow, 398
  mobility, 278
  paper, 247
  two-dimensional, 249, 353
  urea gradient, 123
  zone, 278, 348
Electrophoretic mobility, 530
Electroviscous effect, 460, 494
Elution gradient, 260, 293
Emulsification, 459
Emulsion, 528
  creaming, 537
  flocculation, 537
  interfacial tension, 529, 533, 539
  isoelectric point, 533
  particle size analysis, 537
  stability, 532
  steric stabilization, 534, 535
  viscosity, 536, 538
  zeta potential, 530

Enzyme, 5, 18, 21
  glycolytic, 31
  hydrolase, 21
  protease, 26
  transferase, 21
Enzyme-linked immunoadsorbent assay (ELISA), 417
Epitope analysis, 423
Excluded volume, 63, 64
Exocytosis, 178

Flagellum, 44
Flory-Huggins theory, 119
  interaction parameter, 119
Fluorescence spectroscopy, 78, 497, 506, 509
  depolarization, 78
  energy transfer, 78
  intrinsic, 497
  life time, 507
  probe, 510
Freeze concentration, 107
Frictional coefficient, 68, 70
Functionality, 15, 22, 492

Gel, 386
  polyacrylamide, 342
Gelation
  heat, 459, 493
  impurity effects, 506
  network formation, 527
Genetic engineering, 478
  insulin, 478
Glycoprotein, 12, 198, 202

Halophile, 31
Hofmeister series, *see* Lyotropic series
Homology, 28, 30, 40, 484
  globin, 40

interferon, 484
trypsin inhibitor (PTI), 41
Hormone, 233
  feedback control, 161, 181
  luteinizing, 235
  peptide, 234
  phosphodiesterase, 227, 228
  release/inhibition, 234
  secretion, 160, 161
Hormone–receptor complex, 172, 479
Hybridoma, 422
Hydration
  dynamics, 142
  geometry, 130, 142
  hydrophobic, 115, 128, 132
  infrared spectral determination, 77
  ionic, 128
  number, 145
  protein crystal, 139
  solvent accessibility, 134
  stabilizing role, 127
Hydration sphere, 129, 144
  collagen, 141
  halide ion, 130
  neutron diffraction, 129
Hydrogen bond, 31, 34, 47
Hydrophilic hydration, 130
  peptide bond, 138
  polyproline stabilization, 131
Hydrophilic-lipophilic balance (HLB), 535
Hydrophobic hydration, 115, 128
  cavity model, 132
Hydrophobic interaction, 220, 292, 293, 296, 493, 517, 535
  emulsion stability, 535
  entropic origin, 133, 493
  gelation, 493
  protein stabilization, 117
  range, 135
  thermodynamics, 132
Hydrophobic site, 220, 294

Hydrophobicity, 135, 136, 365, 518
Hypothalamus, 162

Immobiline, 347
Immunization, 416
Immunoblotting, 418
Immunocomplex, 417
Immunoelectrophoresis, 499
Immunoglobulin, 50
Infrared spectroscopy, 76, 471
  amide bands, 76, 150, 471
  hydrogen bond effect, 76
  secondary structure determination, 76
Interferon, 487
  action, 487
  chromatographic separation, 484
  electrophoresis, 485
  homology, 484
Ion channel, 174, 175
Ionic strength, 55, 56
Isoelectric focusing, 309, 340
  ampholyte, 341, 402
  antibody, 424
  field strength, 341
  flat-bed, 418
  pH gradient, 341, 386
  preparative scale, 406
  ultra-thin-layer, 344
  zone diffusion, 344, 411
Isoelectric point, 56, 298, 310
  solubility minimum, 58
Isotachophoresis (ITP), 275, 352
  annular, 392
  capillary, 386
  column, 391
  flat-bed, 389
  heat dissipation, 391
  leading electrolyte, 277
  terminating electrolyte, 277

Joule heat, 406, 408

Subject Index

Lymphocyte, 419, 422
Lyotropic series, 59, 115
  salting in/out, 59, 60, 64, 113, 432, 460

Membrane, 189, 196, 291
  isoelectric, 406
  semipermeable, 65
Microbial proteins, 6
Microtubule, 44, 178
Moffit equation, 80, 470
Molecular weight determination, 49, 63
  aggregation effect, 65
  osmotic pressure, 65, 66
  scattering, 73, 74
  sedimentation, 69
  viscosity, 70
Mutant protein, 5, 31, 42, 105
Myeloma cell, 419, 420

Neuropeptide, 157, 165
Neurotransmitter, 157, 158, 163, 179, 209, 217, 219, 227
Newtonian fluid, 71
Nuclear magnetic resonance (NMR), 83, 472
  active site mapping, 482
  chemical exchange, 84, 474
  chemical shift, 84, 90
  denaturation, 90
  interpretation of spectra, 87
  lanthanide shift, 482
  ligand binding, 88, 92
  nuclear magnetic relaxation, 86, 143
  protein dynamics, 142, 473
  random coil spectrum, 472
  ring current shift, 474
  spin–spin coupling, 86, 88, 472
  torsional angle, 86
Nutritional value, 519
  net protein utilization, 521
  protein efficiency ratio, 521

Optical activity, 79
Optical rotatory dispersion (ORD), 79, 119, 470
  Kronig-Kramer equation, 79
Osmotic pressure, 65

Partial specific volume, 69, 70
Peptide bond, 31
  geometry, 31
  hydration, 138
  torsional angle, 32, 86
Peptide hormone, 234
  synthesis, 161, 233
Photoaffinity labeling, 229
Pituitary gland, 158, 177
Polyproline helix, 34
Precipitation
  exclusion, 439
  isoelectric, 432, 460, 462
  kinetics, 439
  receptor, 207
Production processes, 429
  centrifugation, 442, 460
  concentration, 448, 457
  dewatering, 441
  disruption, 428
  extraction, 430, 455, 458
  filtration, 440
  freeze drying, 448, 501
  interferon isolation, 483
  mixing shear, 431
  osmotic shock, 430
  spray drying, 448, 461
  ultrafiltration, 444, 460
Prosthetic group, 12
Protein efficiency ratio, 521
Protein engineering, 4, 124
Protein folding, 38
Protein function, 18
  blood clotting, 50
  chromatin, 49
  contractile, 45
  enzyme, 18
  histone, 49
  immune response, 48

storage, 18
structural (fibrous), 32, 45, 48
technological (functionality), 22
transducer, 20
transport, 20
Protein production technology, 426
  blood serum, 436
  economics, 2
  isolate, 458
  markets, 5, 7
  raw materials, 2, 6, 428, 453, 454, 501
  yields, 426, 496
Protein refolding, 123
Protein thermodynamics, 95, 135, 514
  denaturation, 97
  free energy of transfer, 100
  ionic strength effects, 113
  two-state model, 102
Proton transfer, 141, 143

Radioimmunoassay, 424
Radius of gyration, 469
Random coil, 63, 472
Rapeseed, 454, 501
Receptor, 169, 170, 175
  benzodiazapine, 225
  binding, 211
  cell surface, 217, 222
  dopamine, 221
  estrogen, 205
  insulin, 223, 224
  intracellular, 213
  membrane, 189, 196
  nicotinic, 222
  nuclear, 205
  opioid, 224
  steroid complex, 176, 207, 208
Retardation factor, 247, 381
Rheology, 460, 527
  consistency index, 51
  flow behavior index, 51
  interfacial, 538

pseudoplasticity, 57, 545
shear stress, 51
shear thinning, 460
textural properties, 57, 526, 541
thixotropy, 545
turbulence, 110
visoelasticity, 57

Scatchard plot, 192, 194
Scattering, 72, 509
  scattering ratio (Rayleigh), 73, 509
  second virial coefficient, 73
  small angle X-ray, 74, 469
Second virial coefficient, 63, 73
  effect of Donnan equilibrium, 67
  excluded volume effect, 63
Sedimentation
  coefficient, 68
  equilibrium, 68
  velocity, 68
Sex hormone, 183
Sexual cycle, 181
  differentiation, 183
Shear thinning, 460
Small angle X-ray scattering (SAXRS), 469
  radius of gyration, 469
Solubility, 54, 432, 459
  classification of protein type, 57
  hydrophobic effects, 432
  nitrogen solubility index (NSI), 492
  organic solvents, 436, 457
  pH effect, 58, 433
  polyelectrolyte character, 54
  protein dispersion index (PDI), 492
  salting out, 59, 432, 460
Sorption, 120, 295
  dilatational modulus, 121
  energy, 146
  hysteresis, 146
  monolayer, 121

Subject Index

relative humidity, 148
water vapor, 146
Soya bean, 453
Spacer molecule, 314
Specific heat, 150
Stability
  emulsion, 518
  foam, 518
  folded structure, 137
  thermal, 518
Stationary phase, 289, 298
Stokes radius, 353
Structure, 25
  chiral, 271
  determination, 26
  domain, 479
  hierarchy, 25
  polyproline helix, 34
  primary, 27
  quaternary, 42, 47
  random coil, 63, 472
  reverse turn, 36
  secondary, 32
  symmetry index, 33
  tertiary, 37
Subunit assembly, 42, 49, 51
  microtubule, 44, 104, 178
  propeptide, 43, 46
  virus coat protein, 42, 104
Support matrix, 314
Svedberg equation, 69
Symmetry index, 33
Synapse stabilization, 225
Synaptic (storage) vesicle, 179

Theoretical plate equivalent, 280, 320, 375
Theta solvent, 64, 72
Thyroid hormone, 213
Titration profile, 55, 495
  ionic strength effect, 55
  pH, 54
  spectrophotometric, 55
Toxin, 20
Tubulin, 44

Ultracentrifuge, 68, 205
Ultrafiltration, 444
  concentration polarization, 445
  membrane, 444
  selectivity, 445
  shear inactivation, 110
Ultraviolet spectroscopy, 77, 470, 499, 505, 508
  aromatic amino acids, 77, 500
  hyperchroism, 470
  hypochroism, 470
  plant protein absorbance, 506
  secondary structure, 470

Van der Waals interactions, 494, 535
Van der Waals radius, 139
  solvent accessibility, 139
Viscosity, 495
  intrinsic, 71, 72
  random coil polymer, 71

Water
  as metal ligand, 140
  bound, 145
  bridge, 140, 142

X-ray diffraction, 471

Zeta potential, 494, 530, 542
Zimm plot, 73